Sustainable Summer Fodder

Increasing human population pressure, especially in developing countries, indicates a massive increase in the consumption of livestock products. Fodder crops are the primary and cheapest source of feed for livestock, but feed shortages or poor-quality fodder are particular constraints to the world's livestock sector. These shortages affect milk production, adult and calf health, calving rates, and livestock numbers. Summer crops, including cereal, legume, grass, and tree fodder, provide an economical source of nutrients suitable for animal health as well as improve their breeding efficiency.

Sustainable Summer Fodder: Production, Challenges, and Prospects presents the unique opportunities and difficulties of supportable cultivation and growth of summer fodder crops and the various methods for enabling crop preservation. It details conventional modern approaches to sustainable production and advanced molecular approaches for developing high-yielding fodder crops.

FEATURES

- Contains information on cultivation, growth, productivity, and protection from pests and diseases of various summer fodder crops
- Presents approaches to sustainable production, summarizes fodder preservation methods, and details molecular approaches to develop high-yielding fodder crops
- Provides insights into agronomic aspects of cereal, legume, grass, and tree species cultivated as summer fodder worldwide
- Discusses the trends in insect pests and diseases, their field identification, and various control methods

This book is an essential resource for agronomists, horticulturalists, livestock farmers and practitioners, and students working in the field.

Sustainable Summer Fodder

Production, Challenges, and Prospects

Edited by
Imran Ul Haq, PhD
Siddra Ijaz, PhD
Hayssam Mohamad Ali, PhD

CRC Press
Taylor & Francis Group
Boca Raton London New York

CRC Press is an imprint of the
Taylor & Francis Group, an **informa** business

First edition published 2024
by CRC Press
2385 NW Executive Center Drive, Suite 320, Boca Raton, FL 33431

and by CRC Press
4 Park Square, Milton Park, Abingdon, Oxon, OX14 4RN

CRC Press is an imprint of Taylor & Francis Group, LLC

ISBN: 978-1-032-20833-6 (hbk)
ISBN: 978-1-032-20899-2 (pbk)
ISBN: 978-1-003-26580-1 (ebk)

DOI: 10.1201/b23394

Typeset in Times
by Apex CoVantage, LLC

Contents

Preface

The livestock sector is one of the critical components of gross domestic product and contributes significantly to most economies of agricultural countries. It plays a vital role in sustainable livelihood, poverty alleviation, and generating year-round income for rural and urban populations from regional and global perspectives. Increasing human population pressure, especially in developing countries, indicates a massive increase in the consumption of livestock products. Fodder crops are the primary and cheapest source of feed for livestock. A regular supply of adequate and nutritious fodder is essential to strengthening this sector. An extensive literature review revealed that feed shortages or poor-quality fodder are particular constraints to the world's livestock sector. These shortages affect milk production, adult and calf health, calving rates, and livestock numbers.

Green fodder, an economical source of nutrients, is highly palatable and digestible. It is suitable for animal health as well as improves their breeding efficiency. We have various summer crops with an excellent nutritional profile, classified as cereal, legume, grass, and tree fodder. Cereal grains are highly rich in carbohydrates and crude fiber contents and moderately rich in phosphorus and vitamins. The **cereal fodder**, sorghum (*Sorghum bicolor*), maize (*Zea mays*), and millet spp., contains high fiber content composed of cellulose, a complex carbohydrate polysaccharide that is indigestible for humans but is a significant source of energy for animals, particularly ruminants. The **grass fodder**, fiorin grass, bluegrass, Columbus grass, fescue, Napier, elephant grass, orchard grass, Rhodes grass, Sudan grass, Timothy grass, and Guinea grass, contains crude fibers, crude proteins, and essential minerals. **Legume fodder**, cowpea, guar, soybean, lablab, and burgundy bean, is particularly rich in proteins and minerals. **Tree fodder**, sesbania, *Glyricidia* spp., subabul, Mott grass, *and Stylosanthes guianensis*, contains essential animal feed components, especially during dry periods when there is a shortage of other green fodder. Moreover, tree fodder can provide 15–29% crude proteins and is considered high milk-producing forages.

This book provides a comprehensive insight into summer fodder crops. This book aims to collect current knowledge on summer fodders for academicians, breeders, researchers, and students. We have tried to make the book a scientific knowledge platform equipped with recent scientific information on winter fodders with practical knowledge for biologists dealing with breeding fodder crops' idiosyncrasies. This book focuses on bioecology, cropping patterns, nutritional profiling, biotic and abiotic stresses, etiology and integrated disease management, biotechnological interventions, and socioeconomic perspectives of summer fodder crops.

Imran Ul Haq
Siddra Ijaz
Hayssam Mohamad Ali

Editors

Dr. Imran Ul Haq, with a bright career in agriculture, plant pathology, and fungal molecular biology, has a postdoc from University of California Davis, USA. He is currently serving as Associate Professor in the Department of Plant Pathology, University of Agriculture Faisalabad, Pakistan. He has supervised more than 40 graduate students and established the Fungal Molecular Biology Laboratory Culture Collection (FMB-CC-UAF) and is an affiliated member of the World Federation for Culture. Collections (WFCC). He has published more than 60 articles, seven books, four laboratory manuals, and several book chapters. He has made colossal contributions in fungal taxonomy by reporting novel species of fungal pathogen in plants. His research interests are fungal molecular taxonomy and nanotechnology integration with other control strategies for sustainable plant disease management.

Dr. Siddra Ijaz, with a vibrant career in agriculture and biotechnology, has a post doc from the Plant Reproductive Biology Laboratory, University of California Davis, USA. She is currently serving as Assistant Professor in the Center of Agricultural Biochemistry and Biotechnology, University of Agriculture, Faisalabad, Pakistan. She has supervised more than 50 M.Phil and PhD students. She has published more than 60 articles, eight books, and several book chapters. Her research focus includes plant genome engineering using transgenic technologies, genome editing through CRISPR/Cas systems, nanobiotechnology, and exploration of genetic pathways in plant–fungus interactions.

Dr. Hayssam Mohamad Ali, is working at the Botany and Microbiology Department, College of Science, King Saud University. Dr. Hayssam Mohamad Ali has a “World Ranking of Top 2% Scientists” according to the Stanford Scientists’ Ranking released by Stanford University in 2021 and 2022. He has more than 300 published papers, 10 published books, and he has reviewed more than 1000 papers in highly impacted journals. He is interested in studying plant physiology, biotic and abiotic stress, sustainability, molecular biology, environmental stresses, plant biotechnology, crop improvement, plant breeding, forestry, and plant biology.

Contributors

Aqleem Abbas
Department of Agriculture and Food Technology
Karakoram International University
Gilgit, Pakistan

Muhammad Abdullah
Department of Animal Nutrition
University of Veterinary and Animal Science
Lahore, Pakistan

Jamshaid Ahmad
University of Veterinary and Animal Sciences, Ravi Campus
Pattoki, Pakistan

Usama Ahmad
Department of Plant Pathology
University of Agriculture
Faisalabad, Pakistan

Manzoom Akhtar
University Institute of Management Sciences
PirMehr Ali Shah Arid Agriculture University
Rawalpindi, Pakistan

Abid Ali
Department of Entomology
University of Agriculture
Faisalabad, Pakistan

Amjad Ali
Department of Agriculture and Food Technology
Karakoram International University
Gilgit, Pakistan

Dr. Hayssam Mohamad Ali
Department of Botany and Microbiology
College of Science
King Saud University
Riyadh, Saudi Arabia

Safdar Ali
Department of Agronomy
University of Agriculture
Faisalabad, Pakistan

Sajjad Ali
Department of Entomology
The Islamia University of Bahawalpur
Bahawalpur, Punjab, Pakistan

Usman Ali
Department of Animal Nutrition
University of Veterinary and Animal Science
Lahore, Pakistan

Tuba Amjad
Centre of Agricultural Biochemistry and Biotechnology (CABB)
University of Agriculture
Faisalabad, Pakistan

Muhammad Anjum Aqueel
Department of Entomology
The Islamia University of Bahawalpur
Bahawalpur, Punjab, Pakistan

Muhammad Jalal Arif
Department of Entomology
University of Agriculture
Faisalabad, Pakistan

Muhammad Arshad
Department of Entomology
University of Agriculture
Faisalabad, Pakistan

Hafiz Muhammad Usman Aslam
Department of Plant Pathology
Institute of Plant Protection (IPP)
MNS–University of Agriculture
Multan, Pakistan

Saba Aslam
Department of Plant Pathology and Microbiology
Iowa State University
Ames, Iowa

Tahir Hussain Awan
Rice Research Institute Kala Shah Kaku, Pakistan

Rabia Tahir Bajwa
Cholistan Institute of Desert Studies
The Islamia University of Bahawalpur
Bahawalpur, Punjab, Pakistan

Muhammad Hamid Bashir
Department of Entomology
University of Agriculture
Faisalabad, Pakistan

Anjum Faraz
Department of Plant Pathology
University of Agriculture
Faisalabad, Pakistan

Muhammad Dildar Gogi
Department of Entomology
University of Agriculture
Faisalabad, Pakistan

Amer Habib
Department of Plant Pathology
University of Agriculture
Faisalabad, Pakistan

Akhtar Hameed
Department of Plant Pathology
Institute of Plant Protection (IPP)
MNS–University of Agriculture
Multan, Pakistan

Faiza Hassan
Institute of Physiology and Pharmacology
Faculty of Veterinary Sciences
University of Agriculture
Faisalabad, Pakistan

Azhar Hussain
Department of Agriculture and Food Technology
Karakoram International University
Gilgit, Pakistan

Yasir Iftikhar
Department of Plant Pathology
College of Agriculture
University of Sargodha
Sargodha, Pakistan

Misbah Ijaz
Faculty of Veterinary Sciences
University of Agriculture
Faisalabad, Pakistan Muhammad

Siddra Ijaz
Centre of Agricultural Biochemistry and Biotechnology (CABB)
University of Agriculture
Faisalabad, Pakistan

Asif Iqbal
Department of Agronomy
University of Agriculture
Faisalabad, Pakistan

Muhammad Aamir Iqbal
Department of Agronomy
University of Poonch
Rawalakot, Pakistan

Maheera Khaliq
Institute of Physiology and Pharmacology
Faculty of Veterinary Sciences
University of Agriculture
Faisalabad, Pakistan

Nasir Ahmad Khan
Department of Plant Pathology
University of Agriculture
Faisalabad, Pakistan

Nabeeha Aslam Khan
Department of Plant Pathology
University of Agriculture
Faisalabad, Pakistan

Rashad Rasool Khan
Department of Entomology
University of Agriculture
Faisalabad, Pakistan

Nabil Killiny
Department of Plant Pathology
University of Florida
Gainesville, Florida

Tehrim Liaqat
Department of Zoology, Wildlife and Fisheries
University of Agriculture
Faisalabad, Pakistan

Oscar Emanuel Liburd
Department of Entomology and Nematology
University of Florida
Gainesville, Florida

Shahid Majeed
Department of Entomology
University of Agriculture
Faisalabad, Pakistan

Riffat Malik
Centre of Agricultural Biochemistry and Biotechnology (CABB)
University of Agriculture
Faisalabad, Pakistan

Asad Manzoor
Faculty of Veterinary Sciences
University of Agriculture
Faisalabad, Pakistan

Muhammad Tahir Mohy-ud-din
Faculty of Veterinary Sciences
University of Agriculture
Faisalabad, Pakistan

Mustansar Mubeen
Department of Plant Pathology
College of Agriculture
University of Sargodha
Sarhodha, Pakistan

Ghalia Nadeem
Centre of Agricultural Biochemistry and Biotechnology (CABB)
University of Agriculture
Faisalabad, Pakistan

Ahmad Nawaz
Department of Entomology
University of Agriculture
Faisalabad, Pakistan

Ahmad Nisar
Department of Plant Pathology
University of Agriculture
Faisalabad, Pakistan

Ayesha Parveen
Department of Zoology, Wildlife and Fisheries
University of Agriculture
Faisalabad, Pakistan

Nick T. Peters
Department of Plant Pathology and Microbiology
Iowa State University
Ames, Iowa

Rabia Ramzan
Department of Entomology
University of Agriculture
Faisalabad, Pakistan

Ifrah Rashid
Cholistan Institute of Desert Studies
The Islamia University of Bahawalpur
Bahawalpur, Punjab, Pakistan

Aqsa Riaz
Department of Zoology, Wildlife and Fisheries
University of Agriculture
Faisalabad, Pakistan

Hasan Riaz
Department of Plant Pathology
Institute of Plant Protection (IPP)
MNS–University of Agriculture
Multan, Pakistan

Muhammad Shahid Rizwan
Cholistan Institute of Desert Studies
The Islamia University of Bahawalpur
Bahawalpur, Punjab, Pakistan

Saima
Department of Animal Nutrition
University of Veterinary and Animal Science
Lahore, Pakistan

Muhammad Rehan Sajid
Institute of Physiology and Pharmacology
Faculty of Veterinary Sciences
University of Agriculture
Faisalabad, Pakistan

Muhammad Abdullah Saleem
Department of Agronomy
University of Agriculture
Faisalabad, Pakistan

Muhammad Kaleem Sarwar
Department of Plant Pathology
University of Agriculture
Faisalabad, Pakistan

Qaiser Shakeel
Cholistan Institute of Desert Studies
The Islamia University of Bahawalpur
Bahawalpur, Punjab, Pakistan

Ashfaq A. Sial
Department of Entomology
University of Georgia
Athens, Georgia

Muhammad Umair Sial
Department of Entomology
University of Agriculture
Faisalabad, Pakistan

Muhammad Sufian
Department of Entomology
University of Agriculture
Faisalabad, Pakistan

Imran Ul Haq
Department of Plant Pathology
University of Agriculture
Faisalabad, Pakistan

Umm E. Ummara
Department of Zoology, Wildlife and Fisheries
University of Agriculture
Faisalabad, Pakistan

1 Conventional and Biotechnological Interventions in Summer Fodders

Siddra Ijaz, Imran Ul Haq, Ghalia Nadeem, Riffat Malik, Tuba Amjad, and Nasir Ahmad Khan

1.1 INTRODUCTION

Pakistan is an agricultural country whose economy relies heavily on crop varieties, contributing about 18.5% to the annual gross domestic product (GDP) value. Fodder crops occupy the major share of livestock feed, with an average cultivated area of 2.24 million hectares. In Punjab, fodder crops have an average yield of 21.6t/ha and are the third major crop variety after wheat and cotton. Fodder crops are agricultural foodstuffs mainly used as animal feed (Ijaz, 2021). Different types of fodder crops vary with seasons. Summer fodders are grown as the season starts and can be stored as the source of animal feed until the beginning of winter. Major summer fodder crops in Pakistan include maize, sorghum, pearl millet, cowpeas, and guar (Rehman et al., 2015). Maize is the third-largest crop in Pakistan after wheat and rice, and its optimum growth temperature ranges from 25–28°C. Maize contains about 70–80% carbohydrates and 6–13% protein, contributing about 1.3 million tonnes to the annual production value. Approximately 60% of its production is used in the animal feed industry in Pakistan. It is a non-legume fodder that serves a major role in livestock and poultry and is typically harvested for 60–65 days. The high fodder quality of maize is attributed to its palatability and the short growing period (Tariq & Iqbal, 2010). Sorghum (*Sorghum bicolor* L.) is a well-known forage crop known as "Jawar" in Pakistan. It can grow well in a dry climate and requires moderate rainfall. The optimum growth temperature of sorghum ranges from 25–32°C. Sorghum contains about 69% carbohydrates and 10% protein. A large cultivated area of sorghum is in Punjab and Sindh Provinces, producing about 47% and 26% of the total land, respectively. It is extensively grown as fodder in Mianwali, Rawalpindi, Jhelum, and Dera Ghazi Khan in Punjab; Dera Ismail Khan and Mardan in Khyber Pakhtunkhwa; Sukkur and Dadu in Sindh; and Sibi and Loralai in Baluchistan (Tonapi et al., 2020). Pearl millet (*Pennisetum glaucum*), locally known as Bajra, is the sixth-largest crop in Pakistan. The optimum growth temperature of pearl millet ranges from 26–29°C. It contains about 41.2% carbohydrates and 6% protein. It is grown largely in arid regions and is utilized as fodder and grain. It is grown in Gujranwala, Rawalpindi, Jhelum, Bahawalpur, Attock, and Mianwali in Punjab; Sibbi and Khuzdar in Baluchistan; Bunnu and Karak in Khyber Pakhtunkhwa; and Hyderabad, Khairpur, and Dadu in Sindh (Khanum et al., 2020). Cowpea (*Vigna unguiculata*) is an important legume of Pakistan and has a high resilience for sandy soil. Locally known as lobia or chunra, this fodder crop has 21% carbohydrates and a high protein content of 20–25% and is cultivated

DOI: 10.1201/b23394-1

in the Barani areas of Pakistan. Cowpea has an optimum growth temperature of 20–35°C. It is a multipurpose crop that can be used as pulses, vegetables, or fodder for livestock. Guar (*Cyamopsis tetragonoloba* L.) is another essential legume with more than 0.181 million hectares of area under cultivation in Pakistan. It is a drought-tolerant and proteinaceous crop and is extensively grown in arid and semiarid regions of Pakistan. Guar contains about 22% carbohydrates and 32% protein. It is used largely as animal feed due to its high protein quality. It has an optimum growth temperature of 28–32°C. In Pakistan, it is mainly grown in Mianwali, Multan, Muzaffargarh, Sargodha, Bahawalpur, Bahawalnagar, and Sindh, covering an almost 0.1925-million-hectare area. Pakistan produces about 20% of guar production worldwide (Kuravadi et al., 2013).

Fodder crops and their products are used as a diet component due to their medicinal importance and in curing and preventing certain human diseases. Maize contains anti-HIV activity phytochemicals because of *Galanthus nivalis* agglutinin (GNA) lectin (Rouf Shah et al., 2016). Maize oil contains essential fatty acids that are required for health. Liquid from maize roots, leaves, and silk treats nausea, vomiting, bladder problems, and stomach complications. Resistance starch in maize minimizes the risks of obesity, atherosclerosis, and cecal cancer (Huma et al., 2019). Sorghum has anticancer effects, reduces weight, is safe for persons suffering from celiac disease, and has antioxidant effects due to phenolic compounds (Xiong et al., 2019).

Seeds and leaves of sorghum are used in medication. Sorghum treats diseases like AIDS, diabetes, obesity, and digestive problems (McGinnis & Painter, 2020). Pearl millet reduces the levels of blood glucose and cholesterol. It treats iron deficiency in anemia, relieves constipation, prevents insomnia, and helps in weight loss (Dias et al., 2018). Cowpea lowers blood cholesterol and improves the circulation of blood and the health of the heart and cancer (da Silva et al., 2021). Guar gum lowers glucose and cholesterol due to its gelling activity. Guar gum reduces weight loss, appetite, and obesity. When hydrolyzed guar gum is added to a diet supplement, it reduces the signs of irritable bowel disease, the need for laxatives, and the frequency of diarrhea (Mudgil et al., 2014). Oat bran reduces blood cholesterol and glucose level. Oat is used for anticancer activity. It relieves itching and swelling on the skin (Sang & Chu, 2017). Consumption of oats for the long term in the diet reduces the risk of celiac disease (Comino et al., 2015).

1.2 ECONOMIC IMPORTANCE AND CONSTRAINTS IN FODDER PRODUCTION IN PAKISTAN

Fodder crops have great economic importance because they are the major source for feeding livestock and have many industrial applications. According to the harvesting and production area, sorghum is the fifth-most significant crop in Pakistan. More than 100 countries grow this dryland crop. It is a staple food in some countries. It is used as feed for cattle, pigs, and poultry. Sorghum is used in industries for the production of starch and ethanol. Guar is a food product for many cattle, fish, poultry, and swine. Guar gum and its derivatives have many applications in industry, such as ore flotation, the production of explosives, and the fracking of oil and gas formation. Guar gum is used in beverage processing, bakery products, processed milk, dairy products, sauces, and salad dressings (Mudgil et al., 2014). Maize is a fodder crop with many minerals, fiber, oil, and vitamins. Its oil is used in cooking, cosmetics, printing ink, and manufacturing soaps. Its stem is used for manufacturing paper, while its seeds are used for making alcohol (Orhun, 2013). Pearl millet is used as a cooking fuel. It has a tolerance against heat, so it can be grown in drylands (Gulia et al., 2007).

Livestock is the major contributor to the country's economic growth, with a significant share of 11.11% of the GDP value. Fodder is preferred over other cereal grains or feed crops due to its high nutritional value. Feed requirements are increasing alongside the livestock population, but in Pakistan, fodder production is not widely recognized, leading to its decline. The significant problem in decreased fodder production in Pakistan is the less area under cultivation and poor management practices. Fodder crops are cultivated only in an area of 2.0 million hectares in Pakistan, which cannot fulfill the country's average fodder requirements. During 1976–1977, the total area under cropped fodder was 2.6 million hectares which dropped to 2.31 million hectares in 2009–2010. Punjab has a significant share of 82.56%, while Sindh, Khyber Pakhtunkhwa, and Baluchistan contribute 11.50%, 4.48%, and 1.46%, respectively, to fodder production (Ijaz, 2021). The unavailability of improved seed is the major factor contributing to low forage production in Pakistan. There is a lack of private and government sectors in quality fodder seed production. For compensating the loss, the seed is imported from other countries or produced by farmers as low-quality fee substituents. Various insect pests and diseases, that is, red leaf spot, root rot, stem rot, wilt, and others, also contributed to the decline of forage production, resulting in significant losses of 40–50% or sometimes even more (Jackai et al., 1986). About 90–95% of sorghum crops are damaged due to serious red leaf spot disease. Increased livestock population, reduced agricultural land, a lack of knowledge about recent innovations and research, and the improper use of fertilizers and pesticides are the other limiting factors to low fodder production in the country (Pawar et al., 2019).

1.3 CONVENTIONAL METHODS OF BREEDING IN CROPS

Mass selection: Mass selection is widely used in fodder crop breeding. Seeds from desirable plants are collected and grown in the next season based on phenotypic selection. Undesirable plants are eliminated from the field. Best plants are harvested, and their progenies are compared. The poor progenies are eliminated, and the remainder is selected (Baenziger et al., 2006). Mass selection is applied in self- and cross-pollinated plants (Allard, 1960). Since mass selection is based on plant phenotype, it works best when the desired trait is highly heritable (Brown & Caligari, 2011).

Pure line selection: Pure line cultivars are vulnerable to disease outbreaks and lack yield stability in various situations (Allard, 1960). From a genetically diverse population, superior plants are selected, and their progenies are evaluated by simple observation based on phenotypic characteristics after several next generations over many years. Individuals of the pure line are homozygous, containing identical alleles on all the loci. Genotypes from different populations are selected and selfed repeatedly until there is no segregation in the progeny (Poehlman & Sleper, 1995). The selection is suitable based on progeny performance during the improvements of low heritable traits.

Hybridization: In hybridization, desirable traits are combined from two or more different varieties of plants. In controlled hybridization, the pollens from the desired male plant pollinate the female plant. If plants are self-pollinated, the male-part anthers are removed from plants selected as females by three different methods: hot water emasculation, open pollination, and hand emasculation. Hand emasculation is most commonly used and has a higher success rate. In the hand emasculation method, the male part is removed by forceps or scissors. The stigma of the female part is covered with a plastic or paper bag to avoid pollination from foreign pollen. When stigma becomes receptive, then desired pollens are transferred. It is tedious, time-consuming, and needs experts (Stetter et al., 2016).

The purpose of hybridization is to combine the desired genes from different varieties. If all the desired traits are not present in parents, each of which has one or more desired traits not present in the other, and they are crossed, giving the hybrid as the first filial generation (F1) by a genetic phenomenon known as heterosis (Wu et al., 2021). Individuals of the F1 generation are crossed, giving the F2 generation, and selection is made by eliminating the undesired individuals. F1 hybrids are more vigorous than their parents. Hybrid vigor has faster growth rate, earlier flowering, high uniformity, and increased yield. Different types of hybrids, such as the single cross, three-way cross, and two-way cross, are obtained in maize. Seeds of single-cross hybrids have higher yields but are costly. Seeds of two-way crosses are the least expensive and have lower yields than the other two crosses (MacRobert et al., 2014). Drought-tolerance varieties of pearl millet have been developed by hybridization (Srivastava et al., 2022).

Backcrossing: Backcrossing transfers the desired character from the donor plant to the cultivar plant while retaining all other desirable characteristics. Two different parents are selected and then crossed. One parent (adapted parent variety) lacking the desired trait is crossed to another parent containing the desired trait, producing F1 hybrid plants. F1 hybrid material with the desired character is backcrossed with an adapted parent variety for generations. An adapted parent that enters into each backcross is known as a recurrent parent, while the other parent containing the desired character enters into the first cross but not into the backcross and is known as a nonrecurrent or donor plant. The purpose of backcrossing is to retrieve the genome of recurrent plant variety, excluding the addition of the desired trait from a noncurrent parent (Briggs et al., 1953). The backcross method requires a limited number of plants and has predictable results (Vogel, 2009). The backcross method is time-consuming and expensive (Sekerci et al., 2016).

Synthetic varieties: Synthetic varieties are produced by intercrossing the superior genomes to produce a hybrid with superior performance in all combinations. Hayas and Garber developed synthetic varieties in maize in 1919. No primary test is required to determine the hybrid's performance compared to mass selection. Synthetic varieties are well known due to their hybrid vigor and capacity to produce seeds for usage in the next generations. For developing synthetic varieties, first breeders produce "breeder seeds" at a small-scale level. In the next stage, government agencies and institutes produce the foundation seed by multiplication. In the third step, the certified seed is produced and distributed to farmers and gardeners (Acquaah, 2015). The varieties of maize, such as MERAJ-2019, MAHZAIB-2019, NOOR-2019, PAGHUNDA-2019, SILVER-2019, and SAR-SUBZ-2019, have been developed in Pakistan. They have excellent productive potential, are early to mature, and can withstand drought. For farmers in the Province of Baluchistan, drought stress is a significant problem (Manès et al., 2012). Several synthetic varieties of fodder crops have been developed and are available in the market.

1.4 CONVENTIONAL BREEDING IN FODDER CROPS

Livestock farming and fodder production are the major sources of income in developing countries like Pakistan. Plant breeding has been an ancient practice from the beginning of agriculture. Breeding involves the production of genetically diverse populations of adaptable plants by new combinations having desired characteristics. Plant breeding aims to obtain plants with the maximum number of desirable characteristics, such as resistance against diseases and insects, tolerance against salinity, soil, temperature, adaptation to the environment, and greater yield and quality. Farmers collected the seeds from the excellent variety that grows well in the field for the next season. This type of selection served as a basis for early plant-breeding techniques. The outcomes of plant breeding were obvious. New, modern varieties are so modified that their wild varieties cannot exist today, or they have been changed so that it is too difficult to find their ancestors.

High yield, superior quality, early maturity, wider adaptation, and tolerance to biotic and abiotic challenges are the key breeding objectives in grain sorghum (Rakshit & Bellundagi, 2019). Breeding strategies in maize have increased yield, tolerance against pests, and oil content (Gedil & Menkir, 2019). The nutritional quality of plants has been increased by breeding, such as corn (maize) varieties containing higher lysine content than previous varieties developed by breeding (Pajić, 2007). Breeding in ryegrasses has improved the production of dry matter, tolerance against freezing, digestibility, and heat or drought tolerance (Wilkins, 1991). The yield and quality of cowpea grains have been increased by breeding (Boukar et al., 2016). The resistance genes against the weed *Striga gesnerioides* from parents are introduced in cultivated cowpeas from breeding strategies (Omoigui et al., 2017).

1.5 NON-CONVENTIONAL BREEDING IN FODDER CROPS

The primary object of plant breeding is to increase yield, and the secondary objectives are to enhance tolerance against biotic and biotech stresses, quality of plants, synchronous maturity, efficient use of water and nutrients, and removal of toxic substances. The enhanced understanding of genetics has made it possible to increase the yield of crops. The molecular markers have helped enhance the yield of crops by improving the degree of precision and selection efficiency. There are four categories of markers, morphological, biochemical, cytological, and molecular.

Morphological markers are also known as phenotypic markers based on visual characteristics of plants such as flower color, shape, growth habitat, seed structure, and other agronomic traits. These markers are simple to use, do not need any special equipment, and are eco-friendly, but they are influenced by the environment (Chesnokov et al., 2020).

Biochemical markers are also known as isozymes and depend on allelic variations in enzymes encoded by different genes, but their function is the same. They are mostly used to study population structure, gene flow, and genetic diversity. They are simple to use, cost-effective, and co-dominant, but they show less polymorphism and are affected by plant age, plant tissue material to be used, and extraction method (Ismail et al., 2020).

The ***cytological marker*** determines the specific characteristics of chromosomes, such as satellite, knob, and number of nucleoli. Cytological markers are based on chromosomal variations such as size, shape, position, and banding pattern, as the Giemsa stain produces a G pattern, the reverse of the G band produces an R band, and quinacrine hydrochloride produces a Q band. These variations are distributed in heterochromatin and euchromatin regions. They are widely used for physical mapping and linkage group identification. They are readily available and require small equipment but show less polymorphism, need experts, and are fewer in number (Nadeem et al., 2018).

Molecular markers are DNA-based markers and are now the most commonly used. They are co-dominant, highly polymorphic, high resolution, highly reproducible, whole genome coverage, low cost, and have no environmental effect. They are categorized by method of detection.

1.6 HYBRIDIZATION-BASED MARKERS

DNA is first digested with restriction enzymes in hybridization-based markers and separated on an agarose gel. The separated fragments of DNA are then transferred to the nitrocellulose membrane by the Southern blotting technique. Fragments are hybridized with radioactively labeled probes and then visualized by autoradiography. Restriction length polymorphism (RFLP) is a hybridization-based marker. It is co-dominant and eco-friendly and has high reproducibility, but it is time-consuming and tedious (Amom et al., 2017). The high grain yield of sorghum varieties has been identified using the RFLP marker (Jordan et al., 2003). The RFLP marker was used to search desirable alleles at the quantitative trait loci in the maize (Zehr et al., 1992).

1.7 POLYMERASE CHAIN REACTION (PCR)-BASED MARKERS

PCR-based markers involve primer-dependent amplification or hybridization followed by gel electrophoresis. PCR-based markers are regarded as second-generation markers based on polymorphism in DNA sequences detected by PCR amplification (Singh et al., 2015). Polymorphism is detected based on band size, absence or presence of amplicons, and mobility. It includes amplified fragment length polymorphism (AFLP), random amplified polymorphic DNA (RAPD), simple sequence repeats or microsatellites (SSRs), inter-simple sequence repeat (ISSR), sequence-related amplified polymorphism (SRAP), single nucleotide polymorphism (SNP), and sequence characterized amplified region (SCAR) (Kordrostami & Rahimi, 2015). Resistance genes against cowpea varieties' aphids have been analyzed using the RFLP marker (Myers et al., 1996). AFLP markers were used to identify the resistant varieties of pearl millet against downy mildew disease (Singru et al., 2003).

1.8 MARKER-ASSISTED SELECTION

Marker-assisted selection (MAS) is the technique of choosing agriculturally significant features in crop breeding by employing morphological, biochemical, or molecular markers as indirect selection criteria (Ashraf et al., 2012). It overcomes all the challenges and difficulties that arise during conventional breeding programs. Plant breeders use it to find dominant and recessive alleles of interest throughout generations. MAS selects genes involved in economically important traits such as forage yield, tolerance against biotic and abiotic stresses, forage quality, and other agronomic traits. Some complex traits, such as forage yield, are controlled by many genes and are known as quantitative trait loci (QTLs). Different molecular markers are associated with a gene or QTL. Some traits, such as resistance and susceptibility, are related to a single major gene having quantitative trait characteristics. The location of this single gene can be found by genetic map using closely linked markers.

In marker-assisted selection, a population of desired traits is developed by hybridizing superior parents. A QTL map is constructed to locate the position of genes controlling the desired traits using different types of molecular markers (Nadeem et al., 2018). For QTL mapping, the selected parents should be diverse and show sufficient polymorphism. F2 populations, double hybrid cross, near-isogenic lines, and backcross populations are commonly used in QTL mapping (Paterson, 1996). QTL mapping is based on the segregation of markers through chromosomal recombination in meiosis. During recombination, the markers linked firmly are transferred together as compared to others far away (Collard et al., 2005). QTLs in different populations are verified to check the validation of QTLs, and different molecular markers are used in germplasm to validate molecular markers. In this way, the marker associated with that particular desired trait is identified (Kumawat et al., 2020).

The plant breeder uses MAS to select genes associated with QTLs or major genes in progeny in fodder crops. After selecting favorable genes, the time from starting a trait-targeted breeding strategy to creating improved germplasm can be greatly shortened (Collard & Mackill, 2008). QTL against diseases such as the crown, leaf dollar spot, and bacterial wilt has been identified in fodder crops (Kumawat et al., 2020). Performing an MAS evaluation at the seedling stage for a trait that often requires identification in 2- or 3-year-old plants is a clear example of using perennial forages to reduce the time and area required to create plant populations with desired characteristics. MAS reduces the environmental influence and determines allele dosage at significant loci, making an easier selection of plants for breeding programs. Recessive genes can be maintained easily, and selection can be made on a single plan (Lema, 2018).

Drought tolerance and mildew resistance genotypes of maize have been developed using QTLS mapping and marker-assisted selection (George et al., 2003). MAS with QTLs in sorghum revealed that the stay-green drought component is associated with stove yield and grain stability and maintains the nutritional value under drought stress (Rama et al., 2014). The sorghum yield has been increased using the single nucleotide polymorphism (SNP) PCR-based marker (Burow et al., 2019).

1.9 GENETIC ENGINEERING AND TRANSGENIC DEVELOPMENT

Genetic modification in crops, also known as genetic manipulation or genetic engineering, is a technique of altering an organism's characteristics by a change in its genetic makeup. An organism's genome contains DNA, which is the determinant of all its characteristics. This technique involves changing the crop genome by adding a new piece of DNA. The change could be how to make the crop resistant to insects, pests, or adverse environmental conditions or changing how a crop grows. The resulting forage cultivars will now contain the new desired DNA (Raman et al., 2017). The key steps in producing a GM crop include the choice of a target gene and isolating and introducing a specific gene into the desired organism. DNA transfer into a plant cell is important in producing genetically modified forage crops. This is done by coating the desired DNA fragment onto the metal particles and targeting specific plant cells. Another method is to use a virus or bacterium for efficient DNA transfer.

In Pakistan, modern biotechnology was applied in 1985, and since then, various genetically modified crops have been developed with the potential of biotic and abiotic stress tolerance. The gene of interest is introduced into a virus or bacterium, which transfers the new DNA into the target host. This technique is one of the most effective ways to improve the quality of forage crops through targeted improvement. It produces unique genetic variants with desirable traits (Matres et al., 2021). The new forage varieties are adapted to various environmental conditions and pest and disease resistance, have increased crop yield and nutritional value in animal feed, and could overcome the increased feed requirements (Nagamine et al., 2022).

Various genetically modified maize varieties have been developed with desirable traits, that is, insect resistance, insect tolerance, male corn sterility, increased lysine level for use in animal feed, and alpha-amylase expression. Bt maize (corn) is the most common genetically modified maize variant, released in 1997. This cultivar variety encodes proteins from the bacterium *Bacillus thuringiensis*, which offers resistance against certain insect pests (Hasan et al., 2021). Genomic and biotechnology tools have improved sorghum's nutritional value by improving the digestibility of kafirins (protein), increased lysine content, increased pro–vitamin A (beta carotene) accumulation, and zinc and iron bioavailability. Agrobacterium-mediated transformation is the most frequently used method in the development of genetically modified sorghum (Elkonin et al., 2018). Genetically modified (GM) millet has been developed with improved nutritional value, resistance to fungal infection, and biotic and abiotic stresses. Biolistic-mediated gene delivery is widely used in transforming millet crops (Ceasar et al., 2009). This technique has also resulted in GM cowpea, which has drought tolerance and insect resistance. It is a staple food to combat hunger and poverty (Ezezika et al., 2012). The genetic manipulation technique has resulted in increasing the overall yield of guar. There are some safety considerations associated with genetic engineering techniques. According to the United States Department of Agriculture, the Environmental Protection Agency, and Food and Drug Administration, genetically engineered crops are assessed and tested to avoid any harm to the consumer and the environment (Caradus et al., 2022).

1.10 GENOME EDITING TECHNOLOGY

Genome editing is another technique to modify crop genomes. DNA is added, removed, or modified at specific locations, thus resulting in the customized genetic makeup of a crop. Gene editing has emerged as a new breeding tool to achieve high yield and desirable traits with greater accuracy and efficiency (Kamthan et al., 2016). Various genome editing tools have been employed to edit the crops' DNA, that is, zinc finger nucleases (ZFNs), transcription activator-like effector nucleases (TALENs), and clustered regularly interspaced short palindromic repeats (CRISPR-Cas). ZFNs and TALENs are artificial DNA-binding proteins that are used for the targeted editing of genomes at site-specific locations. ZFN-based genome editing is a technique based on ZFNs used for targeted genome modification. The linking of the zinc finger domain and FokI endonuclease domain forms ZFN molecules. ZFNs are the most effective targeting reagents, which are used to introduce double-stranded breaks at specific sites and hence result in the precise modification of the genome (Carroll et al., 2011). TALENs have engineered restriction enzymes that can be synthesized to edit genomes at specific sites. These are highly specific and efficient enzymes that allow for precisely altering the DNA sequence (Joung & Sander, 2013).

Nuclease enzyme is used to cut both strands of the DNA, which produces double-strand breaks in them, which are then fixed by the plant's internal DNA repair systems, either through non-homologous end joining, which results in nucleotides insertion or deletion or through homologous directed repair (Shan et al., 2013). CRISPR is a more efficient genome editing tool resulting in multiple gene mutations within a crop. The function of CRISPR in gene editing was first described by two female scientists, Jennifer A. Doudna and Emanuelle Charpentier, for which they both received the Noble Prize in Chemistry in 2020 (Shah et al., 2018).

The potential for crop improvement using gene editing technology has become increasingly popular in recent years. Nutritional and functional attributes are the most targeted traits regarding animal feed production. Maize is the most widely cultivated crop in the world. Using CRISPR, maize (corn) is genetically edited to attain desirable traits such as resistance to drought and high yield. Sorghum is the second most planted silage source because of its high biomass and quality. Successful genome editing has been achieved in sorghum by CRISPR/Cas9 through particle bombardment (Liu et al., 2019). It has also emerged as a promising technique for obtaining stress-tolerant and high-yielding millet varieties (Pati et al., 2022). The developing countries use the leaves of cowpea as animal feed. Gene manipulation by CRISPR has been achieved in cowpea to efficiently disrupt the SNF gene (Ji et al., 2019). Genome editing has produced guar hybrids with an efficient nitrogen-fixing ability (Malik, 2013).

1.11 BIOSAFETY CONCERNS RELATED TO BIOTECHNOLOGY

Biotechnology has played a major role in livestock feed improvement through cheaper and more manageable production. This technology has enhanced the nutritional quality and yields of fodder crops, thus making farming more profitable. However, food safety and quality may be affected by biotechnological approaches either directly or indirectly. Biotechnological methods such as genetic engineering, mutagenesis, and breeding alter the genetic composition of plants, animals, and microorganisms. Genetic engineering is the targeted modification of genes, while mutagenesis and breeding are the non-targeted methods. The biggest challenge regarding GM crops and foods is the possibility of having adverse health effects on the human body. The creation of GM crops requires the use of antibiotic resistance marker genes (ARMGs) for the selection of transformed cells. Cells containing the antibiotic resistance genes and the desired gene will survive in the antibiotic-containing medium. A transgenic crop with desired characteristics still contains antibiotic-resistance genes (Read & Zealand, 2000). It is assumed that GM foods consumption could result in diseases resistant to antibiotic treatment. These technologies also can introduce unwanted compositional alterations that might harm human health (Ahmad et al., 2021). People choose to avoid GM foods as little is known about their potential risks and concerns regarding human safety. Many cultural and religious groups have opposed the new biotechnological approaches, and they consider it an unethical way of food production.

There are a variety of risks associated with plants which would be primarily determined by the new traits incorporated into the organism. It could be the production of new allergens or toxins in the food or increased antibiotic resistance. Some of the newly introduced genes can also result in heavy metal accumulation in the plant tissue. These technologies can also cause potential environmental harm, such as increased weediness and gene transfer to wild relatives. There must be a risk assessment for various biotechnological approaches to determine their potential harms (Giraldo et al., 2019).

REFERENCES

Acquaah, G. Conventional Plant Breeding Principles and Techniques. In *Advances in Plant Breeding Strategies: Breeding, Biotechnology and Molecular Tools*; Springer, **2015**; pp. 115–158.

Ahmad, N.; Raza, G.; Waheed, T.; Mukhtar, Z. Food Safety Issues and Challenges of GM Crops. *Policy Issues Genet. Modif. Crop*. **2021**, 355–369.

Allard, R.W. *Principles of Plant Breeding*; John Wiley & Sons, **1960**; p. 485.

Amom, T.; Nongdam, P. The Use of Molecular Marker Methods in Plants: A Review. *Int. J. Curr. Res. Rev.* **2017**, *9*, 1–7.

Ashraf, M.; Akram, N.A.; Foolad, M.R. Marker-Assisted Selection in Plant Breeding for Salinity Tolerance. *Plant Salt Toler. Methods Protoc.* **2012**, 305–333.

Baenziger, P.S.; Russell, W.K.; Graef, G.L.; Campbell, B.T. Improving Lives: 50 Years of Crop Breeding, Genetics, and Cytology (C-1). *Crop Sci.* **2006**, *46*, 2230–2244.

Boukar, O.; Fatokun, C.A.; Huynh, B.-L.; Roberts, P.A.; Close, T.J. Genomic Tools in Cowpea Breeding Programs: Status and Perspectives. *Front. Plant Sci.* **2016**, *7*, 757.

Briggs, F.N.; Allard, R.W. The Current Status of the Backcross Method of Plant Breeding 1. *Agron. J.* **1953**, *45*, 131–138.

Brown, J.; Caligari, P. *An Introduction to Plant Breeding*; John Wiley & Sons, **2011**; ISBN 1444357700.

Burow, G.; Chopra, R.; Hughes, H.; Xin, Z.; Burke, J. Marker Assisted Selection in Sorghum Using KASP Assay for the Detection of Single Nucleotide Polymorphism/Insertion Deletion. In *Sorghum*; Springer, **2019**; pp. 75–84.

Caradus, J.R. Intended and Unintended Consequences of Genetically Modified Crops—Myth, Fact and/or Manageable Outcomes? *New Zeal. J. Agric. Res.* **2022**, 1–101.

Carroll, D. Genome Engineering with Zinc-Finger Nucleases. *Genetics*. **2011**, *188*, 773–782.

Ceasar, S.A.; Ignacimuthu, S. Genetic Engineering of Millets: Current Status and Future Prospects. *Biotechnol. Lett.* **2009**, *31*, 779–788.

Chesnokov, Y. V; Kosolapov, V.M.; Savchenko, I.V. Morphological Genetic Markers in Plants. *Russ. J. Genet.* **2020**, *56*, 1406–1415.

Collard, B.C.Y.; Jahufer, M.Z.Z.; Brouwer, J.B.; Pang, E.C.K. An Introduction to Markers, Quantitative Trait Loci (QTL) Mapping and Marker-Assisted Selection for Crop Improvement: The Basic Concepts. *Euphytica* **2005**, *142*, 169–196.

Collard, B.C.Y.; Mackill, D.J. Marker-Assisted Selection: An Approach for Precision Plant Breeding in the Twenty-First Century. *Philos. Trans. R. Soc. B Biol. Sci.* **2008**, *363*, 557–572.

Comino, I.; de Lourdes Moreno, M.; Sousa, C. Role of Oats in Celiac Disease. *World J. Gastroenterol.* **2015**, *21*, 11825.

da Silva, A.C.; de Freitas Barbosa, M.; da Silva, P.B.; de Oliveira, J.P.; da Silva, T.L.; Junior, D.L.T.; de Moura Rocha, M. Health Benefits and Industrial Applications of Functional Cowpea Seed Proteins. *Grain Seed Proteins Funct.* **2021**, 1–12.

Dias-Martins, A.M.; Pessanha, K.L.F.; Pacheco, S.; Rodrigues, J.A.S.; Carvalho, C.W.P. Potential Use of Pearl Millet (Pennisetum Glaucum (L.) R. Br.) in Brazil: Food Security, Processing, Health Benefits and Nutritional Products. *Food Res. Int.* **2018**, *109*, 175–186.

Elkonin, L.; Italyanskaya, J.; Panin, V. Genetic Modification of Sorghum for Improved Nutritional Value: State of the Problem and Current Approaches. *J. Investig. Genom* **2018**, *5*, 39–48.

Ezezika, O.C.; Daar, A.S. Overcoming Barriers to Trust in Agricultural Biotechnology Projects: A Case Study of Bt Cowpea in Nigeria. *Agric. Food Secur.* **2012**, *1*, 1–8.

Gedil, M.; Menkir, A. An Integrated Molecular and Conventional Breeding Scheme for Enhancing Genetic Gain in Maize in Africa. *Front. Plant Sci.* **2019**, *10*, 1430.

George, M.L.C.; Prasanna, B.M.; Rathore, R.S.; Setty, T.A.S.; Kasim, F.; Azrai, M.; Vasal, S.; Balla, O.; Hautea, D.; Canama, A. Identification of QTLs Conferring Resistance to Downy Mildews of Maize in Asia. *Theor. Appl. Genet.* **2003**, *107*, 544–551.

Giraldo, P.A.; Shinozuka, H.; Spangenberg, G.C.; Cogan, N.O.I.; Smith, K.F. Safety Assessment of Genetically Modified Feed: Is There Any Difference from Food? *Front. Plant Sci.* **2019**, *10*, 1592.

Gulia, S.K.; Wilson, J.P.; Carter, J.; Singh, B.P. Progress in Grain Pearl Millet Research and Market Development. *Issues New Crop. New Uses.* **2007**, 196–203.

Hasan, M.; Rima, R. Genetic Engineering to Improve Essential and Conditionally Essential Amino Acids in Maize: Transporter Engineering as a Reference. *Transgenic Res.* **2021**, *30*, 207–220.

Huma, B.; Hussain, M.; Ning, C.; Yuesuo, Y. Human Benefits from Maize. *Sch. J. Appl. Sci. Res.* **2019**, *2*, 4–7.

Ijaz, R. Role of Good Quality Fodder in Animal Production. *Agrospheres.* **2021**, *2*, 10–12.

Ismail, A.; Mosa, K.A.; Ali, M.A.; Helmy, M. Biochemical and Molecular Markers: Unraveling Their Potential Role in Screening Germplasm for Thermotolerance. *Heat Stress Toler. Plants Physiol. Mol. Genet. Perspect.* **2020**, 47–76.

Jackai, L.E.N.; Daoust, R.A. Insect Pests of Cowpeas. *Annu. Rev. Entomol.* **1986**, *31*, 95–119.

Ji, J.; Zhang, C.; Sun, Z.; Wang, L.; Duanmu, D.; Fan, Q. Genome Editing in Cowpea Vigna Unguiculata Using CRISPR-Cas9. *Int. J. Mol. Sci.* **2019**, *20*, 2471.

Jordan, D.; Tao, Y.; Godwin, I.; Henzell, R.; Cooper, M.; McIntyre, C. Prediction of Hybrid Performance in Grain Sorghum Using RFLP Markers. *Theor. Appl. Genet.* **2003**, *106*, 559–567.

Joung, J.K.; Sander, J.D. TALENs: A Widely Applicable Technology for Targeted Genome Editing. *Nat. Rev. Mol. cell Biol.* **2013**, *14*, 49–55.

Kamthan, A.; Chaudhuri, A.; Kamthan, M.; Datta, A. Genetically Modified (GM) Crops: Milestones and New Advances in Crop Improvement. *Theor. Appl. Genet.* **2016**, *129*, 1639–1655.

Khanum, S.; Siddique, M.; Kamal, N.; Khanam, B. Study of Correlation, Heritability and Genetic Advance in Pearl Millet. *Cross. Current. Int. J. Agri. Vet. Sci.* **2020**, *2*, 47–60.

Kordrostami, M.; Rahimi, M. Molecular Markers in Plants: Concepts and Applications. *Genet. 3rd Millenn.* **2015**, *13*, 4024–4031.

Kumawat, G.; Kumawat, C.K.; Chandra, K.; Pandey, S.; Chand, S.; Mishra, U.N.; Lenka, D.; Sharma, R. Insights into Marker Assisted Selection and Its Applications in Plant Breeding. In *Plant Breeding-Current and Future Views*; IntechOpen, **2020**; ISBN 1839683104

Kuravadi, N.A.; Verma, S.; Pareek, S.; Gahlot, P.; Kumari, S.; Tanwar, U.K.; Bhatele, P.; Choudhary, M.; Gill, K.S.; Pruthi, V. Guar: An Industrial Crop from Marginal Farms. *Agric. Sustain. Prog. Prospect. Crop Res.* **2013**, 47–60.

Lema, M. Marker Assisted Selection in Comparison to Conventional Plant Breeding. *Agric Res Technol.* **2018**, *14*, 555914.

Liu, G.; Li, J.; Godwin, I.D. Genome Editing by CRISPR/Cas9 in Sorghum through Biolistic Bombardment. In *Sorghum*; Springer, **2019**; pp. 169–183.

MacRobert, J.F.; Setimela, P.S.; Gethi, J.; Regasa, M.W. *Maize Hybrid Seed Production Manual*; CIMMYT, **2014**; pp. 1–36.

Malik, V.S. Use of Genome Editing Technologies for Improving Castor Bean and Guar. *J. Plant Biochem. Biotechnol.* **2013**, *22*, 357–358.

Manès, Y.; Gomez, H.F.; Puhl, L.; Reynolds, M.; Braun, H.J.; Trethowan, R. Genetic Yield Gains of the CIMMYT International Semi-arid Wheat Yield Trials from 1994 to 2010. *Crop Sci.* **2012**, *52*, 1543–1552.

Matres, J.M.; Hilscher, J.; Datta, A.; Armario-Nájera, V.; Baysal, C.; He, W.; Huang, X.; Zhu, C.; Valizadeh-Kamran, R.; Trijatmiko, K.R. Genome Editing in Cereal Crops: An Overview. *Transgenic Res.* **2021**, *30*, 461–498.

McGinnis, M.J.; Painter, J.E. Sorghum: History, Use, and Health Benefits. *Nutr. Today.* **2020**, *55*, 38–44.

Mudgil, D.; Barak, S.; Khatkar, B.S. Guar Gum: Processing, Properties and Food Applications—a Review. *J. Food Sci. Technol.* **2014**, *51*, 409–418.

Myers, G.O.; Fatokun, C.A.; Young, N.D. RFLP Mapping of an Aphid Resistance Gene in Cowpea (Vigna Unguiculata L. Walp). *Euphytica.* **1996**, *91*, 181–187.

Nadeem, M.A.; Nawaz, M.A.; Shahid, M.Q.; Doğan, Y.; Comertpay, G.; Yıldız, M.; Hatipoğlu, R.; Ahmad, F.; Alsaleh, A.; Labhane, N. DNA Molecular Markers in Plant Breeding: Current Status and Recent Advancements in Genomic Selection and Genome Editing. *Biotechnol. Biotechnol. Equip.* **2018**, *32*, 261–285.

Nagamine, A.; Ezura, H. Genome Editing for Improving Crop Nutrition. *Front. Genome Ed.* **2022**, *4*.

Omoigui, L.O.; Kamara, A.Y.; Moukoumbi, Y.D.; Ogunkanmi, L.A.; Timko, M.P. Breeding Cowpea for Resistance to Striga Gesnerioides in the Nigerian Dry Savannas Using Marker-Assisted Selection. *Plant Breed.* **2017**, *136*, 393–399.

Orhun, G.E. Maize for Life. *Int. J. Food Sci. Nutr. Eng.* **2013**, *3*, 13–16.

Pajić, Z. Breeding of Maize Types with Specific Traits at the Maize Research Institute, Zemun Polje. *Genetika*. **2007**, *39*, 169–180.

Paterson, A.H. Making Genetic Maps. *Genome Mapp. Plants*. **1996**, *194*, 23–39.

Pati, D.; Kesh, R.; Mohanta, V.; Pudake, R.N.; Sevanthi, A.M.; Sahu, B.B. Genome-Editing Approaches for Abiotic Stress Tolerance in Small Millets. In *Omics of Climate Resilient Small Millets*; Springer, **2022**; pp. 259–273.

Pawar, M.M.; Ashwar, B.K.; Joshi, P.C.; Patil, S.S.; Madhavatar, M.P.; Thakkar, N.K.; Patel, J. V; Gupta, J.P. Constraints Perceived about Fodder Production by the Dairy Farmers of North Gujarat. *Indian J. Dairy Sci*. **2019**, *72*, 565–568.

Poehlman, J.M.; Sleper, D.A. *Breeding Field Crops*. 4th ed. Iowa State University Press, **1995**; p. 494.

Rakshit, S.; Bellundagi, A. Conventional Breeding Techniques in Sorghum. In *Breeding Sorghum for Diverse End Uses*; Elsevier, **2019**; pp. 77–91.

Rama Reddy, N.R.; Ragimasalawada, M.; Sabbavarapu, M.M.; Nadoor, S.; Patil, J.V. Detection and Validation of Stay-Green QTL in Post-Rainy Sorghum Involving Widely Adapted Cultivar, M35-1 and a Popular Stay-Green Genotype B35. *BMC Genomics*. **2014**, *15*, 1–16.

Raman, R. The Impact of Genetically Modified (GM) Crops in Modern Agriculture: A Review. *GM Crops Food*. **2017**, *8*, 195–208.

Read, D.; Zealand, E.N. *Use of Antibiotic Resistance Marker Genes in Genetically Modified Organisms*; Citeseer, **2000**; ISBN 0478215150.

Rehman, A.; Jingdong, L.; Shahzad, B.; Chandio, A.A.; Hussain, I.; Nabi, G.; Iqbal, M.S. Economic Perspectives of Major Field Crops of Pakistan: An Empirical Study. *Pacific Sci. Rev. Humanit. Soc. Sci*. **2015**, *1*, 145–158.

Rouf Shah, T.; Prasad, K.; Kumar, P. Maize—A Potential Source of Human Nutrition and Health: A Review. *Cogent Food Agric*. **2016**, *2*, 1166995.

Sang, S.; Chu, Y. Whole Grain Oats, More Than Just a Fiber: Role of Unique Phytochemicals. *Mol. Nutr. Food Res*. **2017**, *61*, 1600715.

Şekerci, A.D.; Gülşen, O. Overview of Dahlia Breeding. *Sci. Pap. B Hortic*. **2016**, *60*, 199–204.

Shah, T.; Andleeb, T.; Lateef, S.; Noor, M.A. Genome Editing in Plants: Advancing Crop Transformation and Overview of Tools. *Plant Physiol. Biochem*. **2018**, *131*, 12–21.

Shan, Q.; Wang, Y.; Li, J.; Zhang, Y.; Chen, K.; Liang, Z.; Zhang, K.; Liu, J.; Xi, J.J.; Qiu, J.-L. Targeted Genome Modification of Crop Plants Using a CRISPR-Cas System. *Nat. Biotechnol*. **2013**, *31*, 686–688.

Singh, B.D.; Singh, A.K.; Singh, B.D.; Singh, A.K. Polymerase Chain Reaction-Based Markers. *Marker-Assisted Plant Breed. Princ. Pract*. **2015**, 47–75.

Singru, R.; Sivaramakrishnan, S.; Thakur, R.P.; Gupta, V.S.; Ranjekar, P.K. Detection of Genetic Variability in Pearl Millet Downy Mildew (*Sclerospora graminicola*) by AFLP. *Biochem. Genet*. **2003**, *41*, 361–374.

Srivastava, R.K.; Yadav, O.P.; Kaliamoorthy, S.; Gupta, S.K.; Serba, D.D.; Choudhary, S.; Govindaraj, M.; Kholová, J.; Murugesan, T.; Satyavathi, C.T. Breeding Drought-Tolerant Pearl Millet Using Conventional and Genomic Approaches: Achievements and Prospects. *Front. Plant Sci*. **2022**, *13*, Art-781524.

Stetter, M.G.; Zeitler, L.; Steinhaus, A.; Kroener, K.; Biljecki, M.; Schmid, K.J. Crossing Methods and Cultivation Conditions for Rapid Production of Segregating Populations in Three Grain *Amaranth* Species. *Front. Plant Sci*. **2016**, *7*, 816.

Tariq, M.; Iqbal, H. Maize in Pakistan–an Overview. *Agric. Nat. Resour*. **2010**, *44*, 757–763.

Tonapi, V.A.; Talwar, H.S.; Are, A.K.; Bhat, B.V.; Reddy, C.R.; Dalton, T.J. *Sorghum in the 21st Century: Food, Fodder, Feed, Fuel for a Rapidly Changing World*; Springer, **2020**; ISBN 9811582483.

Vogel, K.E. Backcross Breeding. In *Transgenic Maize*; Springer, **2009**; pp. 161–169.

Wilkins, P.W. Breeding Perennial Ryegrass for Agriculture. *Euphytica*. **1991**, *52*, 201–214.

Wu, X.; Liu, Y.; Zhang, Y.; Gu, R. Advances in Research on the Mechanism of Heterosis in Plants. *Front. Plant Sci*. **2021**, 2124.

Xiong, Y.; Zhang, P.; Warner, R.D.; Fang, Z. Sorghum Grain: From Genotype, Nutrition, and Phenolic Profile to Its Health Benefits and Food Applications. *Compr. Rev. Food Sci. Food Saf*. **2019**, *18*, 2025–2046.

Zehr, B.E.; Dudley, J.W.; Chojecki, J.; Saghai Maroof, M.A.; Mowers, R.P. Use of RFLP Markers to Search for Alleles in a Maize Population for Improvement of an Elite Hybrid. *Theor. Appl. Genet*. **1992**, *83*, 903–911.

2 Nutritional Profiling of Summer Fodder Crops

Saima, Usman Ali, Muhammad Abdullah, Usama Ahmad, Jamshaid Ahmad, and Manzoom Akhtar

2.1 INTRODUCTION

Globally, livestock is one of the fastest-progressing agricultural subsectors vital to human food chains and livelihoods (Robinson et al., 2014). By 2050, rapid growth in the human population is expected, which will lead the intensified interspecies competition for food due to the same food chains in generations. Developing nations are expected to have higher growth rates than developed ones which might result in an inadequate food supply. However, ethnobotanical knowledge of indigenous resources can help better plan their utilization (Thappa et al., 1997; Geng et al., 2017).

In Pakistan, the people mostly raise their livestock on fodders, but the availability of the desired fodder depends on its productivity, which is largely dependent on soil and climatic conditions of the region and the socioeconomic status of the farmers.

2.2 FODDER: A SOURCE OF NUTRITION

Forages, shrubs, and agricultural waste nourish domesticated animals. Animals are fed either by pasturing or through fodders and grains. Generally, "fodder is any agricultural feedstuff offered to domesticated animals to fulfill their nutritional requirements" (Saima et al., 2021). This is the most valued and effective source of feeding, providing the necessary nutrients to animals for different physiological functions (Mishra and Pathak, 2015).

2.3 QUALITIES OF FODDER

The fodder that provides optimum animal growth and production is called excellent quality fodder. The quality of fodder varies due to different factors like the level and stage of maturity of the plant, the nutritional composition of the soil, the irrigation system, and the climatic conditions of the region. The animal's response to the feed is one of the best indicators of its quality. The growth, production, and reproduction of animals are highly dependent on the consumption of the type of fodder. Thus, only good-quality fodder can assure a high yield of animals. Therefore, fodders with the following characteristics should be fed to farm animals for good earnings:

- The fodder should be palatable and have high digestibility.
- Fodder should have a sustainable level of digestible crude protein to meet the animal body's requirements.
- The presence of high mineral content and lack of anti-nutritional substances in the fodder for optimum animal growth and performance.
- Dry matter of fodder should be in the desired amount to fulfill the nutritional requirements of animals.

DOI: 10.1201/b23394-2

2.4 CLASSIFICATION OF FODDERS

Fodders can be classified on a different basis. Some of the major categories are as follows:

1. **Based on family**
   - Poaceae: wheat, maize, jawar, and bajra
   - Leguminosae/Fabaceae: cowpea, lucerne
   - Cruciferae: Chinese cabbage, Japanese rape
2. **Based on maintenance**
   - Maintenance crops include maize, bajra, and sorghum
   - Non-maintenance crops include wheat, rice, straw, sorghum, and maize stover
3. **Based on protein content**
   - Cereals and grasses are categorized into low-protein forage crops
   - Legumes are categorized into high-protein forage crops
4. **Based on season**
   - Rabi crops are winter crops
   - Kharif crops are summer crops
5. **Based on origin**
   - Indigenous species include marvel grass and anjan grass
   - Nonindigenous species include signal grass and timothy grass
6. **Based on the life cycle**
   - Annual or seasonal
     - Legumes: cowpea, berseem
     - Non-legumes: maize, sorghum
   - Perennial forage crops
     - Non-legumes include guinea grass
     - Legumes include subabul and lucerne
7. **Based on habit**
   - Forage crops are broadly categorized into herbs, shrubs, and trees
8. **Based on habitat**

   Following are the categories of fodders based on habitat:
   - Cultivated fodders
   - Wasteland fodders
   - Marshland fodders
   - Aquatic fodders

For animal feeding, we usually use the fodder classification *based on family* or the *basis of the life cycle.*

2.5 CLASSIFICATION OF FODDER CROPS BASED ON FAMILY

The crops are categorized as leguminous (all from the Leguminosae/Fabaceae family) and non-leguminous (crops other than the Leguminosae/Fabaceae family). Some of the basic differences between leguminous and non-leguminous fodders are tabulated in Table 2.1.

2.6 CLASSIFICATION OF FODDER CROPS BASED ON LIFE CYCLE

In Pakistan, domestic fodder (forage) is usually categorized based on the life cycle of plants, that is, annual or seasonal.

TABLE 2.1
Comparison between Leguminous and Non-Leguminous Fodder

Leguminous Fodder	Non-Leguminous Fodder
Nitrogen-fixing bacteria are from the genus *Rhizobium*.	Nitrogen-fixing bacteria are from the genus *Frankia*.
The protein content is high in leguminous fodder.	The carbohydrate content is high in non-leguminous fodder.
Leaves are compound and pinnate.	Leaves can be simple or compound.
These are flowering plants that belong to the family Fabaceae.	These are flowering plants that belong to different classes apart from Fabaceae.
Fruit is a legume or pod.	Fruit is of different types.
This is the best for haymaking.	This is the best for silage making.
Examples are cowpeas, guar, and moth.	Examples are maize, sorghum, and millet.

The annual type is *perennial fodder*. Herbaceous plants that have multi-cut, and these fodders are planted once they become dormant but can recover from rhizomes, tubers, or stolon and have a life span of more than one year. These include ryegrass, mott grass, and lucerne.

The seasonal fodder crops are divided into two cropping seasons, that is, the **kharif/monsoon crops**, as the first germinating season begins in April-June, and harvesting is done in October-December. Typical examples of Kharif fodder are cotton, maize, sorghum, rice, sugarcane, bajra, jowar, and Rhodes grass. The "rabi/winter crops" as the second germinating season starts in October–December and harvesting is done in April–May. Typical examples of rabi fodder are lucerne, berseem, wheat, rapeseed, oats, mustard, and lentil (masoor).

2.6.1 Fodder Plants

Several plants are grown for fodder, but some of the most common plants are the following:

- Millet
- Sorghum
- Maize
- Sadabahar
- Cowpeas
- Guar
- Moth
- Jantar

Some of the essential fodders used for feeding livestock in Pakistan are tabulated along with their germination and harvesting time in Table 2.2.

2.7 SORGHUM (*Sorghum bicolor* L.)

Sorghum is the fifth-most important cereal crop and is a major grain crop for feeding animals. It is a crop of the summer season and is used as forage as well as grains. A prominent feature of this crop is that it can tolerate heat and drought conditions, mainly due to its efficient root system. That is why it can be cultivated in both rain-fed and irrigated areas. It can be used in feeding forms like hay silage, green chop, and crop residues. It provides an excellent forage source in dryland areas. It has four desirable qualities, that is, high dry-matter yield, use of water and efficient use of

TABLE 2.2
Cultivation and Harvesting Time of Some of the Essential Fodders

Fodder	Cultivation	Harvesting
Maize	Mar–Sep	May–Dec
Sorghum	Mar–Jul	Jun–Nov
Millet	April–Aug	Jun–Nov
Cowpeas	Mar–Jul	Jul–Oct
Guar	Apr–Jul	Jun–Nov

nitrogen source, nutritious and juicy, and the ability to withstand high temperatures and minimal soil moisture requirement. It is used in a variety of beverages after going through fermentation, malting, and other processes. It can be used in making hay or silage to feed animals during the fodder shortage season.

2.7.1 Varieties

The different varieties of sorghum are sorghum-2011, JS.2002, hegari, JS-263.

2.7.2 Sowing Time

Sowing time for fodder is March–August, and seed is sowed in July for all the varieties.

2.7.3 Harvesting

Sorghum yield per hectare is about 20 tons, but up to 75 tons per hectare has also been reported depending on optimal growth conditions (Balole and Legwaila, 2006). Sorghum can be harvested at the medium dough stage when ensiling is the primary objective. A height of 80 cm is desirable when hay formation is objective because it is easy to be cured at that stage.

2.7.4 Qualities of Soil

Soils having a pH of 5.5–7.5, loamy clay properties, and suitable water drainage properties are best suited for their cultivation. Areas of the world having an average annual rainfall of less than 400–750 mm are considered suitable for its cultivation.

2.7.5 Salient Features

- Tall height
- Low hydrocyanic acid (HCN) content
- High-yielding fodder
- Long duration and wider adaptability

2.7.6 Nutritional Aspects

The plant's nutrient composition depends on its maturity stage, climatic conditions, variety, and other factors (Table 2.3).

TABLE 2.3
Nutritional Composition of Sorghum

Nutrients	Percentage
Dry Matter (as feed)	11.82–38.19
Crude Protein (DM)	7.0–8.5
Ash (DM)	6.8–7.2
Ether Extract (DM)	1.5–2.0
Crude Fiber (DM)	30–32
NDF	55–57.9
ADF	33–35

Source: Aguiar et al. (2006).
Note: DM = Dry Matter; NDF = Neutral Detergent Fiber; ADF = Acid Detergent Fiber.

2.8 MILLET (*Pennisetum glaucum* L.)

These are the group of small-seeded annual grasses that are mainly grown in the Asia and African regions. It is a very important summer crop and can be harvested in irrigated and rain-fed areas. It has a higher tolerance than other crops regarding a low nutrient profile, drought tolerance, and temperature fluctuations. African countries account for 60% of the global area and 55% of the worldwide millet production. Asian countries are the second-most producers, counting for 39% of the worldwide area and 42% of the global output. It is a highly nutritious, high-energy food and is consumed as food in most developing countries. It is also used for bird feeding but only in developing countries. Millet fodder is an important feed resource, particularly in Africa and Asia, in the post–monsoon season when other feed resources are less available.

2.8.1 Varieties Name

Sgd. Bajra 2011 and MB-87 are its two varieties.

2.8.2 Sowing Time

The sowing time for millet crops is April–August.

2.8.3 Harvesting

Harvesting is usually performed in July.

2.8.4 Qualities of Soil

These can survive in high saline soils and maintain good nutritive value.

2.8.5 Salient Features

- Tall
- Single cut

- Much resistant to lodging
- Long duration and tolerant to insects and pests
- Better digestibility and palatability

2.8.6 Nutritional Aspects

The pearl millet is quite palatable, but the nutritive value may vary according to the environmental conditions and variety of plants (Table 2.4).

TABLE 2.4
Nutritional Composition of Pearl Millet

Nutrients	Percentage
Dry Matter (as feed)	14.1–28.5
Crude Protein (DM)	6.6–17.0
Ash (DM)	9.2–15.8
Ether Extract (DM)	1.4–3.0
Crude Fiber (DM)	21.1–39.7
NDF	46.1–64.8
ADF	30.7–45.1

Source: Aguiar et al. (2006).
Note: DM = Dry Matter; NDF = Neutral Detergent Fiber; ADF = Acid Detergent Fiber

2.9 MAIZE (*Zea mays* L.)

The maize crop is used both for fodder and seed purposes. It is an important crop of the summer season, especially for milch animals. It has the potential to overcome shortages during lean periods. It is also named "gacha" and can be preserved as silage. Its grains are also a rich carbohydrate source and are used as feed for birds. Out of the total maize production, about 60% is used in poultry feeds, 25% in industries, and the remaining part is used for food for animals and humans. Maize is ranked as the third major crop in Pakistan after wheat and rice in all provinces of Pakistan.

2.9.1 Varieties

Super green maize and sgd.2002 are the two varieties of maize used in Pakistan.

2.9.2 Sowing Time

It is cultivated two times a year: mid-February–mid-March and then July–August.

2.9.3 Harvesting

Maize fodder is harvested when most of the cobs are immature or milky. Harvesting the crop at the immature cob stage is desirable because of its shorter duration of plantation and harvesting

(55–66 days). Apart from the shorter time frame, the immature cob stage yields crop residues (stalks, leaves, husks) succulent and highly palatable to livestock. The harvested fodder can either be used to feed livestock or stored as silage.

2.9.4 Potential Yield (Maund/Acre)

The estimated yields for fresh green fodder range between 10 and 50 t/ha.

2.9.5 Qualities of Soil

Soils with a good water supply and optimum fertility are best suited for their growth. Cobs at the milky stage are considered the best stage for its harvesting. Maize fodder crops can yield up to 40–50 t/ha.

2.9.6 Salient Features

- Long duration
- High yield of fodder
- Broad leaves
- Wider adaptability

2.9.7 Nutritional Aspects

The nutrient value of maize fodder depends on the harvesting stage and part of the plant being evaluated. The leaves of the plant are higher in protein content than the whole plant. Organic matter digestibility of 71–72% can be anticipated from flowering to the grain stage (Table 2.5).

TABLE 2.5
Nutritional Composition of Maize

Nutrients	Percentage
Dry Matter (as feed)	12.6–47.6
Crude Protein (DM)	3.0–12.8
Ash (DM)	2.5–11.8
Ether Extract (DM)	0.7–3.1
Crude Fiber (DM)	19.1–36.6
NDF	54.7–81.5
ADF	23.3–45.7

Source: AFZ (2011).
Note: DM = Dry Matter; NDF = Neutral Detergent Fiber; ADF = Acid Detergent Fiber.

2.10 COWPEAS (*Vigna unguiculata* L.)

Cowpea is one of the important forage crops in food legumes and traditional cropping systems, especially in semiarid areas that include Asia, Africa, and Central and South America. Its dual-purpose character makes it an attractive fodder for humans and livestock. Due to its prominent feature of drought tolerance and warm weather, it is mostly cultivated in semiarid regions. This crop has the unique ability to fix nitrogen through its nodules and grows well in poor soils with more than

85% sand, less organic matter, and low levels of phosphorus. Along with these attributes, its rapid growth and ground covering check soil erosion and produce nitrogen-rich residues that increase soil fertility. It is a crop of the summer season and greatly beneficial for milch animals. It is used as intermediate fodder when animals are fed maize, lucerne, berseem, and sorghum. This crop is very much tolerant to the dry climate, which is why mostly grown in barani areas.

2.10.1 Variety

Rawan-2003 is the variety of cowpea forage crops.

2.10.2 Sowing Time

It is cultivated in the months of March–July.

2.10.3 Harvesting

Cowpeas can produce approximately 0.5–4.0 tons of dry matter per hectare. The yield depends mainly on environmental conditions, soil type, and water availability. The intercropping system has also yielded encouraging results. Most of the time, it is intercropped with cereals, especially maize, sorghum, and millet.

2.10.4 Qualities of Soil

Soils with sandy, light, and heavier textures with better drainage are well suited for cultivating cowpeas.

2.10.5 Salient Features

- More protein and wider adaptability
- Lengthy wine and thick stem

2.10.6 Nutritional Aspects

Cowpeas are good quality forage and rich in some essential nutrients (Table 2.6).

TABLE 2.6
Nutritional Composition of Cowpeas

Nutrients	Percentage
Dry Matter (as feed)	11.1–26.4
Crude Protein (DM)	13.5–24.3
Ash (DM)	8.1–14.4
Ether Extract (DM)	1.3–4.1
Crude Fiber (DM)	11.5–35.9
NDF	28.4–55.0
ADF	17.8–40.4

Source: Heinritz et al. (2012).
Note: DM = Dry Matter; NDF = Neutral Detergent Fiber; ADF = Acid Detergent Fiber.

2.11 GUAR (*Cyamopsis tetragonoloba* L.)

Guar is a summer season crop commercially grown in India, Pakistan, the US, and some other countries as seed, fodder, and a green crop. It is also termed a "cluster bean." It is an important drought crop and is grown in the dryland areas of the country. It is also considered a multipurpose crop because it can be used as a vegetable, green manure, fodder, and feed. Its straw is a very good dry feed for the livestock. Fodder of guar and its grain is very nutritious and rich in protein, fats, and minerals. It also increases soil fertility due to its leguminous nature. Its seeds contain specific gums used in the food and textile industry.

2.11.1 Varieties

BR-2017, BR-99, BR-90 & 2/1 are the common guar varieties in Pakistan and worldwide.

2.11.2 Sowing Time

It is cultivated from April–July.

2.11.3 Harvesting

Between June and November.

2.11.4 Qualities of Soil

Like other legume crops, guar improves the availability of nitrogen in the soil and is also used for reclamation of the high salinity and low fertility of soils (Ecoport, 2010).

2.11.5 Salient Features

- Hairy and erect structure
- Early maturing phase
- High protein and galactomannan content
- Heavy fruiting
- Suitable for vegetable, fodder, and grain purposes
- Single stemmed
- Short stature

2.11.6 Nutritional Aspects

Leguminous fodders generally have higher crude protein and lower fiber. The digestibility of cowpeas depends on the plant part consumed, the crop's age, and environmental alterations. The digestibility ranges from 50–56% when a whole plant is offered to animals. However, digestibility is 60–75% in the case of leafy parts. In the case of grazing, animals usually prefer to consume leafy portions of the plant (Table 2.7).

TABLE 2.7
Nutritional Composition of Guar

Nutrients	Percentage
Dry Matter (as feed)	93.4–96.7
Crude Protein (DM)	40.0–44.3
Ash (DM)	4.7–8.8
Ether Extract (DM)	4.2–7.5
Crude Fiber (DM)	7.2–16.1
NDF	20.5–21.0

Source: AFZ (2011).
Note: DM = Dry Matter; NDF = Neutral Detergent Fiber.

2.12 ANTI-NUTRITIONAL FACTORS IN SUMMER FODDERS

Anti-nutritional factors (ANFs) are phytochemicals produced from the feedstuff by normal metabolism or secondary metabolites, can be formed by the inactivation of nutrients, can have an effect on nutrient and feed utilization, and can interfere with the metabolism, health, and productivity of animals by altering the nutrient intake, digestibility, absorption, and utilization. All anti-quality substances in summer feedstuff are toxic but not lethal to animals, including phytic acid, tannins, trypsin inhibitors, lectins, haemagglutinins, saponins, cyanogenic glycosides, dihydroxyphenylalanine, oxalic acid, and proanthocynidins. These substances have deleterious effects on the compound stomach and monogastric animals when consumed (Kumar et al., 2017; Akande and Fabiyi, 2010; Cheeke and Shull, 1985).

ANFs can be classified based on the utilization of nutrients and their chemical properties.

2.12.1 Utilization of Nutrients

- Substances that affect the digestion or utilization of proteins, for example, protease inhibitors, lectins, saponins, and polyphenolic compounds
- Substances decrease the solubility and utilization of minerals, for example, gossypol, phytic acid, oxalic acid, and glucosinolates
- Substances decrease the absorption of vitamins, for example, antivitamins (A, D, E, K, B1, B6, B12)

2.12.2 Chemical Properties

- Proteins, for example, protease inhibitors, lectins
- Glycosides, for example, saponins, cyanogens, and glucosinolates
- Phenols, for instance, gossypol, tannins
- Miscellaneous, for example, anti-metal and antivitamins (Ramteke et al., 2019)

A description of some of the important ANFs follows.

2.12.2.1 Tannins

Sequin first introduced the term *tannin* in 1976. Tannins are primarily present in guar, cowpeas, millet, sadabahar, and sorghum (2–10%), especially condensed tannin; these are water-soluble polyphenolic compounds that have a molecular weight of more than 500. Hydrolyzable and condensed tannins are two distinct types of tannins. Hydrolyzable tannins are quickly hydrolyzed by enzymes, bases, acids, and water to produce gallotannins and ellagitannins (Ramteke et al., 2019). Condensed tannins are polymers of flavonol (flavonoids) and have a profound capability to decrease nutrient digestibility compared to hydrolyzable tannins, which are toxic for ruminants due to their hydrolysis in the rumen (Smitha Patel et al., 2013). Both types of tannins are widely distributed in nature. Astringents in nature cause a dry sensation in the mouth due to the reduced lubricant action of saliva glycoproteins.

Moreover, tannins decrease mucus production and edema of epithelial cells and damage the gastrointestinal tract (GIT), directly or indirectly promoting tannin absorption and leading to toxicity. Due to phenolic–OH groups, tannins can bind and precipitate protein and various amino acids. The high tannin content in fodder disturbs fiber digestion, especially cellulose, by binding with a bacterial enzyme in the rumen and acting as an inhibitor of protein-digesting enzymes like trypsin, chymotrypsin, amylase, and lipase. The presence of tannin feedstuff decreases the quality of forage protein, feed intake, dietary iron absorption, and feed efficiency (Pooja et al., 2020; Kumar et al., 2017; Smitha et al., 2013; Cheeke, 1995).

2.12.2.2 Cyanogens

A minute quantity of cyanide is present in each summer forage. Cyanogens are cyanogenic glycosides consisting of aglycone cyanide and one or more sugar molecules. Three different types of glucosides present in nature are dhurrin in jowar, millet, sorghum, cowpeas, grasses, and linamarin in pulses, cassava, and linseed. On hydrolysis, cyanogen yields hydrocyanic acid and prussic acid in the cytosol, and this reaction occurs in the presence of an enzyme produced by the same plant. However, sometimes, this reaction takes place in the rumen by microbes. It indicated that compound stomach animals are more prone to the toxicity of cyanogens (CN) than simple stomach animals (Smitha et al., 2013). Because swine and equines can destroy HCN by gastric hydrochloric acid, which is absent in the rumen of ruminants. Normally HCN rapid detoxification occurs in the liver after its absorption in the presence of an enzyme rhodanese, which converts CN into Thiocyanate (SCN). Extra cyanide ions prohibit the activity of cytochrome oxidase enzyme and lead to a block in adenosine triphosphate (ATP) formation; ultimately, death occurs due to lack of energy to cells and tissue. In cattle, 2 mg/kg of the body weight is lethal, whereas in sheep, 4 mg/kg is (Sarah, 2007). Bharani and Deghani (2004) documented that in fodder crops concentration of prussic acid was significantly higher in the first cut than the second, possibly due to increased metabolic activity and environmental temperature degradation of prussic acid that occur during the growth and development process. HCN comprises 5–10% of the molecular weight of CN, and its presence in forages results in the simultaneous release of HCN. HCN concentration must be 0–500 ppm on dry matter (DM) basis. HCN toxicity causes retarded growth and death if consumed in a high amount. Sheep is less susceptible than buffalo and cattle (Pooja et al., 2020; Ramteke et al., 2019).

2.12.2.3 Oxalates

Mostly summer fodder crops (guar, cowpeas, sadabahar, sorghum) have oxalate anti-nutritional content in oxalic acid, while animal-origin feedstuff has a low quantity of oxalate. Mature leaves of plants have high oxalate content as compared to young. In mature plants, oxalates are present in seeds as calcium oxalate and can affect monogastric animals (pig and poultry) and in ruminants (cattle and sheep); oxalate poisoning is characterized by difficult breathing, weakness, depression, syncope, and ultimately death. After entry into the gastrointestinal tract, oxalic acid is produced, which forms a complex with another useful nutrient present in the diet and hinders their availability to the animal body and renders them unabsorbable. Fodder containing a high level of oxalic acid can cause nutritional deficiencies and irritation of intestinal mucosa (Tadele, 2015). Oxalate can form strong chelates with macro-mineral minerals like potassium, sodium, calcium, and magnesium to produce oxalate salts. In the gut, calcium reacts with oxalate to produce insoluble calcium oxalate, which hampers calcium absorption and stimulates bone mineral mobilization to cover hypocalcemia (Rahman and Kawamura., 2011). Reduced calcium absorption through the gut and bone mineral mobilization leads to secondary osteodystrophy fibrosa and hyperparathyroidism (Pooja et al., 2020; Ramteke et al., 2019; Kumar et al., 2017). Some researchers reported that oxalate content is higher in immature plants than in mature plants (Rahman et al., 2006; Marais et al., 1997; Davis, 1981; Jones and Ford, 1972).

2.12.2.4 Saponins

Saponins are surface active nonvolatile compounds with a diverse structure containing polyglycolic aglycone nonpolar molecules with the moiety of steroid triterpenoids (C_{27} or C_{30}) linked with one or more monosaccharides; collectively, they have termed a sapogenin. Saponins are commonly present in guar, millet, cowpeas, and sorghum. Compared to the stem, leaves have two times more saponin content, which declines with maturity. The degradation of saponins occurs in the rumen by microbes, so no growth suppression occurs (Sharma et al., 1969; Molyneux et al., 1980). Feeding of forages containing high saponin content leads to gas accumulation in the rumen by reducing the surface tension of ruminal digesta. They have characteristic foaming and bitter taste. Structural complexity results in various physical, chemical, and biological properties like bitterness and sweetness, frothing and emulsifying, medicinal and pharmaceutical, hemolysis of red blood cells, and antimicrobial and insecticidal activities. Their physiochemical interaction with various macro- and micronutrients in the intestine results in poor absorption and uptake of cholesterol, dietary lipid absorption, fat-soluble vitamins (A and E), and glucose. Saponins have a hypocholesterolemic effect. Toxicity (0.4–0.5%) due to saponins can lead to poor weight gain, poor appetite, retarded growth, reduced feed intake, egg production, and gastroenteritis in monogastric animals (pigs and poultry; Pooja et al., 2020; Ramteke et al., 2019; Kumar et al., 2017).

2.12.2.5 Nitrate

During hot, dry, or drought conditions, the accumulation of nitrates is more due to water deficiency. Summer fodders like sorghum, millet, guar, and sadabhar mostly grow in stressful conditions, so during the early growth period of plants, nitrates are used by the plant for the development of stem and leaves and used by leaves for conversion of nitrates into protein, extra absorbed nitrates are deposited in stem portion of plants, and a low concentration is present in leaves and grains. When ruminants are fed with high-nitrate-containing forages, in the rumen, nitrate is converted into nitrite, and bacteria use nitrite to produce amino acids and proteins (Andrews and Kumar, 1992; Singh et al., 2000). An unusually high concentration of nitrates present in forage leads to partial conversion, and nitrate accumulation occurs. Accumulated nitrate enters the bloodstream through the ruminal wall and binds with hemoglobin, forming methemoglobin that cannot carry oxygen. Blood turns brown-chocolate color rather than the normal bright red. A lack of oxygen leads to the death of animals due to asphyxiation. Forage containing 0–1000 ppm of nitrate is considered safe for animals, and more than 4000 ppm is potentially toxic (Pooja et al., 2020; Ramteke et al., 2019; Kumar et al., 2017).

REFERENCES

AFZ (2011) IO-La Banque de données de Alimentation Animale/French feed database. Association Française de Zootechnie. www.feedbase.com

Aguiar EM, Lima GC, Santos MVF, Carvalho FFR, Guim A, Medeiros HR and Borges AQ (2006) Yield and chemical composition of chopped tropical grass hays. Revista Brasileira de Zootencine 35(6): 2226–2233.

Akande KE and Fabiyi EF (2010) Effect of processing methods on some anti-nutritional factors in legume seeds for poultry feeding. International Journal of Poultry Science 9(10): 996–1001, ISSN 1682–8356.

Andrews DJ and Kumar KA (1992) Pearl millet for food, feed, and forage. Advances in Agronomy 48: 89–139.

Bahrani MJ and Deghani GA (2004) Summer forage sorghum yield, protein and prussic acid contents as affected by plant density and nitrogen topdressing. Journal of Agricultural Science and Technology 6: 73–83.

Balole TV and Legwaila GM (2006) *Sorghum bicolor* (L.) Moench. Record from Protabase. Brink M and Belay G. (eds). PROTA (Plant Resources of Tropical Africa/Ressources végétales de l'Afrique tropicale). Wageningen, Netherlands. www.feedipedia.org/node/379

Cheeke PR (1995) Endogenous toxins and mycotoxins in forage grasses and their effects on livestock. Prated Symposium, Fescue and Other Toxic Grass: Effects on Livestock Production, pp. 909–916.

Cheeke PR and Shull LR (1985) Tannins and Polyphenolic compounds. In: Natural Toxicants in Feeds and Poisonous Plants. Westport, CT: AVI Publishing Company.

Davis AM (1981) The oxalate, tannin, crude fiber, and crude protein composition of young plants of some *Atriplex* species. Journal of Range Management 34: 329–331.

Ecoport (2010) Ecoport database. www.ecoport.org (Accessed on 11 November 2022).

Geng Y, Hu G and Ranjitkar S (2017) Prioritizing fodder species based on traditional knowledge: A case study of mithun (*Bos frontalis*) in Dulongjiang area, Yunnan Province, Southwest China. Journal of Ethnobiology and Ethnomedicine 13: 24.

Heinritz SN, Hoedtke S, Martens SD, Peters M and Zeyner A (2012) Evaluation of ten tropical legume forages for their potential as pig feed supplement. Livestock Research for Rural Development 24(1): 1–3. www.lrrd.org/lrrd24/1/hein24007.htm

Jones RJ and Ford CW (1972) Some factors affecting the oxalate content of the tropical grass (*Setaria sphacelata*) Australian Journal of Experimental Agriculture and Animal Husbandry 12: 400–406.

Kumar B, Tirkey N and Kumar S (2017) Anti-Nutrient in Fodders: A Review. Chemical Science Review and Letter 6(24): 2513–2519.

Marais JP, Barnabas AD and Figenschou DL (1997) Effect of calcium nutrition on the formation of calcium oxalate in kikuyugrass. In: Proceedings of the XVIII International Grassland Congress, Canada, p. 45.

Mishra S and Pathak P (2015) Fodder production and conservation: A potential source of livelihood for women. In: Vasudevan P, Sharma S, et al. (eds), Women, Technology and Development. New Delhi, India: Narosa Publishing House, pp. 136–150.

Molyneux RJ, Steven KL and James LF (1980) Chemistry of toxic range plants. Volatile constituents of broomweed (*Guiterrezia sarothrae*). Journal of Agriculture and Food Chemistry 28: 1332–13333.

Naveed S. Ali U, Haque MN, Abdullah M, Ahmad U, Ahmad J, Akhtar M (2021) Winter fodder opportunities and challenges in livestock. In: Haq M and Ijaz S (eds), Sustainable Winter Fodder: Production, Challenges, and Prospects. London, UK: CRC Press, Taylor & Francis Group, p. 273.

Pooja AP, Arunjith P and Jacob G (2020) Anti-nutritional factors in fodder crops. Agriculture &. Environment 1(4): 71–77.

Rahman MM and Kawamura O (2011) Oxalate accumulation in forage plants: Some agronomic, climatic and genetic aspects. Asian-Australian Journal of Animal Science 24(3): 439–448.

Rahman MM, Niimi M, Ishii Y and Kawamura O (2006) Effects of season, variety and botanical fractions on oxalate content of napiergrass (*Pennisetum purpureum* Schumach). Grassland Science 52(4): 161–166.

Ramteke R, Doneria R and Gendley MK (2019). Antinutritional factors in feed and fodder used for livestock and poultry feeding. Acta Scientific Nutritional Health 3(5): 39–48.

Robinson TP, Wint GW and Conchedda G (2014) Mapping the global distribution of livestock. PLOS One 9(5): 96084–96089.

Sarah R (2007) Prussic acid poisoning in livestock. Prime Facts 417: 1–3. www.dpi.nsw.gov.au/primefacts

Sharma DD, Chandra S and Negi SS (1969) The nutritive value and toxicity of OHI (*Albizzic stipulate* Bovin) treeleaves. Journal of Research Ludhiana 6: 388–393.

Singh ATUS, Tiwans MS and Puri KP (2000) Effect of application method of level of nitrogen fertilizer on nitrate content in oat fodder. Indian Journal of Animal Nutrition 17: 315–319.

Smitha Patel PA, Alagundagi SC and Salakinkop SR (2013) The anti-nutritional factors in forages—A review. Current Biotica 6(4): 516–526, ISSN 0973–4031.

Tadele Y (2015) Important anti-nutritional substances and inherent toxicants of feed. Food Science and Quality Management 36: 40–47.

Thappa B, Walker D and Sinclair F (1997). Indigenous knowledge of the feeding value of tree fodder. Animal Feed Science & Technology 67: 97–114.

3 Aflatoxins of Summer Fodders

Causes and Management

Faiza Hassan, Maheera Khaliq, Muhammad Rehan Sajid, Misbah Ijaz, Muhammad Tahir Mohy-ud-din, and Asad Manzoor

3.1 INTRODUCTION

According to the economic survey of Pakistan 2021–2022, the agriculture sector showed an astonishing increase of 4.40%. The major factor responsible for this growth is high yield and agriculture credit. The crop sector recorded a remarkable growth of 6.58%. Kharif fodder, such as maize, depicted a significant growth of 19.0% during 2021–2022, and its production showed tremendous growth from 8.9 to 10.6 million tonnes. Livestock share in agriculture and gross domestic product (GDP) is 61.89% and 14.04%, respectively (Economic Survey of Pakistan, 2021–2022). However, due to mycotoxins, the production of crops decreased, and it also harms animal health, thus reducing the GDP of Pakistan.

Chemicals that are hazardous and can only be produced by biological organisms are called toxins. Toxins include phytotoxins (from algae), phycotoxins (plants), mycotoxins (fungus), venom (animals), and endotoxins (Madsen, 2005). Mycotoxins are universal adulterants and a massive group of naturally occurring toxins in feed and food. Toxic fungal metabolites like mycotoxins are fed to livestock, pets, and aquaculture animals through contact with diet (i.e., feed). Livestock is usually exposed to different toxins through the feedstuff. Low-molecular-weight natural products, or tiny compounds known as mycotoxins, are created by the filamentous fungus as secondary metabolites. Mycotoxins are toxic secondary metabolic by-products of kingdom fungi, especially by molds (Guchi, 2015).

Mycotoxins include distinctive and heterogenous substances that provoke various harmful effects in humans and animals. Mucotoxigenic fungi are the foremost factor involved in contaminating multiple crops during their pre- and post-harvesting. Mycotoxins usually affect about 25% of the total food supply of the world annually (Ames, 1989). Different factors that affect the intensity of mycotoxin contamination severity include humidity, temperature, excessive moisture, drought, alteration in harvesting practices, and insect infestation. Effects of mycotoxin contamination in domestic animals like poultry and dairy cattle include decreased feed conversion ratio and reproductive ability, retarded growth, resistance to infectious diseases, decrease in vaccine efficacy, and different pathologic effects on the liver, kidney, and other organs. Mycotoxicosis is a disease of animals and humans caused by mycotoxins. Different mycotoxins are involved in the epidemic of diseases in humans. Lesions associated with mycotoxicosis appear in organs like kidneys, liver, central nervous system (CNS), and epithelial tissues of mucous membranes and skin. Some of them are carcinogenic and are potent to humans and animals. The effect of mycotoxins may be acute or chronic. Thus, mycotoxins proposed a major risk to animal and public health (Park and Pohland, 1986).

DOI: 10.1201/b23394-3

3.2 TYPES OF MYCOTOXINS

Ochratoxin A, zearalenone, fumonisins, and aflatoxins are major mycotoxins. A brief description of these mycotoxins follows.

3.2.1 Zearalenone

Zearalenone is phenolic resorcylic acid lactone manufactured by different strains of *Fusarium*, especially by *F. graminearum* and *F. sporotrichiodes*. According to the Agricultural Science and Technology Council, it is a natural contaminant of wheat, corn, oats, barley, hay, and sorghum (CAST, 2003). Its predisposing factors are low temperature and high humidity and present in higher concentration during autumn harvesting of crops. It has an estrogenic effect in different animals (Hidy et al., 1977). Its presence is confirmed by pink on grains contaminated by zearalenone. Zearalenone is ubiquitously present in corn. Usually, the favorable condition for the growth of *Fusarium* species is moist and cool. Zearalenone occurs in a pre-harvesting stage in sorghum, wheat, and corn (HAN et al., 2011). Storing large amounts of grain should be prohibited to avoid fungal contamination. Zearalenone harms poultry and cattle and causes enlargement of the uterus and swelling of the vulva. In ewes, zearalenone (ZEN) causes a decrease in ovulation rate and lengthens the cycle length, and the duration of estrus is increased (Smith et al., 1990). ZEN also decreases egg production. ZEN binds with receptors of estrogen, thus leading to a change in hormonal levels. ZEN is not responsible for mortalities.

3.2.2 Ochratoxin A

Ochratoxin A is a considerable mycotoxin, a fluorescent compound formed chiefly by *Aspergillus ochraceus* and *Penicillium verrucosum* (CAST, 2003). In several geographical areas, it is also produced by some members of *Aspergillus niger* (Tjamos et al., 2004). Ochratoxin is produced in conservation circumstances that are predisposing factors for toxin production and mold growth. It is difficult to observe its presence with the naked eye despite its presence on different commodities. The major species of fungus that produce ochratoxin are differentiated from other species due to their color variation: yellowish tan shows *A. ochraceus* presence, bluish green shows different *Penicillium* species, and black shows *A. niger* presence. For ochratoxin, the visible mold is not observed on grains. The musty odor from different grains may be an important factor for ochratoxin's existence. The presence of "hot spots" in storage can occur, so careful sampling is done. Warm temperatures and high moisture are prerequisites for the contamination of grains in storage conditions. So, the drying and storing of grains in insect-controlled areas should be preferred.

Ochratoxin is primarily nephrotoxic, but in high concentrations, it is also hepatotoxic. It is assumed to be carcinogenic in rats and mice, thus also causing human diseases. Ochratoxin is mainly responsible for Balkan endemic nephropathy disease in humans (Mantle, 2002). Ochratoxin is also present in barley, raisins, soy products, and coffee (CAST, 2003). It may gather in fluids and tissues of animals and human bodies and shows its negative effects due to its slow elimination from the body.

3.2.3 Fumonisins

Nonfluorescent mycotoxins are fumonisins whose important representatives are FB1, FB2, and FB3. *Fusarium verticillioides* and *F. proliferatum* significantly produce them. The difference in the ability of toxin production is dependent on strains. It is mostly in corn but appears on rice and sorghum (CAST, 2003). Drought stress after warm weather and wet weather later in the growing season is considered the predisposing factor for fumonisins growth. Leukoencephalomalacia (white matter softening), a major disease in horses, is caused by Fumonisins (Marasas et al., 1988). Esophageal tumors are one of the disorders associated with fumonisins contamination (Marasas, 1996). Its hepatoxic action is due to interference with sphingolipid metabolism and inhibition of sphingosine synthesis. The absence of fumonisins was observed in cattle milk (Richard et al., 1996). FB1 is the most potent fumonisin and is considered to be carcinogenic. FB1 has tumor-promoting and tumor-initiating activities (Jaskiewicz et al., 1987).

3.2.4 Aflatoxins

Aflatoxins are bifurcation products of polyketide-derived carcinogenic and toxic secondary metabolites. Aflatoxins are primely produced by *A. parasiticus* and *A. flavus*. Seventeen were reported. Major aflatoxins include B_1, B_2, G_1, G_2, M_1, and M_2. Due to the carcinogenic effects of aflatoxin B_1 in animals and humans, it gained considerable importance. The worldwide crop threat is aflatoxin contamination, affecting the country's food, feed, and economic status. The contamination of crops occurs during storing, harvesting, and transportation, leading to mycotoxin production. It eradicates 25% of crops worldwide per year. The name 'Aflatoxin' was given to this mycotoxin in 1960 after observing the involvement of aflatoxin in causing Turkey X disease (Awuchi et al., 2020).

Aflatoxin M1 (AFM1) is a hydroxylated metabolite found in fluids and animal tissues as a biotransformational product of aflatoxin B1 (AFB1). Aflatoxins are mutagenic, teratogenic, carcinogenic, and immunosuppressive. AFM1 is not considered a fungal contaminant. Aflatoxins are commonly present in millet, sesame seed, chili, tree nuts, sorghum, rice, wheat, and sunflower seed crops. The grains are exposed to aflatoxin contamination, especially in drought situations (Shotwell et al., 1974). They also affected the food chain and were observed in animal feedstuff. Aflatoxins are also transferred from fungal-contaminated food to milk, eggs, and meat in humans after consumption (Awuchi et al., 2020).

3.3 OTHER MYCOTOXINS

T2 toxin belongs to trichothecenes, a mycotoxin family primarily produced by *F. sporotrichioides*. It is in corn, barley, rice, and wheat. T2 affects the skin, gastrointestinal, erythroid, and lymphoid cells (Niyo et al., 1988; Richard et al., 1991). Its mechanism of action involves protein synthesis inhibition, thus leads disturbance in RNA and DNA synthesis (Cast, 2003).

Deoxynivalenol (DON), commonly known as vomitoxin, belongs to the Trichothecenes family and is primarily manufactured by *Fusarium* species, especially *F. culmorum* and *F. graminearum* (Richard, 2000). DON has immunosuppressive activity and is nephrotoxic in animals. Vomition syndrome occurs in humans after ingesting DON-contaminated grains (Bhat et al., 1989).

Other mycotoxins like penitrem, cyclopiazonic acid, patulin, and citrinin are pathogenic for humans and animals (Baker et al., 2003).

3.4 IMPORTANCE OF MYCOTOXINS

The effects of mycotoxicosis may be classified as acute or chronic. Reye's syndrome and Kwashiorkor have been linked with aflatoxins (Blunden et al., 1991). Fewer effects on ruminants as compared to nonruminants were observed. Hyperestrogenism is the side effect of zearalenone toxins. Its presence is confirmed by different clinical signs such as infertility, pseudopregnancy, prolonged estrus period, enlargement of the uterus and hyperplasia of mammary glands, stillbirths, decreased libido, and vaginal prolapse (Gupta, 2012). The toxic metabolites of fumonisins (B1 and B2) appear to be associated with pulmonary edema and leukoencephalomalacia (Wilson et al., 1990). It also has a hepatotoxic effect in cattle, horses, primates, and rabbits (Gumprecht et al., 1995). Ochratoxin A is nephrotoxic, teratogenic, carcinogenic, immunotoxic, mutagenic, and genotoxic (Sava et al., 2006).

3.5 FACTORS INVOLVED IN MYCOTOXIN PRODUCTION

Physical, biological, and chemical factors related to the synthesis of mycotoxins. Physical factors comprise environmental conditions like humidity and temperature, which are predisposing factors that favor fungus growth and mycotoxins production. The use of fertilizers and fungicides contributes to chemical factors. Biological factors show the relationship between the substrate and toxic species of fungus (D'Mello et al., 1997).

3.6 CIRCUMSTANCES FOR AFLATOXIN PRODUCTION

The fungi involved in aflatoxin production include *A. nomius*, *A. flavus*, and *A. parasiticus*. These are commonly present in most soils. There are frequently involved in the deterioration of plant materials (Acur et al., 2019). The general conditions required for the growth of aflatoxigenic species are 80% to 85% moisture contents and 13 to 42°C temperature, and optimal growth at 25 to 37°C. For example, for developing *A. flavus* in cereal grains rich in starch, the destructive moisture content is 17% to 18% in peanuts and 9% to 10.5% for soybeans. The higher moisture content limit is 30% for the growth of *A. flavus* and to produce aflatoxins. Below the temperature of 13°C, *A. flavus* growth will be very slow, and at 37°C, it will grow more rapidly. However, there is no production of aflatoxins below 13°C or above 42°C. Factors other than temperature and relative humidity are carbon, plant metabolites, nitrogen, and sugars in the substrates that will influence the production of the aflatoxin. The upper and lower range of temperature and net evaporation is more significant than temperature and humidity alone. Whenever there is the growth of fungus in the seed embryo, more production of aflatoxins will occur. Epigenetics and genetic factors will also affect the production of aflatoxins. The isolates of aflatoxigenic *A. flavus* range from 40% to 70% of the *A. flavus* total population, and this percentage of aflatoxigenic isolates can differ between ingredients and finished feeds. The seeds damaged during the harvesting and handling are more vulnerable to the attack of *Aspergillus* and other fungi. In the damaged seeds with high moisture, the development of aflatoxigenic fungi and aflatoxin production occurs rapidly.

3.7 *IN VIVO* PRODUCTION OF AFLATOXINS

In human beings and animals, *A. flavus* and *A. fumigatus* have been recognized as pathogens. Toxigenic fungi can also produce aflatoxins in tissues. The culturing of *A. flavus* and *A. fumigatus* and isolation of their assays from different tissues have displayed that these fungi can produce aflatoxins, and the presence of aflatoxins was shown by chemical analyses of the infected tissues (Pepeljnjak and Šegvć, 2004). Naturally, aflatoxigenic fungi that infect animals and humans are immunocompromised. Different conditions or factors that are responsible for aflatoxin production are discussed in the following subsections.

3.7.1 Physical Factors

Aflatoxin production mainly depends on water, temperature, atmospheric gases, pH, relative humidity, moisture, and temperature. pH values range from 1.7–9.3 is suitable for mold growth, but 3–7 is the optimum pH for aflatoxins production (Yoshinari et al., 2010). A pH = 7 is favorable for AFG production, while a pH = 5 is favorable for AFB production. Lower pH decreases the incidence of fungal growth, and a pH above 6 is suitable for aflatoxin production (Eshelli et al., 2015). Light is also a prime factor for controlling aflatoxin production. Production of aflatoxin peaks in darkness compared to sunlight (Rushing and Selim, 2019). The optimum relative humidity value is 85%, which is suitable for aflatoxins growth, while its production shows a tremendous increase when relative humidity reaches 95%. High temperatures enhance the growth of ABF, and its value occurs between 12 and 48°C, showing aflatoxin production, but 25–35°C is the optimum temperature. Oxygen and carbon dioxide availability also affect aflatoxin production (Mahbobinejhad et al., 2019).

3.7.2 Nutritional Factors

The presence of substrate and nutritional substances like lipids, amino acids, nitrogen, and carbon affects aflatoxin production. The production of aflatoxins is greatly affected by carbohydrates due to the carbon presence that favors fungal growth (Ma et al., 2014). Aflatoxin production is promoted by sucrose, glycerol, glucose, xylose, and ribose, while sorbose, peptone, and lactose do not affect aflatoxin production (Liu et al., 2016). The level of aflatoxin from *A. flavus* has been increased due to nitrite and nitrate presence. Glutamate, glycine, alanine, magnesium, and zinc metals increase aflatoxin production. The production of Aflatoxin B1 (AFB_1) was increased by using glutamic acid, arginine, aspartic acid, and glycine at 0.5% concentration. Lipids also influenced aflatoxin production (Scarpari et al., 2014).

3.7.3 Biological Factors

The type of fungus is the prime factor involved in aflatoxin production (Kinyungu et al., 2019). Weeds, fungal species, and insect injuries are included in biological factors. Plant stress greatly influences aflatoxin production, and weeds cause plant stress. The output of aflatoxins is also affected by strains of fungus. *A. flavus* produces fewer toxins as compared to *A. parasiticus*. *A. flavus* is mainly related to aflatoxin contamination and growth because it is present in soil and possesses saprobic nature, due to which it can easily grow on field crops, plant debris, piles, and stored grains (Kakde, 2012). Due to the natural inhabitant of soil, *A. flavus* causes aflatoxin production during the pre-harvesting stages of crops.

3.8 ETIOLOGY OF AFLATOXINS

Aflatoxins naturally occur on various matrices like meat, milk, spices, vegetables, fruits, cereals, and oils. Twenty aflatoxins were discovered (Molyneux et al., 2007). The foremost aflatoxins are produced primarily by *A. flavus* and *A. parasiticus*. The chief aflatoxin formed by *A. flavus* is AFB1, whereas its toxicity varies from nontoxic to toxigenic. L strain of *A. flavus* is mostly pathogenic. Variation in toxicity caused by different strains of *A. parasiticus is* not pathogenic and produces different aflatoxins like aflatoxin G1 (AFG1), aflatoxin B2 (AFB2), and aflatoxin G2 (AFG2; Coppock et al., 2018). Some fungus species accountable for producing aflatoxin include *A. pseudotamari*, *A. bombycis*, *A. rambelli*, *A. ochraceoroseus*, *A. nomius*, *Eimericella astellata*, and *E. venezuelensis* (Klich et al., 2000; Mishra and Das, 2003). *A. flavus* and *A. parasiticus* are proved to be pathogenic for both animals and humans (Zain, 2011).

The toxicity level difference is due to the epoxidation of a double bond at the 8,9 position, and their potency is associated with the cyclopentenone ring. Hydroxylated products of AFB1 and AFB2 are AFM1 and aflatoxin M2 (AFM2) (McLean and Dutton, 1995; Wogan, 1966). The order of toxicity is

$$AFB1 > AFG1 > AFB2 > AFG2.$$

Aflatoxin production depends on water activity, substrate composition, temperature, and pH. These conditions favor the growth of different species of *Aspergillus*. Hence, they play an important role in the etiology of aflatoxin (Schmidt-Heydt et al., 2009).

3.9 TYPES OF AFLATOXINS

Aflatoxins are mainly synthesized from various fungi species including *Fusarium*, *Aspergillus*, *Alternaria*, and *Penicillium*. Most toxigenic aflatoxin strains are produced from *Aspergillus parasiticus* and *Aspergillus flavus*. The various types of aflatoxin include AFB1, AFG1, AFM1, AFB2, AFG2, and AFM2. The main aflatoxins in food crops and their by-products are B1, G1, B2, and G2; however, the aflatoxins in animal by-products (dairy products) are M1 and M2. AFB1 and AFB2 are produced due to *A. flavus*. AFG1 and AFG2 are produced due to *A. parasiticus*. Mainly it infects an inclusive range of food possessions including oilseeds (peanut, cotton, sunflower, soybean), cereals (wheat, maize, rice, pearl millet, sorghum), nuts (coconut, Brazil nut, walnut, pistachio, almond), spices (ginger, turmeric, black pepper, coriander, chilies), yam, and different dairy products (Rajarajan et al., 2013). The green appearance is due to *A. flavus* fungus. It can upsurge its population even under full stress (Jeyaramraja et al., 2018).

3.10 AFB1

The chief and foremost harmful aflatoxin for both humans and animals is AFB1. It is cancer-causing because it is associated with hepatocellular carcinoma, ultimately leading to liver cancer. The carcinogenic potential of AFB1 is due to its capability to damage the DNA through the peroxidation of lipids (Zhang et al., 2015). Interfering with the inconstancy of the cells, which are accountable for the uplift of immunity, will overwhelm the immune systems of animals and humans. On the one hand, bulky doses of aflatoxins will result in direct damage and lead to death, while on the other hand, small but long-lasting doses will lead to nutritional and immunologic effects. Because there is an accumulation of aflatoxin, low and high doses will lead to liver cancer. Due to decreased immunization, children's risk of early infection increases, and young children have more chances of developing aflatoxin toxicity.

AFB1 is found in animal feeds, biotransformed in the liver, and transported into milk by AFM1. The CYP450 activates AFB1, which is present in the liver, and converts the activated AFB1 into aflatoxin B1–8,9-epoxide. This aflatoxin B1–8,9-epoxide is accountable for different carcinogenic properties. Other than the cancer-causing effects, it also shows undesirable effects on the heart, testes, brain, kidney, and liver. Due to the lack of food regulation acts, there are more aflatoxin toxicity outbreaks in underdeveloped and developing countries.

The conversion factor of the AFB1 into AFM1 was estimated as 0.3% to 6% in an animal model (Var and Kabak, 2009). Hence, AFB1 level in the feed of dairy animals is positively associated with AFM1 in milk. As milk consumption is higher in infants and aged people with low immunity, different health agencies of the world monitor the existence of AFM1 in milk and dairy products, due to the existence of mycotoxins, especially AFB1, in the feed of animals, the dairy sector suffers from abortion, a decrease in milk production, and the death of animals. Pakistan is among the leading milk producers (38 million tonnes/year), although Pakistan's dairy industry faces huge losses due to aflatoxins in feed. The higher level of toxins in the feed is due to environmental conditions, a lack of awareness, improper feed processing techniques, and ineffective monitoring agencies. Supplying toxin-free feed to different animals is necessary to produce toxin-free milk (Ismail et al., 2017).

3.11 AFLATOXIN B2

Aspergillus flavus fungus has two metabolites, aflatoxins B and G, recently isolated and structure elucidated. The metabolites are associated with animal feed toxicity (Allcroft and Carnaghan, 1963). When culturing was done under laboratory conditions, the fungus produced a multifaceted blend compound of blue and yellowish-green fluorescent. This compound was separated from *A. flavus* culture, which was grown at 30°C for 7 days on sterilized crushed wheat. A nomenclature system was suggested in which the two main components, B and G, would become B1 and G1. The data indicated that aflatoxin B2 is dihydro-aflatoxin B1. AFB2 (isolated from culture) toxic properties were likened to AFB1 by a biological assay procedure (Nesbitt et al., 1962).

3.12 AFM1

AFM1 is stable to the temperature in various milk processing techniques such as pasteurization and ultra-high temperature. It is a principal metabolite of AFB1 hydroxylation formed in the liver. AFM1 is secreted in the milk of human and dairy animals to whom AFB1-dirtied feedstuff was given (Hussain and Anwar, 2008). The conversion of AFB1 in the dirtied feed to the AFM1 is influenced by various factors like breed, milk yield, digestion, diet, and ingestion rate (Duarte et al., 2013). However, about 0.3–6.2% of AFB1 that livestock ingests in the form of contaminated feed was reported to be converted into AFM1 in milk. After the consumption of AFB1-dirtied feed, within 12–24 h, AFM1 in milk appears at a detectable level, and with time, its concentration starts to decrease, and it is undetectable after 72 h (Van Egmond, 1989). AFM1 has the same damage to the liver as AFB1, but the cancer-causing potential is 2–10% of AFB1 (Sun et al., 2002). The carcinogenic potential of AFM1 is ten times less than AFB1; however, due to its higher consumption in children, lower body weight, and less capacity to excrete toxins, it can be a probable health threat to humans. AFM1 can cause DNA damage, abnormalities of the chromosome, a mutation in genes, and the transformation of cells depending on exposure (Van Egmond, 1989). However, milk is well-thought-out as a faultless food for consumers of all ages because of its good nutritious value (increased vitamins, proteins, calcium, antioxidants, etc.). It is considered to have a high potential to transfer AFM1 into the human diet.

Considering the cancer-causing ability of AFM1 and milk and dairy products as a foodstuff in daily diet, most countries set up the maximum acceptable limit of AFM1 in milk. The European Commission set a maximum acceptable limit of 0.05 mg/L AFM1 in milk, while the United States Food and Drug Administration (FDA) set a maximum acceptable limit of 0.5 mg/L AFM1. The United States Food and Drug Administration maximum acceptable limit for aflatoxins in food products, excluding milk, is 20 mg/kg. Small farmers have farms which are not managed properly. Because of poor knowledge about farming and financial restrictions, storage, and management of feed in small farms are not suitable, resulting in the incidence of aflatoxins in animal feedstuff.

3.13 PATHOGENESIS

A large number of factors are involved in the etiology of aflatoxins. The *aflR* gene is involved in the pathway to producing aflatoxins (Chang et al., 1993; Woloshuk et al., 1994). The transcription of this gene is controlled by G-protein coupled pathway with the help of FadA and through cyclic adenosine monophosphate (AMP)-dependent protein kinase enzyme, protein kinase A (PKA). The activity of PKA shows antagonistic activity with *aflR* (Shimizu and Keller, 2001). Aflatoxin production is also affected by Ca^{+2}/calmodulin pathway. The function of the *aflR* gene after transcription is also associated with the *aflJ* gene and involved in aflatoxin synthesis (Chang and Yu, 2002).

3.14 AFLATOXINS IN HUMANS

A. parasiticus and *A. flavus* are two main fungi mostly concerned with manufacturing aflatoxins for health concerns. These fungi are mostly seen in putrefying and dead vegetation with favorable conditions, as in subtropical and tropical regions (increased humidity and temperatures). Scarcity stress, damage due to insects, and poor storage can significantly contribute to the enhanced occurrence of molds. AFM1, a metabolite of AFB1, can be sensed in milk in those areas where aflatoxin exposure is high (WHO, 2018). By consuming milk (breast milk), dairy products, and yogurt, humans could be exposed to aflatoxin. The poor storage of food crops in warm or humid environments, in which mold growth is favored, can result in higher contamination levels of aflatoxins. Before harvesting and storage, crops could be infected with aflatoxins. Before harvesting, food crops could be infested with aflatoxin, most likely in cottonseed, tree nuts, peanuts, acha, wheat, maize, and others. After the harvesting, food crops could be infected with aflatoxin, most likely in rice, coffee, and spices (Awuchi et al., 2020).

3.15 AFLATOXINS IN ANIMALS

Increased vulnerability to disease, reduced productivity, damage to the liver, poor eggshell quality, reproductive efficiency, reduced egg production, and inferior carcass quality are the effects of aflatoxins seen in poultry (WHO, 2018). Damaged liver leading to hepatic disease is manifested in aflatoxicosis in pigs. The chances of aflatoxin infection are higher in pigs. Reduced weight gain, along with impairment of kidney and liver functions and reduced milk secretion, is mostly seen in cattle infected with aflatoxins can be decreased. The susceptibility and vulnerability of the animals to the toxic effect of aflatoxins depend on different enzymes that metabolize the aflatoxins, such as cytochrome P450s and glutathione S-transferases. Aflatoxins can cause hepatic damage in dogs. If the exposure is small, then consistent intake of aflatoxins is required for many weeks, even months, for the manifestation of liver disease (Bingham et al., 2003). According to previous publications, toxic levels of aflatoxins in dog food range from 100 to 300 ppb and require long exposure for weeks or months to develop aflatoxicosis. Turkeys are very vulnerable to the toxic effects of aflatoxicosis. According to Goldblatt, an important relationship has been found between AFB1 exposure (4 mg per kg of body weight) and the onset of growing abnormalities in descendants. In different countries, pet food has been detected with aflatoxin (Awuchi et al., 2020).

3.16 TYPES OF FODDERS

3.16.1 Rabi/Winter Fodders

Fodders are grown at the beginning of winter to meet the needs of animals until the start of the summer season. For example, oats, ryegrass, berseem, and wheat are important rabi fodders.

3.16.2 Kharif/Summer Fodders

These are grown in the summer season for the provision of feed and food until the beginning of the winter season. For example, maize, millet, sorghum, and gwar are important kharif fodders.

3.16.3 Multicut/Perennial Fodders

These are grown only once and cut at different times of the year for prolonged utilization. For example, lucern, mott grass, and Sudan grass are important perennial fodders. They are also known as evergreen fodders.

3.17 PREVALENCE OF AFLATOXINS IN SUMMER FODDERS

A wide variety of regional fodders and food items are affected by aflatoxin contamination. Feed and food such as spices, nuts, dried fruits, corn, maize, sorghum, and millet are of prime importance for aflatoxin contamination (Martinez-Miranda et al., 2019). Contamination mostly occurs in post-harvested crops under optimum storage conditions for fungal growth. Fifty percent of sorghum, 92.9% of millet, and 67.9% of maize were affected by aflatoxin contamination in storage conditions (Sirma et al., 2015).

The effects of aflatoxin include retarded seedling growth, inhibition of seed germination, affected root elongation and synthesis of carotenoid, inhibited nucleic acid and protein synthesis, and affected chlorophyll and enzyme synthesis in plants.

3.18 SORGHUM

Sorghum bicolor is a chief crop grown for fodder and grain in semiarid areas, especially in African and Asian countries. It is cultivated from March to May and harvested from June to November. Its protein level ranges from 7–9% (Katile et al., 2010). Due to its cultivation in the kharif season, it is greatly affected by molds, especially *Aspergillus*, *Curvul aspergillusaria*, and *Fusarium*. The major aflatoxin present in sorghum is AFB1. Aflatoxins mainly affect the seeds of sorghum. The quality of sorghum is affected due to the mycotoxin production in sorghum; thus, reducing the production of sorghum. *A. flavus*, *A. tamari*, *A. fumigatus*, *A. niger*, and *A. terreus* affect different sorghum species (Divakara et al., 2014).

3.19 MAIZE

Maize (*Zea mays* L.) is one of the major commodities of agriculture globally. It is cultivated from February to September and harvested from May to December. It has 8–10% protein. *Aspergillus flavus* and *A. parasiticus* are involved in aflatoxin contamination, but *A. flavus* is the main fungus in aflatoxin contamination in maize and affects the pre-harvesting stage of maize (Lewis et al., 2005).

3.20 MILLET

Pearl millet (*Pennisetum typhoides*) is a feed crop and essential food in African and South Asian countries. Usually, it is grown for forage and hay production. It is cultivated from April to September and harvested from June to November. It has 7–9% protein. *Aspergillus fasciculatus, A. flavus, A. fumigatus, A. terries, A. repens, A. flaviceps, A. niger, A. giganteus, A. versicolor* were main fungi observed from millet fodder. It is easily grown in sandy and acidic soils because of the intensive root system (Raghavender et al., 2007).

3.21 GWAR

Gwar (*Cyamopsis tetragonoloba*) is mostly grown in dry conditions. Its cultivation is from April to July, and harvesting is from June to November. Most seedlings are affected by aflatoxins.

Mostly *Aspergillus flavus* affects seeds of gwar. Gwar seeds were contaminated by *A. nidulellus*, *A. flavus*, *A. fumigatus*, *A. unguis*, *A. parasiticus*, and *A. niger* (El-Nagerabi and Elshafie, 2001).

3.22 COWPEAS

Cowpea (*Vigna unguiculata*) is present in semiarid areas and less affected by aflatoxin contamination (Houssou et al., 2008). Resistance to aflatoxin contamination is due to beta and alpha proteins and tannins. Mostly the growth of cowpea is influenced by aflatoxin during their storage period. Approximately 30% of the cowpea crop is infected with aflatoxin after post-harvesting. It is mostly affected by *Aspergillus flavus* and *A. parasiticus*. AFB1, AFB2, AG1, and AFG2 are present in cowpeas.

3.23 TOXICITY OF AFLATOXINS

Biotransformation of AFB1 contains prime by-products such as AFM1 and aflatoxin Q. M1 and Q1 aflatoxins are toxic; however, other molecules are less reactive and easily eradicated in urine. However, AFB1 is a mutagen and enormously electrophilic; it reacts with the nucleophilic end of DNA or ribonucleic acid RNA or proteins (Oliveira et al., 1997), thus affecting the function of cells through mutation. The AFB1-DNA adduct formation is tremendously linked with the carcinogenicity of AFB1. Classically, AFB1 counters with DNA and results in G→T transversion mutation. Such a type of mutation is related to hepatocellular carcinoma.

Nucleic acids and proteins bond covalently with aflatoxins, resulting in variations in sequences of the base in nucleic acids and protein structures, leading to the impairment of activity. The protein synthesis is inhibited by the covalently binding of very reactive AFB1–8,9-exo-epoxide and Dihydrodiol with DNA, RNA, and proteins. Epoxide reacts with guanine of DNA and RNA at position N7, and dihydrodiol reacts to amino groups of the bases forming a Schiff base (McLean and Dutton, 1995). AFB1 also negatively impacts carbohydrate metabolism, which will cause a decrease in glycogen in the liver and increased glucose in the blood.

The enzyme phosphoglucomutase catalyzes the conversion of glucose-6-phosphate into glucose-1-phosphate, and AFB1 inhibits this enzyme resulting in the buildup of glucose 6-phosphate and a reduction in glycogen synthesis. Furthermore, AFB1, AFG1, and AFM1 inhibit the electron transport system and cytochrome oxidase activity (Kiessling, 1986). As a result of this surplus, glucose will not be stored as animal starch (glycogen) but either accumulate in the blood or transform into glucose-6-phosphate, resulting in metabolic intermediates synthesis through the pentose phosphate pathway.

3.24 ACUTE TOXICITY

The process by which acute toxicity of aflatoxins develops is not well known. The adduct of aflatoxin with protein is linked with acute intoxication, blockage of protein synthesis, and blockage of enzymes involved in vibrant functions like repair and replication of DNA and immune response. Acute aflatoxicosis results in DNA fragmentation when exposed to a high level of aflatoxins, as observed in the mice when AFB1 was injected at 0.02 mg/kg for 21 days. According to the latest study on AFB1, acute toxicity in poultry is responsible for acute aflatoxicosis and takes the lead to the development of the adduct of aflatoxin–albumin. Metabolite of AFB1-phase I was considered to be involved in acute toxicity. Lipids and protein adduct formation result from AFB2a binding with cellular proteins and phospholipids through a covalent bond. This is also a reason for acute aflatoxicosis.

Consequently, acute aflatoxicosis may develop from a sudden emphasis on all the damages mentioned above at a very high dose in a very short time. At this high dose, aflatoxins can overwhelm the depolluting ability of the cell, resulting in the creation of toxic metabolites which cause severe damage to DNA, disturbance in the progression of the cell cycle, fragmentation of the DNA, metabolic disorders, necrosis of tissue, and cytotoxicity, ultimately resulting in organ failure. However, these things are amassed gradually with permanent exposure at low doses, eventually resulting in cancer of the liver.

3.25 CHRONIC DISEASES

Repetitive contact with aflatoxins with low doses for a long period causes prolonged diseases, of which cancer is the most recurrent and severe. Intake of aflatoxins in the diet has been linked with liver cancer like hepatocellular carcinoma (HCC) and bile duct hyperplasia (McGlynn and London, 2005). The repetitive exposure of aflatoxins to organs like bone, the pancreas, the viscera, the kidney, and the bladder is described as developing cancer (Fouad et al., 2019). Aflatoxins cause work-related lung cancer through inhalation and skin cancer via direct contact in mammals with severe diseases like immunosuppression, cytotoxicity, mutagenicity, and estrogenic and teratogenic effects produced in long-lasting exposure of aflatoxins (Klvana and Bren, 2019). Aflatoxins are involved in nutritional disorders like growth faltering and kwashiorkor by disturbing micronutrient absorption (vitamins, iron, zinc) and protein synthesis (Turner, 2013). In animals, aflatoxin-contaminated feeds induce impairment of production and reproduction, increase vulnerability to diseases, and make the quality of food they produce poor. The aflatoxicosis outbreaks cause hundreds of deaths and could be prevented by investigating suspicious crops/foods, obvious mold growth, and disposal of such food if the level of aflatoxin surpasses the regulatory standards. Exposure to aflatoxins in the diet is associated with cancer of the liver. The influence of AFB1 is increased by more than 60 times through long-lasting infections with hepatitis virus B (HB) (Henry et al., 2002). In 2018, 781,631 deaths from liver cancer were recorded, according to the global cancer observatory of the International Agency for Research on Cancer (IARC).

3.26 AFLATOXICOSIS

The consumption of aflatoxin-contaminated food or any other substance by humans and animals is known as aflatoxicosis.

3.27 EXPOSURE OF HUMANS TO AFLATOXINS

Humans could be exposed to aflatoxins in two ways:

1. Humans directly ingest foods of plant origin contaminated with aflatoxins, especially AFB1.
2. Aflatoxin is directly ingested from milk and other dairy products. AFM1 is the main aflatoxin present in milk products. Aflatoxin residual may be observed in animal tissues, eggs, and meat.

3.28 AFLATOXINS AND GROWTH DISTURBANCES

Maternal consumed food is a prime factor involved in the exposure of children to aflatoxins by ingestion of aflatoxin-contaminated food. Exposure may occur through breastfeeding and eating aflatoxin-contaminated foods. Children may also be exposed to aflatoxin in utero, which can retard the growth of infants.

3.29 AFLATOXINS AND KWASHIORKOR

Kwashiorkor is defined as a disease caused by the intake of a low-protein and high-carbohydrate diet. Kwashiorkor is due to zinc and niacin deficiencies, free radical production, depletion of antioxidants, and abnormal microbiota, but the consumption of aflatoxins is also associated with kwashiorkor (Tesfamariam et al., 2020). AFB1, AFB2, AFG1, AFG2, AFM1, and AFM2 were detected in the brain, kidney, stool, heart, urine, and serum of affected people. Females are at a greater risk of being attacked by this disease (Soriano et al., 2020).

3.30 HEPATOCELLULAR CARCINOMA

Metabolism of biomolecules occurs in the liver, which is considered a core part of the body, and thus takes part in maintaining blood glucose. AFB1 plays a critical role in the development of many diseases. Broad exposure to mycotoxins in individuals showed a high level of AFG1 (Leong et al., 2012). Fatty change, necrosis, megalocytosis, congestion, and fibrosis were histopathological examinations of the liver in individuals affected with aflatoxin contamination (Zuberi et al., 2019). Fat droplets and vacuolation are also associated with histopathology due to the liver's production of AFB1 metabolite (AFG1). Thus, aflatoxins cause liver degeneration (Kaleibar and Helan, 2013).

AFB1 is biotransformed into metabolic products due to hydroxylation, reduction, epoxidation, and demethylation (Yiannikouris and Jouany, 2002). Due to these processes, different pathways will occur, thus producing various metabolites like AFM1, AFP1, and AFB1–8,9-epoxide (AFBO; Wu et al., 2009). CYP450 enzymes convert aflatoxin into aflatoxin-8,9-endo-oxide and aflatoxin-8,9-exo-epoxide, which is highly reactive and reacts with RNA, proteins, and DNA and makes derivatives. Aflatoxin can also respond with tumor-suppressor genes (p53). Aflatoxin-8,9-exo-epoxide binds with DNA and converts into pro-mutagenic 8,9 dihydro8-(N7 guanyl)-9-hydroxy AFB1 adduct (Wild and Turner, 2002). The conversion of AFB1-N7-Gua adduct into the AFB1–formamido–pyrimidine adduct and the apurinic site can occur, and the former is more stable than the latter in secondary form (Hussain et al., 2007). The AFB1–FABY adduct can cause transversion mutations between guanine and thymine. Mutagenesis has occurred due to its incorporation in DNA. These mutations are important in producing malignancy disorders (Brown et al., 1999). AFP1 is catalyzed by nicotinamide adenine dinucleotide phosphate (NADPH) (Zhang et al., 2021). Formation of AFB1–8,9-exo-epoxide after AFB1 oxidation will cause mutagenic effects after reacting with DNA (Kensler et al., 2011). In humans, CYP3A4 and CYP2A13 enzymes in the liver and lung have a tremendous role in the biotransformation of AFB1 to AFBO.

The inhibition of lipid and carbohydrate metabolism can also be due to AFM1. The level of antioxidants was also decreased due to AFM1. Inflammatory responses can also be intensified by aflatoxin poisoning due to an increase in the level of pro-inflammatory cytokines (IL-6, TNF-α, IL-8) and a decrease in anti-inflammatory cytokine (IL-4) levels (Marchese et al., 2018).

The occurrence of type 2 diabetes (T2D) is due to excessive exposure to aflatoxins that result in a disturbance in normal endocrine functions. The level of glucose, acetoacetate, glycine, citrate, betaine, and acetate is remarkably increased after aflatoxin exposure, thus leading to metabolic disorders. Aflatoxin significantly decreased serum ketone bodies, valine, choline, leucine, lactate, lipoprotein, creatinine, and citrulline levels (Cheng et al., 2017). The hepatocarcinogenic effect of aflatoxins increases the level of hepatic enzymes and causes liver damage. A downregulation in cholesterol, hepatic triglyceride, and phospholipids levels and upregulation in free acids levels was induced due to acute exposure to aflatoxins (Rotimi et al., 2018)

3.31 TERATOGENESIS

The embryos of pregnant birds and animals can be exposed to aflatoxins in the form of fertilized eggs or embryonic forms, leading to different pathological disorders and harmful effects on health (Benkerroum, 2020). Highly affected mothers from aflatoxins can easily transfer aflatoxins and their related toxic secondary metabolites to fetuses through systemic circulation pathways in all mammals, including animals and humans. Thus, metabolites of aflatoxin, DNA aflatoxin, and albumin adducts were identified from their mothers' fetal cord and blood samples (Abdulrazzaq et al., 2002; Hsieh and Hsieh, 1993). The metabolism of aflatoxin and related metabolites has occurred through the same mechanisms as in adults (Fink-Gremmels and Malekinejad, 2007). Different fetal issues like fetal loss, premature birth, and retarded fetal growth have been recorded in highly exposed mothers. A restricted growth rate has been observed both in animals and humans. An indirect relationship between birth weight and acceptable biomarkers or aflatoxin in cord blood has been documented (Vries et al., 1989; Abdulrazzaq et al., 2004). Daily intake of aflatoxins from 0 to 100 μg/kg has shown different harmful effects like skeletal anomalies, a reduction in litter size and live birth, and a deterioration in organ development. These harmful effects are due to inhibition in protein synthesis and aflatoxins binding to DNA (El-Nahla et al., 2013).

Congenital anomalies and other consequences have also been reported due to the ingestion of aflatoxin-rich feed in cases of pregnant animals. Systemic inflammation induced due to up and downregulation of maternal pro and anti-inflammatory cytokines leads to placental growth impairment that negatively impacts fetal growth. Other congenital anomalies include miscarriage, premature birth, and stillbirth (Vries et al., 1989). Due to the cytotoxic effect of aflatoxins, mothers have induced anemia due to erythrocytes' lysis and disturbance in nutrient metabolism (selenium, iron, vitamins; Yousef et al., 2003). The relation of anemia due to aflatoxin ingestion was evaluated in vitro or in vivo, taking 0.5–1 mg/kg of body weight.

3.32 OTHER HARMFUL EFFECTS RELATED TO AFLATOXINS

Due to different mechanisms involved in aflatoxin metabolism, various harmful health illnesses have been observed. Reproductive issues, neurodegenerative disorders, retarded growth, stunting, retarded physical maturity, and sexuality problems are included in health illnesses (Williams et al., 2004).

3.33 NEURODEGENERATIVE DISEASES

Neurodegenerative disorders also develop due to ongoing exposure to aflatoxins compared to other harmful effects. When functional macro-molecules of neuronal brain cells interact with reactive oxygen species (ROS) and AFBO, they prevent lipid and protein synthesis, leading to the degeneration of those cells. The function and structure of mitochondrial brain cells were also disrupted, causing oxidative phosphorylation, thus guiding their apoptosis (Verma, 2004). Histopathological examination revealed astrocytes gliosis, necrosis, and vasodilation, and disrupted oxidative stress parameters were also observed, thus leading to neurodegenerative disorder in rats taken 1/600th of aflatoxin level (Alsayyah et al., 2019). Neurodegenerative disorders also occur due to the dysregulation of immune responses to immune-competent cells and the creation of pro-inflammatory cytokines in CNS (Mehrzad et al., 2017).

3.34 MALNUTRITION

Malnutrition is a significant aspect under consideration in developing countries due to its association with other harmful diseases in a child facing drought conditions. A balanced diet should be provided for the production of healthy individuals. The ingestion of aflatoxins interferes with minerals and vitamin absorption, thus aggravating children's nutritional condition. Absorption of vitamins E, A, and C and selenium are mostly affected (Obuseh et al., 2011). This increases sensitivity to aflatoxins, and children become deficient in these vitamins. Increased susceptibility is due to inhibiting the CYP450 enzyme complex (Sun et al., 2015). Chronic exposure also leads to kwashiorkor and marasmus diseases.

3.35 RELATION BETWEEN CLIMATE CHANGE AND AFLATOXINS

Humidity and temperature are the main environmental factors responsible for aflatoxin production. Consequently, climate change inherently imposes a new dynamic in naturally occurring pollutants. Relative humidity and temperature increase due to dry season rainfall, leading to aflatoxin production, mostly in tropical areas. Sterigmatocystin, a metabolite of *Aspergillus*, was present at 20 μg/concentration in southern Norway. The physiology of plants is also changed due to temperature and photoperiod management. Foodborne risks include aflatoxins most sensitive to climate change (Nesic et al., 2015).

3.36 EFFECT OF AFLATOXINS ON ANIMALS

3.36.1 Routes of Animal Exposure

1. Ingestion of aflatoxins harms the production and health of animals. For example, the production of toxins in cereals (corn) under optimum conditions for fungal growth can occur, thus reaching animal farms and causing contamination.
2. Other ways of exposure are inhalation and dermal contact.

Animals may be affected by only one aflatoxin or a combination of aflatoxins simultaneously, affecting different organs (Murugesan et al., 2015). The negative influence of aflatoxins includes decreased livestock productivity, increased cost of veterinary-related care, and a continuous harmful effect on feed and food. Public health is also significantly affected owing to aflatoxin contamination.

The susceptibility of aflatoxins in animals varies due to different factors like gender, species, nutrition, and species due to the difference in the initiation of the aflatoxin metabolism rate. The CYP450 enzyme complex activates AFB1. Different enzymes are responsible for AFB1 biotransformation in different species like in poultry; CYP1A5, CYP2A6, CYP1A1, and CYP3A37 are involved in biotransformation (Monson et al., 2015; Yarru et al., 2009). Rate of aflatoxin-8,9-epoxide (AFBO) production and its combination with (GSH) glutathione to decrease glutathione-S-transferase toxicity is different in various species (Dohnal et al., 2014).

Among animals, rabbits are more significantly affected by the detrimental effects of aflatoxins than ducks, turkeys, and chickens. The order of susceptibility is

Rabbits > Ducks > Turkeys > Chickens

Fish are less susceptible to toxic effects, and cattle and sheep are least susceptible. Male birds are more sensitive than female birds, and ducks are also largely more affected than chickens and quails (Lozano and Diaz, 2006). Young animals are more vulnerable to aflatoxins than older animals.

Aflatoxin toxicity is positively and adversely affected by diet. Negative results may be due to the frequency of contaminated feed used, gender, species, animal age, inclusion rate in feed, and exposition period. Different ingredients may also reduce aflatoxins through reduction, exclusion, and sorbent mechanisms in the gastrointestinal tract (Yarru et al., 2009).

3.37 CALVES

LD_{50} of AFB_1 in calves is 0.5–1.5 mg/kg. Anorexia, submandibular edema, diarrhea, dysentery, photosensitization, depression, and anorexia were observed in affected calves. Death may also be seen. In necropsy findings, hemorrhages in the pericardium, skeletal muscles, alimentary tract serosa, lymph nodes, and subcutaneous tissues were observed in the necropsy findings. Histopathological examination revealed enlarged and vacuolated hepatocytes (Dhanasekaran et al., 2011).

3.38 DAIRY AND BEEF CATTLE

Liver damage is the most prominent effect in cattle and beef cattle. Fatty degeneration, biliary proliferation, fibrosis, megalocytosis, and veno-occlusive lesions were observed in histopathological examination. Animals may become immunosuppressive. Cell-mediated immunity may be linked with the immunotoxic effect of AFB1.

In dairy cattle, conversion of AFB1 into AFM1in milk has occurred. Ninety-six hours are required to remove aflatoxins from milk by giving an uncontaminated feed. Milk production will be decreased due to a high aflatoxin level (4 ppm) in 1 week, and a low aflatoxin level (0.4 ppm) can lower the output within 3–4 weeks. AFM1 concentration depends on the intake of AFB1 through contaminated feed. Milk becomes more toxic as the level of aflatoxin increase. The toxic level in milk is also influenced by liver metabolism and its route of excretion (Dhanasekaran et al., 2011).

3.39 EFFECTS ON RUMINANTS

Aflatoxicol is a carcinogenic derivative generated by the partial degradation of aflatoxins under the influence of ruminal flora in ruminants. The degradation of toxins has been done by rumen microbiota present in ruminants, so they are more resistant than nonruminants. Intake of aflatoxins may lead to reproductive problems, decreased feed utilization, decreased milk production, immune suppression, and reduced wool and wool yield in goats, cattle, deer, and sheep. In cattle, reduced feed efficiency has been observed. Consumption of contaminated feed for a long period can impair the growth rate in cattle. Intake of 600 μg kg^{-1} of aflatoxin can reduce feed efficiency and weight gain in steers (Zain, 2011). Lethargy and inappetence can also occur. The level of aflatoxin ranges between 100 and 1000 μg kg^{-1} can decrease ruminal motility and feed efficiency, inhibit growth, and increase kidney and liver weight. Embryotoxicity has been reported due to aflatoxin consumption (Zain, 2011).

In cattle, the function of the immune system has been affected by suppression in the blastogenesis of lymphocytes because aflatoxin B1 inhibits the mitogen-induced activation of lymphocytes. In sheep, increased concentration of aflatoxins leads to nephritic lesions, disturbance in mineral metabolism, and hepato-toxicosis. In lambs, 2.5 mg/kg of aflatoxin consumption decreased feed intake and changed blood parameters (Gallo et al., 2016).

3.40 EFFECTS ON POULTRY

Teratogenicity, mutagenicity, hematological problems, hepato-toxicity, immune suppression, and carcinogenicity are unfavorable effects noted in poultry as a result of intake of AFB1. Poultry is greatly affected by AFB1, even in low concentrations. The order of vulnerability in poultry is

Ducks > Turkeys > Japanese > Quail > Chicken (Monson et al., 2015)

Immune response has been depressed due to aflatoxin exposure. Reduced phagocytosis and apoptotic process in the thymus, bursa of Fabricius, and spleen have been observed by Rawal and co-workers (Rawal et al., 2010). Enlarged and congested kidneys, enlarged, pallor liver, and depletion in erythrocyte production and spleen lymphoid were observed in broilers with 1 mg/kg^{-1} of AFB1 (Kumar & Balachandran, 2009).

Exposure to aflatoxins can become a hazardous risk for animal health, increasing the sensitivity to infections and decreasing the effectuality of vaccination. Newcastle disease outbreaks correlate with aflatoxins in broiler feed (Yunus et al., 2009).

The 36.9–95.2 µg/kg concentration of AFB1 in broiler feed showed biliary hyperplasia, apoptosis, hepatocytic necrosis, hypertrophy of Kupffer cell, micro-vesicular fatty degeneration, reduced antioxidant levels, and changes in biochemical serum parameters (Yang et al., 2012).

Blood coagulation disorders may also be caused by aflatoxins in broilers and characterized by massive hemorrhages in kidneys, stomach, intestines, muscles, lungs, and heart that ultimately lead to death. The activity of blood coagulation factors like prothrombin, X, VII, IX, V, and fibrinogen factor is indicated by prothrombin time, and these factors can diagnose liver lesions in poultry. A direct relation is observed between prothrombin time and the dose of aflatoxin.

Layers are also affected by AFB1, resulting in decreased egg production and inferior egg quality, and the presence of residues is also observed in muscles and eggs. The transmission of AFB1 from feed to egg is about 5000:1 (Hossein and Gürbüz, 2015). Decreased body weight has been observed due to disturbance in maintaining hypothalamic neuropeptides due to aflatoxin (Trebak et al., 2015). Lesions in the heart, liver, ovaries, and kidneys may also be observed in laying hens.

Nutrient absorption is also affected by aflatoxins. Reduced surface area, disturbance in nutrient transporters, barrier function loss, and inflammation could be the main effects on the gastrointestinal tract (Grenier and Applegate, 2013).

3.41 EFFECTS ON OTHER ANIMALS

Exposure of aflatoxins to marine animals via the feed chain can also occur. Decreased weight gain, reduced feed conversion ratio (FCR), and decreased production are the main consequences of fish exposed to aflatoxins. Histopathological examination reveals liver and kidney lesions (Anater et al., 2016). The average weight and length of Nile tilapia were not significantly affected by aflatoxin contamination, but fingerlings are greatly affected by aflatoxins (Cagauan et al., 2004). Eye opacity, blindness, decreased appetite, static movement, abnormal swimming, and yellowed body surfaces are outcomes of aflatoxin exposure.

Jaundice, death, depression, lameness, and anorexia are consequences of aflatoxin ingestion (58.4 µg/kg) in horses. Heart and skeletal muscles are greatly affected in ponies. Histopathological findings in horses are hyperplasia of the bile duct, enlarged livers, and damaged kidneys (Zain et al., 2011).

Mycotoxins in companion animals lead to death due to their severity. Anorexia, weakness, hepatitis, and severe depression are observed in dogs due to aflatoxin (AFB1) contamination. The presence of mycotoxins in pet food has been examined. Aflatoxins contaminated 88% of dog food at a concentration of 5 g/kg (Gazzotti et al., 2015).

3.42 DETECTION OF AFLATOXINS

For the detection of aflatoxins, two main techniques are used. The first is the patient's urine analysis to detect the AFB1-guanine adduct. If the patient has been exposed to AFB1 recently, then AFB1–guanine adduct will be present. This technique measures present exposure only. AFB1 might differ daily; hence, it is not an effective technique for prolonged exposure (Hussein and Brasel, 2001). The second technique is the quantification of AFB1–albumin adduct in blood serum. This method measures the exposure to aflatoxins over weeks, even months. Due to the difference in clinical signs and immune suppression, which results from infectious diseases, identifying aflatoxins in humans and animals is challenging. Among the techniques frequently used for detecting aflatoxin in humans, one measures breakdown product in urine, and another measures the AFB1–albumin level in the blood. Different procedures have been developed to identify aflatoxin's presence in foods. Aflatoxins are significant, and techniques for their detection have been researched broadly.

3.43 CHROMATOGRAPHIC METHODS

The sample to be separated is distributed between the stationary and mobile phases (Braithwaite and Smith, 1985). It is based on the interaction of the mobile phase with the stationary phases. The mobile phase is fluid, while the stationary phase might be liquid or solid.

3.43.1 Thin-Layer Chromatography

This technique was first used by scientist de Iongh along with his co-workers. It is the most widely used parting method in the detection of aflatoxins. It comprises two phases. Stationary phase made of either cellulose or silica or alumina stopped an inactive substance like glass/plastic, known as matrix. The mobile phase contains a combination of acetonitrile, water, and methanol (Betina, 1985), which transfers the sample to a solid stationary phase. In thin-layer chromatography (TLC), aflatoxins are allocated on mobile and stationary phases depending on the difference in sample solubility. Based on their molecular structures and association with two phases, diverse samples either stick to the stationary phase or endure in the mobile phase, allowing quick and efficient separation. In different foods, TLC is commonly used for aflatoxin detection. The benefit of TLC over other methods is that in a single test, it can determine numerous types of mycotoxins (Younis and Malik, 2003). Whereas TLC is very sensitive, it requires the pretreatment of samples, trained technicians, and high-priced equipment (Wacoo et al., 2014). High-performance liquid chromatography (HPLC) has lessened the challenges related to TLC procedures due to mechanization, sample application, and plate analysis. Currently, it is the most effective and accurate method for detecting aflatoxin. However, the need for trained operators, equipment cost, and broad sample pretreatment limits the HPTLC to the laboratory.

3.43.2 HPLC

HPLC is prevalent among all chromatographic techniques for organic compound parting and calculation. HPLC is used for the determination of nearly 80% carbon-based composites (Li et al., 2011). It uses a stationary phase limited to either glass or plastic tube and a mobile phase consisting of organic solvents that flow over the solid adsorptive substance. When a sample to be determined is coated on the column's top, it flows and circulates between the two phases. This is achieved as the elements in the samples have various attraction for both phases, hence moving over the column at changed rates.

3.43.3 Gas Chromatography

As per other chromatographic techniques, the sample analyzed through gas chromatography (GC) depends on the sample's differential partitioning between two phases. In this type of chromatography, gas is used as a mobile phase, and liquid layered over the inactive solid particles is used as a stationary phase. The stationary phase comprises inactive particles coated with fluid and is normally limited to a lengthy tube of stainless steel called a column, sustained at a suitable temperature. The vaporization of the sample occurs in the gaseous phase, while these vapors are conducted across the stationary phase with the help of gas. The various chemical elements of the sample will allocate themselves among both phases. The elements of the samples having a superior affinity with the stationary phase are stopped in the column, while elements with less affinity will pass across the column less obstructed. Due to this, each element of the sample has its partition coefficient. After that, detection is done through a mass spectrometer (MS) and flame-ionization detector (FID) or an electron-capture detector (ECD) (Pascale, 2009). As it is an expensive technique and other cheaper chromatographic techniques are available commercially, it is less frequently used in the detection of aflatoxins commercially. Moreover, GC also needs an initial cleanup step before the detection; hence, it is limited to examining very few mycotoxins, like A-trichothecenes and B-trichothecenes (Wacoo et al., 2014).

3.44 SPECTROSCOPIC METHODS

3.44.1 Fluorescence Spectrophotometry

Absorption through the ultraviolet (UV)–visible region is an important process for the disentanglement of materials with different molecular structures. However, absorption is followed by light discharge of diverse wavelengths for a few particles. For the particles that release the energy of definite wavelengths, fluorescence is significant for their characterization and examination, and it is utilized to examine aflatoxins in food grains and raw peanuts (Babu, 2010). Within 5 minutes, the fluorometric method can enumerate aflatoxins in the 5–5000-ppb range. However, a derivatization process is needed for improved analysis of aflatoxins and better detection using fluorometry.

3.44.2 Fourier Transform Infrared Spectroscopy

FTIR is a spectroscopic technique that is useful for analyzing aflatoxins. This technique depends on the variation in molecular vibrations on exposure to infrared (IR) light. Molecules having bond vibrations in them can be recorded. As the bond's length and strength differ from one molecule to another molecule, the absorbance rate of IR radiation varies among different bonds. For example, the different bonds should vibrate with changed frequencies. Each carbon-based compound has a different IR spectrum; thus, each pure compound is identified by analyzing its infrared spectra. The spectrum comprises a plot of percentage transmittance, contradicting the wave number. The use of FTIR spectroscopy for the examination of aflatoxins in peanuts, as well as in peanut cake, has been reported. Pearson used transmission and reflectance spectroscopy to determine aflatoxins level in corn kernels. An examination of kernels (approximately 95%) was divided into two concentrations of aflatoxins: high (>100 ppb) and low (<10 ppb) (Wacoo et al., 2014).

3.45 IMMUNOCHEMICAL METHODS

This method depends on the specificity of antigen–antibodies binding. Specific binding and higher affinity depend not on antigens and antibodies; receptors and ligands also show attraction and specificity (Sargent and Sadik, 1999). Antibody–antigen complexes can be calculated by measuring the variation in transmission density of photons spectrophotometrically with different light energy. Furthermore, this technique does not need skilled personnel, and is less laborious and requires less time, making it superior to the chromatographic and spectrophotometric methods. In different illustrations, the events of binding and resultant complexes need amplification to signal gratitude better. This is achieved by employing several labels, including radioisotopes, enzymes, and fluorophores. Because of these advances, these techniques are widely used in fields. For the analysis of aflatoxins, chief immunochemical procedures consist of radioimmunoassay (RIA) and enzyme-linked immune-sorbent assay (ELISA).

3.45.1 RIA

The RIA technique depends on the reasonable binding of radioactive-labeled and nonradioactive antigens. The radioactive-classed antigen contests with unlabeled nonradioactive antigen for an actual number of antibody/antigen attachment locations on the same antibody. An investigated amount of identified antigen and an unspecified amount of unknown antigen respond to a known and restrictive amount of antibody competitively. The dose of tagged antigen is inverse to the quantity of untagged antigen in the sample. Solid phase RIA technique is employed to identify AFB_1 amount in peanuts. The detection limit of 1 μg/kg was achieved using the solid-phase RIA technique. The main advantage of this technique is that it can perform many detections simultaneously with high sensitivity and specificity. The requirement of pure antigen, the radioactive isotope that has probable health dangers, and the storage problem of low-level radioactive wastes are major disadvantages of this technique (Twyman, 2005).

3.45.2 ELISA

The probable health risks associated with using RAI are directed to look for a harmless substitute. This was attained by tagging antigens or antibodies using enzymes instead of isotopes. ELISA has become the choice approach for research institutions, medical diagnostic laboratories, and governing bodies to assess quality. The principle at which ELISA works is fundamentally the same as that of other immunochemical techniques. ELISA depends on the antibodies' specificity for the antigens. The sensitivity of ELISA is improved by tagging either antibodies or antigens by using an enzyme, and specific substrates are used for analysis. Competitive ELISA is easily performed and gives a beneficial quantity antigen and antibody concentration analysis. Nowadays, ELISA is used to determine aflatoxins (Ali, 2016). The advantage of using the ELISA technique is that the ELISA plate has 96 wells, and various samples can be analyzed concurrently. Moreover, ELISA kits are inexpensive and easily used, and no health hazards are linked to using the ELISA technique. However, these are very laborious and time-consuming because of the multiple washing steps.

3.45.3 Lateral Flow Devices (Immunodipsticks)

These are immuno-chromatographic assays. These are also identified as lateral flow devices. It consists of a porous casing that confirms the movement, a sample pad that confirms the contact between the liquid and casing, an absorbent pad for increasing the flowing liquid volume, and an inflexible backing that will give sustenance to the device. Through the capillary action of the absorbent pad, the liquid analyte that was attached to the sample pad travels toward the end and out of the membrane (Li et al., 2009). The mixture of gold and gold layered with the antibody is used as labels in immunodipsticks, which usually give binding zones of red color (Huybrechts, 2011).

AFB1 in pig feed was detected through the lateral flow device that Delmulle developed. The detection threshold of the device was 5 μg/kg in 10 min. Ho and Wauchope established another immuno-chromatographic technique. This technique depends on the competition among free AFB1 and dye-containing liposome-tagged AFB1 for the conforming antibody. In less than 12 minutes, this device detects approximately 18 nanograms of aflatoxins. While the equipment was intended for immediate qualitative graphic reading, this machine has also been modified to be used in an optical density scanning mode, which quantitatively permits the quantitative detection of aflatoxins. Lateral flow devices are easily used. These devices give fast detection of aflatoxins in a few minutes. These devices are cost-effective and can be used for monitoring aflatoxins daily.

3.46 IMMUNOSENSORS

These biosensors utilize Ag/Abs as biological appreciation elements combined with a signal transducer like carbon, graphite, and gold, which helps detect the binding of related species (Ricci et al., 2007). Considering the signal transduction, immunosensors could be classified into optical, electrochemical, and piezoelectric sensors.

3.46.1 Optical Immunosensors

Depending upon the various transduction approaches for the detection of aflatoxins, numerous optical immunosensors devices have been established. One such optical device already developed for recognizing aflatoxins is surface plasmon resonance (SPR). The platform of SRP depends on the amount of variation in refractive index generated by the attachment of the sample with its bio-specific companion, which is immobilized on the sensors' surface. As the sample is streamed across the sensor's surface, there is a change in the echoing SPR wavelength, directly related to the refractile change at the surface (Piliarik et al., 2010). An SPR sensor surface comprises a biorecognition coating that binds to either an antigen or antibody, which results in an equivalent increase in the mass on the sensor's surface that is directly related to an upsurge in refractive index. The upsurge in the refractive index is observed through the alteration in resonance angle. The quantifiable variations in amount are due to the attachment and detachment of antibodies to its aimed antigen. An SPR immunosensor was stopped with monoclonal antibodies, although it combatted revival problems at the sensor surface because of the high attraction for binding of the monoclonal antibodies. Using 1M ethanolamine solution with 20% (v/v) acetonitrile and pH 12, regeneration was accomplished when polyclonal anti-AFB1 antibodies on the sensor's surface were immobilized. The sensor attained a linear recognition range of 3.0–98.0 ng/mL with decent reproducibility. If the recent problems of regeneration are handled, the SPR will offer aflatoxins determination free of labels.

3.46.2 Electrochemical Immunosensors

Many electrochemical immunosensors are being used in the analysis of aflatoxins. The signal is generated in potential membrane form when ions attach to a sensing casing. Then the difference in potential is calculated. The potential difference (pd) and the concentration have logarithmic relation. The measurement of signals could be in the form of linear sweep voltammetry, differential pulse voltammetry, electrochemical impedance spectroscopy, cyclic voltammetry, and chronoamperometry (Välimaa et al., 2010). Several types of nonenzymatic electrochemical immunosensor were also developed to analyze aflatoxins.

3.46.3 Piezoelectric Quartz Crystal Micro-Balances

Quartz crystal micro-balances (QCMs) are devices free from labels and applied to determine antigens directly. The piezoelectric QC depends on variations in the surface mass of the electrode on antigen reaction with a similar antibody that is restrained on the crystal surface of the quartz. Meanwhile, the variation in mass is directly related to the intensity of the antigen–antibody complex, so this method allows the revealing and evaluation of the immune complex (Zhou et al., 2011). Piezoelectric QCM has been used for the analysis of AFB1. During the development, layered both sides of QCM with electrodes of gold. A piezoelectric immunosensor was tried to detect AFB1 by immobilizing the DSP-anti-AFLAB1 antibody over gold-layered crystals of quartz. The sensor could detect AFB1 concentration within the 0.5–10-ppb limit. Jin et al. (2009) and other scientists developed a QCM-dependent sensor to determine AFB1. This device can detect aflatoxins in contaminated milk at 0.01 to 10.0 ng/mL. QCM is considered a technology free from labels since direct detection of mycotoxins is difficult due to their small size.

3.47 MANAGEMENT OF AFLATOXINS

Various types of crops are contaminated by mycotoxins pre- or post-harvesting. Processing, transport, manufacturing, and storage are different stages during which aflatoxins contaminate crops. Contamination of crops is minimized by taking preventive measures pre- and post-harvesting (detoxification) and during storage. Different methods can prevent aflatoxin contamination, but mycotoxin contamination can never be avoided because they are universally present (Kendra and Dyer, 2007).

3.48 HACCP MONITORING DURING PRE- AND POST-HARVESTING STORAGE

In animal feed, management of aflatoxins during the pre-harvesting stage is generally based on excellent agricultural approaches by producers and proper legislation to monitor aflatoxins present in food and feed and adequate control of feeds.

Agronomical practices greatly influence aflatoxin's contamination in crops. The foremost strategy to save crops from aflatoxin contamination is the reduction of mold proliferation during the storage and cultivation of crops. Different approaches should be adopted, like cultivating insusceptible cereals, seed selection, and strong plants that can endure pest attacks. Varieties are selected on molecular techniques that can resist aflatoxin attack (Bhatnagar-Mathur et al., 2015). Ostry and his co-workers (2015) observed that Bt corn has less aflatoxin concentration than non-Bt corn hybrids.

Frequently, mycotoxigenic fungi are first inoculated by crop leftover. The eradication of agricultural wastes significantly prevents cultivated crop contamination (Jard et al., 2011). One of the most critical strategies to prevent contamination is a selection of suitable harvest seasons, like spores and flowering time, because insects greatly attack it. A low level of aflatoxin occurs when groundnuts are harvested earlier than in delayed-harvested groundnuts (Waliyar et al., 2015).

Planting of crops should be timely to avoid crops from stress conditions like rising temperatures and drought during maturation and seed development stages. Limiting molds can be accomplished with factors such as plowing, correct fertilization, managing weeds, proper cultivation of crops, crop rotation, and prevention of growing many plants at once. Fungal infections can be avoided using pesticides (Zain, 2011). Insects are considered vectors for fungal spores and affect kernel's grain integuments to encourage fungi colonization that produces mycotoxin (Jard et al., 2011). Aflatoxin contamination can be minimized by treating the soil with *Aspergillus* (nontoxigenic) and *Trichoderma* strains and using microorganisms like bacteria and competitive exclusion (Dorner and Cole, 2002).

One of the most important factors during contamination is collecting and transporting harvested grains from the field to storage and drying facilities in containers. Containers should be moisture, insect, rodent, and bird free; and fungal growth can be easily visible (Alvarado et al., 2017).

The fungal attack can be avoided by reducing the destruction of grains during and before their storage. Drying cereals should be properly done so that grains are damaged at a minimum, and the moisture level should be lower than the required level to support fungal growth. Storage for a long period and mixing different grains should also be avoided. Aflatoxin-contaminated grains should be buried or burned (Lanyasunya et al., 2005).

Monitoring of grain quality and installation integrity should be done before storage. Storage conditions such as humidity, insect control, temperature, and moisture should be monitored during storage conditions. Moisture content should be less than 15 g/100 g of grains, and low concentration of oxygen and low temperature should be maintained during storage. Subtropical and tropical regions have more susceptibility to contamination than temperate regions due to pH, temperature, and humidity levels favorable for fungal growth. Seventy percent humidity, 10–40°C temperatures, and a pH of 4–8 are good factors for fungal growth.

Due to the activities of insects in improperly stored grains under the influence of moisture, ideal growth conditions for fungal growth have occurred. Essential oils have been used to disinfect stored grains instead of phosphine gas. Although it would be perfect for preventing insects, the extended use of pesticides can lead to the emergence of resistant populations.

Fungal growth may be suppressed by adding different substances like antioxidants, antifungal agents, essential oils, and preservatives and controlling atmospheres during storage. Antioxidants like butylated hydroxy-toluene, selenium, ethoxyquin, and vitamins E, A, and C have been proven to be anti-aflatoxigenic agents (Chulze, 2010). The toxicity of mycotoxins can also be decreased by food additives like piperine and allyl sulfides and by food components like coumarins, fructose, chlorophyll, and phenolic compounds (Devreese et al., 2013). Weak acids such as propionic, sorbic, and benzoic acids have been used extensively in feed to avoid fungal spoilage. Essential oils are also used due to their fungicidal activities: anethole, α-p-cymene, eugenol, and carvacrol. A significant decline in the AFB1 level in corn has been observed, dependent on the concentration of essential oils and incubation periods (Esper et al., 2014).

Lower oxygen and higher carbon dioxide concentrations should be monitored in stored grains for mycotoxin production and fungal growth. Silo bags should be used due to their water-proof quality and gas-tightness ability.

The feed and grain industry has extensively adopted the HACCP system to stop and control risks connected with toxins (Gil et al., 2016). The classification of mycotoxins as a chemical or biological threat may also be done (Chulze, 2010). Mycotoxins can fit in the HAACP system at critical points, and their required thresholds must be determined. The Food and Agriculture Organization also suggested adopting the HACCP system to control mycotoxins systematically via the food chain from field to its consumption, involving all stages involved in animal feed production and ingredients.

In the HACCP program, different measurements should be considered. The foremost consideration to avoid contamination risk in feeds is pre- and post-harvest measurements (Kim et al., 2017). For excellent corn drying for prolonged or medium storage in silos, the measurements are bacterial level, storage humidity, fungal growth, free from insects, storage temperature, grain destruction percentage, and fungal contamination (Lanyasunya et al., 2005).

3.49 DECONTAMINATION/DETOXIFICATION

A detoxifying feed with mycotoxins can be done by applying different approaches like enzymatic, microbial detoxification, and mycotoxin binders' usage in feed. Among chemical substances, ammonia has been evaluated for aflatoxin reduction. Adding sorbent materials to feed is the most important and common approach among post-harvest approaches during their movement through the GIT (gastrointestinal tract) (Jard et al., 2011). Mycotoxin binders are adsorbents, interceptor molecules, entero-sorbents, sequestrants, adsorbents, or tapping agents. Inorganic sorbents, especially organic toxins having microbial origin and clay minerals, are also used as mycotoxin binders because they can reduce the absorption of mycotoxins through the GIT (Murugesan et al., 2015; Pappas et al., 2016).

Physical structure, distribution, surface availability, pore size, and efficacy at gastrointestinal pH values are important considerations when selecting the effectiveness of an adsorbent. Inactive mycotoxin, no toxic product generation, and no effect on the nutritional profile of feed are characteristics of the decontamination process. Meanwhile, due to the variety of mycotoxins' chemical structures, decontaminating feed by a single method is impossible (Murugesan et al., 2015).

The peculiarities of mycotoxins, such as solubility, dissociation constants, shape, polarity, charge distribution, and polarity, must be assessed. *In vitro* and *in vivo* systems are used for testing sorbents. *In vivo* tests are used to check the efficiency of binders. Their efficiency can be evaluated with the help of biological markers like changes in biochemical factors and tissue remains and by performance responses.

The classification of silicate binders is done based on structure. The Phyllosilicate family is identified by sheet-type context (Jard et al., 2011). Hydrated sodium calcium alumino-silicates have frequently been described due to restricted adsorption of aflatoxin during digestion. The production of lactone (β-keto) or lactone's system has occurred.

Other important silicates include zeolites, clinoptillolites, and bentonites. Polyvinylpyrrolidone and cholestyramine belong to synthetic polymers that are mineral binders or sequestrants. Fungal coccidia also can bind with AFB1. Indigested dietary fibers also show an absorbance effect (Mayura et al., 1998). Humic acid, yeast extracts, and yeast also can adhere to aflatoxins and decrease their effects. The potential of lactic acid to adhere to mycotoxins is also observed, but the disadvantage of using lactic acid is reversible adsorption.

Mycotoxins can be adsorbed by calcium di-octahedral smectite clay in the digestive tract; thus, toxin bioavailability is reduced and leads to a decline in AFB1 exposure (Mitchell et al., 2014). The feed and farm industry uses zeolites, mica, charcoal, kaolinite, and mica as sorbents. They are used because of their capability to capture aflatoxins due to the electrical charges present on them. Combining synthetic or natural surfactants with zeolite or smectite minerals can also be used as sorbents. Their combination gives organo-zeolites or organoclays, which are also used (Fruhauf et al., 2012), and plant extracts are also used due to their adsorbent nature.

3.50 INNOVATIVE APPROACHES FOR AFLATOXIN PREVENTION

The use of sorbents in preventing and managing toxins is limited; thus, new approaches have been developed for aflatoxin prevention. Toxin production by *A. parasiticus* has been reduced using zinc chelators (Wee et al., 2016). A remarkable decrease in aflatoxin production was observed, but no detectable changes in gene expression like *aflR* and *ver1* were observed. AFB1 can be degraded by *Pleurotus ostreatus* fungus having ligninolytic enzymes. AFB1 was also produced during the development of *A. flavus* and *P. ostreatus* in rice straw. *Aspergillus oryzae* has the potential to detoxify AFB1 (Lee et al., 2017).

3.51 PHYSICAL METHODS

Different physical methods like irradiation, mechanical sorting, density segregation, and heat inactivation can reduce aflatoxin contamination. Reduction in aflatoxins is directly related to the initial contamination level. Most aflatoxins are heat resistant, but purification with heat has been evaluated in some plants, like burning almonds at 200°C can efficiently decrease aflatoxin levels in it (Yazdanpanah et al., 2005; Hwang and Lee, 2006). Aflatoxins can also be significantly reduced by washing, broiling, cooking, and steaming (Hwang and Lee, 2006). The suppression of sprouting, pathogen elimination, and reduction in aflatoxin levels can be done using ionization and irradiation techniques (Molins, 2001).

UV irradiation can also significantly decrease AFB1, AFM1, and AFB2 in different crops. The detoxification of AFB1 can easily be done by sunlight because sunlight can degrade AFB1 efficiently. In about 80% of maize and 17% of groundnut cake, aflatoxin production was reduced when exposed to 10–12 hours of sunlight (Rushing and Selim, 2019). Gamma radiation can also reduce aflatoxin production.

3.52 CHEMICAL METHODS

Alkalies, aldehydes, acids, oxidizing agents, and different gases are significantly interrupted with aflatoxigenic growth and production of aflatoxins. The degradation of aflatoxins on legumes and cereals can easily be degraded by ozone due to its electrophilic strike on carbon bonds in the furan ring (Udomkun et al., 2017). The ammoniation process can also be used in the prevention of aflatoxin production. Food preservatives like boric acid, crystal violet, sodium acetate, and benzoic acid also suppress the development of *A. flavus* and aflatoxin production.

Antioxidants and chemicals like sodium bisulfite, ammonia, and propionic acid detoxify aflatoxins (Diaz et al., 2004). Approximately 100 different chemicals are identified as having anti-aflatoxigenic activity. Butylated hydroxy-toluene, propylparaben, and butylated hydroxy-anisole are different antioxidants having fungitoxic activity against *A. flavus* (Phillips et al., 1987). Hexane extract and chloroform extracts showed a remarkable decrease in the level of aflatoxins.

3.53 BIOLOGICAL METHODS

Biological techniques, a novel technique against various mycotoxins, can also degrade aflatoxins. It has gained much importance because of its eco-friendly approach. Fungi, enzymes, yeast, bacteria, algae, and fungi are different microbes having anti-aflatoxigenic effects. Microbes show anti-aflatoxigenic activity by antagonistic action. Myco-toxigenic fungal species can also do the elimination of aflatoxins.

Probiotics can also be included in the biological control of aflatoxins. Patulin can be removed by *Enterococcus faecium*. A mixture of *Propionibacterium* with *Lactobacillus* can also significantly reduce aflatoxin levels. A combination of *Streptococcus lactis* and *Bacillus subtillis* efficiently decreased *Aspergillus flavus* transmission. Thyme and O-methoxy-cinnamaldehyde are useful against fungal growth and significantly decrease aflatoxin contamination (Zaika and Buchanan, 1987; Haskard et al., 2004).

3.54 AFLATOXIN MANAGEMENT IN MAIZE

Aflatoxin contamination in maize can be prevented mainly after post-harvesting of maize. Mostly maize is exposed to aflatoxins after the post-harvesting stage. Storage method, storage length, pest management measures, and pest infestation are the most important post-harvest operations. For storage, sun drying, shelling, and bagging of maize must be done, but it may also be stored in husk form in silos. Drying can reduce the level of aflatoxin production by 15%. Storing maize in polypropylene sacs has been preferred due to availability. Maize is mostly stored for 5–8 months, but only 1% of farmers store it beyond 12 months. The most hazardous challenge in maize storage is an infestation caused by rodents, molds, and insects, affecting 44.1–60.2% of stored maize. Pest infestation mostly occurs when storage exceeds more than 8 months. Actellic and bioremethrin have been used to avoid pest infestation because they provide a medium for mild growth, leading to aflatoxin contamination (Sugri et al., 2015). Acetic acid, sodium chloride, and propionic acid are also used for aflatoxin reduction in maize.

Different biological methods have been used in corn to prevent aflatoxin contamination. More than 90% of maize kernels are colonized by nontoxigenic strains due to coccidia or coccidia by-product application using foliar nontoxigenic strains of *A. flavus* (Lyn et al., 2009). Eighty-six percent of aflatoxin contamination may be reduced by using grain affected by nontoxigenic *A. flavus* strains. Glufosinate herbicide can reduce aflatoxin contamination in corn kernels (Tubajika and Damann, 2002).

3.55 AFLATOXIN MANAGEMENT IN SORGHUM

Aflatoxin contamination in sorghum is due to temperature and water activity. So, it is necessary to control these parameters. Aflatoxin can grow at 37°C with 0.99 water activity maximally. The minimum requirements for mycelial growth were 0.91 water activity with 25–37°C. Sorghum stored at less than 0.91 water activity can prevent aflatoxin contamination. An ideal temperature for preventing aflatoxin production is 15°C (Lahouar et al., 2016).

The product's acceptable color, texture, viscosity, and functional properties were obtained using intense extrusion conditions (Méndez-Albores et al., 2009). A significant reduction in aflatoxins was observed using the extrusion cooking temperature technique, moisture content, and citric acid in milled sorghum. The cooking of milled sorghum can degrade 17–92% of aflatoxins.

REFERENCES

Abdulrazzaq, Y. M., Osman, N., & Ibrahim, A. (2002). Fetal exposure to aflatoxins in the United Arab Emirates. *Annals of Tropical Paediatrics*, *22*(1), 3–9.

Abdulrazzaq, Y. M., Osman, N., Yousif, Z. M., & Trad, O. (2004). Morbidity in neonates of mothers who have ingested aflatoxins. *Annals of Tropical Paediatrics*, *24*(2), 145–151.

Acur, A., Arias, R. S., Odongo, S., Tuhaise, S., Ssekandi, J., Muhanguzi, D., . . . & Kiggundu, A. (2019). Genetic diversity of aflatoxin-producing Aspergillus flavus isolated from groundnuts in selected agro-ecological zones of Uganda.doi.org/10.21203/rs.2.13266/v1

Ali, S. (2016). Green analytical methods in analysis of aflatoxins. In *Materials science forum* (Vol. 842, pp. 172–181). Trans Tech Publications Ltd.

Allcroft, R., & Carnaghan, R. B. A. (1963). Groundnut toxicity: An examination for toxin in human food products from animals fed toxic groundnut meal. *Veterinary Record*, *75*, 259–263.

Alsayyah, A., ElMazoudy, R., Al-Namshan, M., Al-Jafary, M., & Alaqeel, N. (2019). Chronic neurodegeneration by aflatoxin B1 depends on alterations of brain enzyme activity and immunoexpression of astrocyte in male rats. *Ecotoxicology and Environmental Safety*, *182*, 109407.

Alvarado, A. M., Zamora-Sanabria, R., & Granados-Chinchilla, F. (2017). A focus on aflatoxins in feedstuffs: Levels of contamination, prevalence, control strategies, and impacts on animal health. *Aflatoxin-Control, Analysis, Detection and Health Risks*, 116–152.

Ames, I. A. (1989). Mycotoxins, economic and health risks. *CAST (Council of Agricultural Science and Technology), Task Force Report, 116.*

Anater, A., Manyes, L., Meca, G., Ferrer, E., Luciano, F. B., Pimpao, C. T., & Font, G. (2016). Mycotoxins and their consequences in aquaculture: A review. *Aquaculture*, *451*, 1–10.

Awuchi, C. G., Amagwula, I. O., Priya, P., Kumar, R., Yezdani, U., & Khan, M. G. (2020). Aflatoxins in foods and feeds: A review on health implications, detection, and control. *Bulletin of Environment, Pharmacology and Life Sciences*, *9*, 149–155.

Babu, D. (2010). *Rapid and sensitive detection of aflatoxin in animal feeds and food grains using immunomagnetic bead based recovery and real-time immuno quantitative PCR (RT-IQPCR) assay.* Oklahoma State University.

Baker, J. L., Bayman, P., Mahoney, N. E., Klich, M. A., Palumbo, J. D., & Campbell, B. C. (2003). Ochratoxigenic Aspergillus lanosus and A. alliaceus isolates from California tree nut orchards. In *Proceedings of the 3rd Fungal Genomics*, 4th Fumonisin, and 16th Aflatoxin Elimination Workshops, Savannah, Georgia.

Benkerroum, N. (2020). Chronic and acute toxicities of aflatoxins: Mechanisms of action. *International journal of environmental research and public health*, *17*(2), 423.

Betina, V. (1985). Thin-layer chromatography of mycotoxins. *Journal of Chromatography A*, *334*, 211–276.

Bhat, R., Ramakrishna, Y., Beedu, S., & Munshi, K. L. (1989). Outbreak of trichothecene mycotoxicosis associated with consumption of mould-damaged wheat products in Kashmir Valley, India. *Lancet*, *333*(8628), 35–37.

Bhatnagar-Mathur, P., Sunkara, S., Bhatnagar-Panwar, M., Waliyar, F., & Sharma, K. K. (2015). Biotechnological advances for combating *Aspergillus flavus* and aflatoxin contamination in crops. *Plant science*, *234*, 119–132.

Bingham, A. K., Phillips, T. D., & Bauer, J. E. (2003). Potential for dietary protection against the effects of aflatoxins in animals. *Journal of the American Veterinary Medical Association*, *222*(5), 591–596.

Blunden, G., Roch, O. G., Rogers, D. J., Coker, R. D., Bradburn, N., & John, A. E. (1991). Mycotoxins in food. *Medical Laboratory Sciences*, *48*(4), 271–282.

Braithwaite, A., & Smith, F. J. (1985). Plane chromatography. In *Chromatographic methods* (pp. 24–84). Springer.

Brown, R. L., Chen, Z. Y., Cleveland, T. E., & Russin, J. S. (1999). Advances in the development of host resistance in corn to aflatoxin contamination by *Aspergillus flavus*. *Phytopathology*, *89*(2), 113–117.

Cagauan, A. G., Tayaban, R. H., Somga, J. R., & Bartolome, R. M. (2004, September). Effect of aflatoxin-contaminated feeds in Nile tilapia (*Oreochromis niloticus* L.). In *Abstract of the 6th International Symposium on Tilapia in Aquaculture (ISTA 6) Section: Health Management and Diseases Manila, Philippines* (Vol. 12, p. 16).

Chang, P. K., Cary, J. W., Bhatnagar, D. E. E. P. A. K., Cleveland, T. E., Bennett, J. W., Linz, J. E., . . . Payne, G. A. (1993). Cloning of the *Aspergillus parasiticus* apa-2 gene associated with the regulation of aflatoxin biosynthesis. *Applied and Environmental Microbiology*, *59*(10), 3273–3279.

Chang, P. K., & Yu, J. (2002). Characterization of a partial duplication of the aflatoxin gene cluster in *Aspergillus parasiticus* ATCC 56775. *Applied Microbiology and Biotechnology*, *58*(5), 632–636.

Cheng, J., Huang, S., Fan, C., Zheng, N., Zhang, Y., Li, S., & Wang, J. (2017). Metabolomic analysis of alterations in lipid oxidation, carbohydrate and amino acid metabolism in dairy goats caused by exposure to Aflotoxin B1. *Journal of Dairy Research*, *84*(4), 401–406.

Chulze, S. N. (2010). Strategies to reduce mycotoxin levels in maize during storage: A review. *Food Additives and Contaminants*, *27*(5), 651–657.

Coppock, R. W., Christian, R. G., & Jacobsen, B. J. (2018). *Veterinary toxicology*.

D'Mello, J. P. F., Macdonald, A. M. C., Postel, D., & Hunter, E. A. (1997). 3-Acetyl deoxynivalenol production in a strain of *Fusarium culmorum* insensitive to the fungicide difenoconazole. *Mycotoxin Research*, *13*(2), 73–80.

Devreese, M., De Backer, P., & Croubels, S. (2013). Different methods to counteract mycotoxin production and its impact on animal health. *Vlaams Diergeneeskundig Tijdschrift*, *82*(4).

Dhanasekaran, D., Shanmugapriya, S., Thajuddin, N., & Panneerselvam, A. (2011). Aflatoxins and aflatoxicosis in human and animals. *Aflatoxins-Biochemistry and Molecular Biology*, *10*(22717), 221–254.

Diaz, D. E., Hagler, W. M., Blackwelder, J. T., Eve, J. A., Hopkins, B. A., Anderson, K. L., . . . Whitlow, L. W. (2004). Aflatoxin binders II: Reduction of aflatoxin M1 in milk by sequestering agents of cows consuming aflatoxin in feed. *Mycopathologia*, *157*(2), 233–241.

Divakara, S. T., Aiyaz, M., Hariprasad, P., Nayaka, S. C., & Niranjana, S. R. (2014). *Aspergillus flavus* infection and aflatoxin contamination in sorghum seeds and their biological management. *Archives of Phytopathology and Plant Protection*, *47*(17), 2141–2156.

Dohnal, V., Wu, Q., & Kuča, K. (2014). Metabolism of aflatoxins: Key enzymes and interindividual as well as interspecies differences. *Archives of Toxicology*, *88*(9), 1635–1644.

Dorner, J. W., & Cole, R. J. (2002). Effect of application of nontoxigenic strains of *Aspergillus flavus* and *A. parasiticus* on subsequent aflatoxin contamination of peanuts in storage. *Journal of Stored Products Research*, *38*(4), 329–339.

Duarte, S. C., Almeida, A. M., Teixeira, A. S., Pereira, A. L., Falcão, A. C., Pena, A., & Lino, C. M. (2013). Aflatoxin M1 in marketed milk in Portugal: Assessment of human and animal exposure. *Food Control*, *30*(2), 411–417.

El-Nagerabi, S. A. F., & Elshafie, A. E. (2001). Incidence of seed-borne fungi and aflatoxins in Sudanese lentil seeds. *Mycopathologia*, *149*(3), 151–156.

El-Nahla, S. M., Imam, H. M., Moussa, E. A., Ibrahim, A. M., & Ghanam, A. R. (2013). Teratogenic effects of aflatoxin in rabbits (*Oryctolagus cuniculus*). *Journal of Veterinary Anatomy*, *6*(2), 67–85.

Eshelli, M., Harvey, L., Edrada-Ebel, R., & McNeil, B. (2015). Metabolomics of the bio-degradation process of aflatoxin B1 by actinomycetes at an initial pH of 6.0. *Toxins*, *7*(2), 439–456.

Esper, R. H., Gonçalez, E., Marques, M. O., Felicio, R. C., & Felicio, J. D. (2014). Potential of essential oils for protection of grains contaminated by aflatoxin produced by *Aspergillus flavus*. *Frontiers in Microbiology*, *5*, 269.

Fink-Gremmels, J., & Malekinejad, H. J. A. F. (2007). Clinical effects and biochemical mechanisms associated with exposure to the mycoestrogen zearalenone. *Animal Feed Science and Technology*, *137*, 326–341.

Fouad, A. M., Ruan, D., El-Senousey, H. K., Chen, W., Jiang, S., & Zheng, C. (2019). Harmful effects and control strategies of aflatoxin b1 produced by *Aspergillus flavus* and *Aspergillus parasiticus* strains on poultry. *Toxins*, *11*(3), 176.

Fruhauf, S., Schwartz, H., Ottner, F., Krska, R., & Vekiru, E. (2012). Yeast cell based feed additives: Studies on aflatoxin B1 and zearalenone. *Food Additives & Contaminants: Part A*, *29*(2), 217–231.

Gallo, A., Solfrizzo, M., Epifani, F., Panzarini, G., & Perrone, G. (2016). Effect of temperature and water activity on gene expression and aflatoxin biosynthesis in *Aspergillus flavus* on almond medium. *International Journal of Food Microbiology*, *217*, 162–169.

Gazzotti, T., Biagi, G., Pagliuca, G., Pinna, C., Scardilli, M., Grandi, M., & Zaghini, G. (2015). Occurrence of mycotoxins in extruded commercial dog food. *Animal Feed Science and Technology*, *202*, 81–89.

Gil, L., Ruiz, P., Font, G., & Manyes, L. (2016). An overview of the applications of hazards analysis and critical control point (HACCP) system to mycotoxins. *Revista de Toxicología*, *33*(1), 50–55.

Grenier, B., & Applegate, T. J. (2013). Modulation of intestinal functions following mycotoxin ingestion: Meta-analysis of published experiments in animals. *Toxins*, *5*(2), 396–430.

Guchi, E. (2015). Implication of aflatoxin contamination in agricultural products. *American Journal of Food and Nutrition*, *3*(1), 12–20.

Gumprecht, L. A., Marcucci, A., Weigel, R. M., Vesonder, R. F., Riley, R. T., Showker, J. L., . . . Haschek, W. M. (1995). Effects of intravenous fumonisin B1 in rabbits: Nephrotoxicityxs and sphingolipid alterations. *Natural Toxins*, *3*(5), 395–403.

Gupta, R. C. (2012). Basic and clinical principles. *Veterinary Toxicology*, 604–608.

Han, Z., Ren, Y., Zhou, H., Luan, L., Cai, Z., & Wu, Y. (2011). WHO food additives series 44, 2000. *Journal of Chromatography. B, Analytical Technologies in the Biomedical and Life Sciences*, *879*(5), 411–420.

Haskard, C., Salminen, S., & Salminen, E. (2004). Lactic acid bacteria as a tool for enhancing food safety by removal of dietary toxins. *Lactic Acid Bacteria: Microbiological and Functional Aspects*, *139*, 397.

Henry, S. H., Bosch, F. X., & Bowers, J. C. (2002). Aflatoxin, hepatitis, and worldwide liver cancer risks. *Mycotoxins and Food Safety*, 229–233.

Hidy, P. H., Baldwin, R. S., Greasham, R. L., Keith, C. L., & McMullen, J. R. (1977). Zearalenone and some derivatives: Production and biological activities. *Advances in Applied Microbiology*, *22*, 59–82.

Hossein, A., & Gürbüz, Y. (2015). Aflatoxins in poultry nutrition. *KSÜ Doğa Bilimleri Dergisi*, *18*(4), 1–5.

Houssou, P. A., Schmidt-Heydt, M., Geisen, R., Fandohan, P., Ahohuendo, B. C., Hounhouigan, D. J., & Jakobsen, M. (2008). Cowpeas as growth substrate do not support the production of aflatoxin by Aspergillus sp. *Mycotoxin Research*, *24*(2), 105–110.

Hsieh, L. L., & Hsieh, T. T. (1993). Detection of aflatoxin B1-DNA adducts in human placenta and cord blood. *Cancer Research*, *53*(6), 1278–1280.

Hussain, I., & Anwar, J. (2008). A study on contamination of aflatoxin M1 in raw milk in the Punjab Province of Pakistan. *Food Control*, *19*(4), 393–395.

Hussain, S. P., Schwank, J., Staib, F., Wang, X. W., & Harris, C. C. (2007). TP53 mutations and hepatocellular carcinoma: Insights into the etiology and pathogenesis of liver cancer. *Oncogene*, *26*(15), 2166–2176.

Hussein, H. S., & Brasel, J. M. (2001). Toxicity, metabolism, and impact of mycotoxins on humans and animals. *Toxicology*, *167*(2), 101–134.

Huybrechts, B. (2011). *Evaluation of immunoassay kits for aflatoxin determination in corn & rice*. CODA-CERVA Veterinary and Agrochemical Research Centre: NRL. Available online: www. favvafsca. fgov. be/laboratories/approvedlaboratories/generalinformation/_documents/2012-05-04_Ev-aluation_immunoassay_kits_aflatoxin. pdf (Accessed on 4 May 2012).

Hwang, J. H., & Lee, K. G. (2006). Reduction of aflatoxin B1 contamination in wheat by various cooking treatments. *Food Chemistry*, *98*(1), 71–75.

Ismail, A., Riaz, M., Akhtar, S., Yoo, S. H., Park, S., Abid, M., . . . Ahmad, Z. (2017). Seasonal variation of aflatoxin B1 content in dairy feed. *Journal of Animal and Feed Sciences*, *26*(1), 33–37.

Jard, G., Liboz, T., Mathieu, F., Guyonvarc'h, A., & Lebrihi, A. (2011). Review of mycotoxin reduction in food and feed: From prevention in the field to detoxification by adsorption or transformation. *Food Additives & Contaminants: Part A, 28*(11), 1590–1609.

Jaskiewicz, K., van Rensburg, S. J., Marasas, W. F., & Gelderblom, W. C. (1987). Carcinogenicity of *Fusarium moniliforme* culture material in rats. *Journal of the National Cancer Institute, 78*(2), 321–325.

Jeyaramraja, P. R., Meenakshi, S. N., & Woldesenbet, F. (2018). Relationship between drought and preharvest aflatoxin contamination in groundnut (*Arachis hypogaea* L.). *World Mycotoxin Journal, 11*(2), 187–199.

Jin, X., Jin, X., Liu, X., Chen, L., Jiang, J., Shen, G., & Yu, R. (2009). Biocatalyzed deposition amplification for detection of aflatoxin B1 based on quartz crystal microbalance. *Analytica Chimica Acta, 645*(1–2), 92–97.

Kakde, U. B. (2012). Fungal bioaerosols: Global diversity, distribution and its impact on human beings and agricultural crops. *Bionano Front, 5*, 323–9.

Kaleibar, M. T., & Helan, J. A. (2013). A field outbreak of aflatoxicosis with high fatality rate in feedlot calves in Iran. *Comparative Clinical Pathology, 22*(6), 1155–1163.

Katile, S. O., Perumal, R., Rooney, W. L., Prom, L. K., & Magill, C. W. (2010). Expression of pathogenesis-related protein PR-10 in sorghum floral tissues in response to inoculation with *Fusarium thapsinum* and *Curvularia lunata*. *Molecular Plant Pathology, 11*(1), 93–103.

Kendra, D. F., & Dyer, R. B. (2007). Opportunities for biotechnology and policy regarding mycotoxin issues in international trade. *International Journal of Food Microbiology, 119*(1–2), 147 151.

Kensler, T. W., Roebuck, B. D., Wogan, G. N., & Groopman, J. D. (2011). Aflatoxin: A 50-year odyssey of mechanistic and translational toxicology. *Toxicological Sciences, 120*(suppl_1), S28–S48.

Kiessling, K. H. (1986). Biochemical mechanism of action of mycotoxins. *Pure and Applied Chemistry, 58*(2), 327–338.

Kim, S., Lee, H., Lee, S., Lee, J., Ha, J., Choi, Y., . . . Choi, K. H. (2017). Invited review: Microbe-mediated aflatoxin decontamination of dairy products and feeds. *Journal of Dairy Science, 100*(2), 871–880.

Kinyungu, S., Isakeit, T., Ojiambo, P. S., & Woloshuk, C. P. (2019). Spread of *Aspergillus flavus* and aflatoxin accumulation in postharvested maize treated with biocontrol products. *Journal of Stored Products Research, 84*, 101519.

Klich, M. A., Mullaney, E. J., Daly, C. B., & Cary, J. W. (2000). Molecular and physiological aspects of aflatoxin and sterigmatocystin biosynthesis by *Aspergillus tamarii* and *A. ochraceoroseus Applied Microbiology and Biotechnology, 53*(5), 605–609.

Klvana, M., & Bren, U. (2019). Aflatoxin B1–formamidopyrimidine DNA adducts: Relationships between structures, free energies, and melting temperatures. *Molecules, 24*(1), 150.

Kumar, R., & Balachandran, C. (2009). Histopathological changes in broiler chickens fed aflatoxin and cyclopiazonic acid. *Veterinarski Arhiv, 79*(1), 31–40.

Lahouar, A., Marin, S., Crespo-Sempere, A., Saïd, S., & Sanchis, V. (2016). Effects of temperature, water activity and incubation time on fungal growth and aflatoxin B1 production by toxinogenic *Aspergillus flavus* isolates on sorghum seeds. *Revista Argentina de microbiologia, 48*(1), 78–85.

Lanyasunya, T. P., Wamae, L. W., Musa, H. H., Olowofeso, O., & Lokwaleput, I. K. (2005). The risk of mycotoxins contamination of dairy feed and milk on smallholder dairy farms in Kenya. *Pakistan Journal of Nutrition, 4*(3), 162–169.

Lee, H. S., Nguyen-Viet, H., Lindahl, J., Thanh, H. M., Khanh, T. N., Hien, L. T. T., & Grace, D. (2017). A survey of aflatoxin B1 in maize and awareness of aflatoxins in Vietnam. *World Mycotoxin Journal, 10*(2), 195–202.

Leong, Y. H., Latiff, A. A., Ahmad, N. I., & Rosma, A. (2012). Exposure measurement of aflatoxins and aflatoxin metabolites in human body fluids. A short review. *Mycotoxin Research, 28*(2), 79–87.

Lewis, L., Onsongo, M., Njapau, H., Schurz-Rogers, H., Luber, G., Kieszak, S., . . . Kenya Aflatoxicosis Investigation Group. (2005). Aflatoxin contamination of commercial maize products during an outbreak of acute aflatoxicosis in eastern and central Kenya. *Environmental Health Perspectives, 113*(12), 1763–1767.

Li, P., Zhang, Q., Zhang, D., Guan, D., Liu, D. X., Fang, S., . . . Zhang, W. (2011). Aflatoxin measurement and analysis. In *Aflatoxins—Detection, measurement and control*. IntechOpen.

Li, P., Zhang, Q., & Zhang, W. (2009). Immunoassays for aflatoxins. *TrAC Trends in Analytical Chemistry, 28*(9), 1115–1126.

Liu, J., Sun, L., Zhang, N., Zhang, J., Guo, J., Li, C., . . . Qi, D. (2016). Effects of nutrients in substrates of different grains on aflatoxin B1 production by *Aspergillus flavus*. *BioMed Research International, 2016*.

Lozano, M. C., & Diaz, G. J. (2006). Microsomal and cytosolic biotransformation of aflatoxin B1 in four poultry species. *British Poultry Science*, *47*(6), 734–741.

Lyn, M. E., Abbas, H. K., Zablotowicz, R. M., & Johnson, B. J. (2009). Delivery systems for biological control agents to manage aflatoxin contamination of pre-harvest maize. *Food Additives & Contaminants: Part A*, *26*(3), 381–387.

Ma, X., Wang, W., Chen, X., Xia, Y., Wu, S., Duan, N., & Wang, Z. (2014). Selection, identification, and application of Aflatoxin B1 aptamer. *European Food Research and Technology*, *238*(6), 919–925.

Madsen, J. M. (2005). *Bio warfare and terrorism: Toxins and other mid-spectrum agents*. Army Medical Research Inst of Chemical Defense Aberdeen Proving Ground MD.

Mahbobinejhad, Z., Aminian, H., Ebrahimi, L., & Vahdati, K. (2019). Reduction of aflatoxin production by exposing *Aspergillus flavus* to CO_2. *Journal of Crop Protection*, *8*(4), 441–448.

Mantle, P. G. (2002). Risk assessment and the importance of ochratoxins. *International Biodeterioration & Biodegradation*, *50*(3–4), 143–146.

Marasas, W. F. O. (1996). Fumonisins: History, world-wide occurrence and impact. *Fumonisins in Food*, 1–17.

Marasas, W. F. O., Kellerman, T. S., Gelderblom, W. C., Thiel, P. G., Van der Lugt, J. J., & Coetzer, J. A. (1988). Leukoencephalomalacia in a horse induced by fumonisin B_1 isolated from *Fusarium moniliforme*.

Marchese, S., Polo, A., Ariano, A., Velotto, S., Costantini, S., & Severino, L. (2018). Aflatoxin B1 and M1: Biological properties and their involvement in cancer development. *Toxins*, *10*(6), 214.

Martinez-Miranda, M. M., Rosero-Moreano, M., & Taborda-Ocampo, G. (2019). Occurrence, dietary exposure and risk assessment of aflatoxins in arepa, bread and rice. *Food Control*, *98*, 359–366.

Mayura, K., Abdel-Wahhab, M. A., McKenzie, K. S., Sarr, A. B., Edwards, J. F., Naguib, K., & Phillips, T. D. (1998). Prevention of maternal and developmental toxicity in rats via dietary inclusion of common aflatoxin sorbents: Potential for hidden risks. *Toxicological Sciences*, *41*(2), 175–182.

McGlynn, K. A., & London, W. T. (2005). Epidemiology and natural history of hepatocellular carcinoma. *Best Practice & Research Clinical Gastroenterology*, *19*(1), 3–23.

McLean, M., & Dutton, M. F. (1995). Cellular interactions and metabolism of aflatoxin: An update. *Pharmacology & Therapeutics*, *65*(2), 163–192.

Mehrzad, J., Malvandi, A. M., Alipour, M., & Hosseinkhani, S. (2017). Environmentally relevant level of aflatoxin B1 elicits toxic pro-inflammatory response in murine CNS-derived cells. *Toxicology Letters*, *279*, 96–106.

Méndez-Albores, A., Veles-Medina, J., Urbina-Álvarez, E., Martínez-Bustos, F., & Moreno-Martínez, E. (2009). Effect of citric acid on aflatoxin degradation and on functional and textural properties of extruded sorghum. *Animal Feed Science and Technology*, *150*(3–4), 316–329.

Mishra, H. N., & Das, C. (2003). A review on biological control and metabolism of aflatoxin. *Critical Reviews in Food Science and Nutrition, 43*, 245.

Mitchell, N. J., Xue, K. S., Lin, S., Marroquin-Cardona, A., Brown, K. A., Elmore, S. E., . . . Phillips, T. D. (2014). Calcium montmorillonite clay reduces AFB1 and FB1 biomarkers in rats exposed to single and co-exposures of aflatoxin and fumonisin. *Journal of Applied Toxicology*, *34*(7), 795–804.

Molins, R. A. (Ed.). (2001). *Food irradiation: Principles and applications*. John Wiley & Sons.

Molyneux, R. J., Mahoney, N., Kim, J. H., & Campbell, B. C. (2007). Mycotoxins in edible tree nuts. *International Journal of Food Microbiology*, *119*(1–2), 72–78.

Monson, M. S., Coulombe, R. A., & Reed, K. M. (2015). Aflatoxicosis: Lessons from toxicity and responses to aflatoxin B1 in poultry. *Agriculture*, *5*(3), 742–777.

Murugesan, G. R., Ledoux, D. R., Naehrer, K., Berthiller, F., Applegate, T. J., Grenier, B., . . . Schatzmayr, G. (2015). Prevalence and effects of mycotoxins on poultry health and performance, and recent development in mycotoxin counteracting strategies. *Poultry Science*, *94*(6), 1298–1315.

Nesbitt, B. F., O'kelly, J., Sargeant, K., & Sheridan, A. N. N. (1962). Toxic metabolites of *Aspergillus flavus*. *Nature, London*, *195*(4846).

Nesic, K., Milicevic, D., Nesic, V., & Ivanovic, S. (2015). Mycotoxins as one of the foodborne risks most susceptible to climatic change. *Procedia Food Science*, *5*, 207–210.

Niyo, K. A., Richard, J. L., Niyo, Y., & Tiffany, L. H. (1988). Effects of T-2 mycotoxin ingestion on phagocytosis of *Aspergillus fumigatus* conidia by rabbit alveolar macrophages and on hematologic, serum biochemical, and pathologic changes in rabbits. *American Journal of Veterinary Research*, *49*(10), 1766–1773.

Obuseh, F. A., Jolly, P. E., Kulczycki, A., Ehiri, J., Waterbor, J., Desmond, R. A., . . . Piyathilake, C. J. (2011). Aflatoxin levels, plasma vitamins A and E concentrations, and their association with HIV and hepatitis B virus infections in Ghanaians: A cross-sectional study. *Journal of the International AIDS Society*, *14*(1), 1–10.

Oliveira, C. A. F. D., & Germano, P. M. L. (1997). Aflatoxinas: Conceitos sobre mecanismos de toxicidade e seu envolvimento na etiologia do câncer hepático celular. *Revista de Saúde Pública*, *31*, 417–424.

Ostrý, V., Malíř, F., & Pfohl-Leszkowicz, A. (2015). Comparative data concerning aflatoxin contents in Bt maize and non-Bt isogenic maize in relation to human and animal health–a review. *Acta Veterinaria Brno*, *84*(1), 47–53.

Pappas, A. C., Tsiplakou, E., Tsitsigiannis, D. I., Georgiadou, M., Iliadi, M. K., Sotirakoglou, K., & Zervas, G. (2016). The role of bentonite binders in single or concomitant mycotoxin contamination of chicken diets. *British Poultry Science*, *57*(4), 551–558.

Park, D. L., & Pohland, A. E. (1986). Rationale for the control of aflatoxin in animal feeds. *Bioactive Molecules*.

Pascale, M. N. (2009). Detection methods for mycotoxins in cereal grains and cereal products. *Zbornik Matice srpske za prirodne nauke* (117), 15–25.

Pepeljnjak, S., & Šegvć, M. (2004). An overview of mycotoxins and toxigenic fungi in Croatia. *An overview on Toxigenic Fungi and Mycotoxins in Europe*, 33–50.

Phillips, T. D., Kubena, L. F., Harvey, R. B., Taylor, D. R., & Heidelbaugh, N. D. (1987, June). Mycotoxin hazards in agriculture-new approach to control. In *Journal of the American Veterinary Medical Association* (Vol. 190, No. 12, pp. 1617–1617).

Piliarik, M., Bocková, M., & Homola, J. (2010). Surface plasmon resonance biosensor for parallelized detection of protein biomarkers in diluted blood plasma. *Biosensors and Bioelectronics*, *26*(4), 1656–1661.

Raghavender, C. R., Reddy, B. N., & Shobharani, G. (2007). Aflatoxin contamination of pearl millet during field and storage conditions with reference to stage of grain maturation and insect damage. *Mycotoxin Research*, *23*(4), 199–209.

Rajarajan, P. N., Rajasekaran, K. M., & Devi, N. A. (2013). Aflatoxin contamination in agricultural commodities. *Indian Journal of Pharmaceutical and Biological Research*, *1*(04), 148–151.

Rawal, S., Kim, J. E., & Coulombe Jr, R. (2010). Aflatoxin B1 in poultry: Toxicology, metabolism and prevention. *Research in Veterinary Science*, *89*(3), 325–331.

Ricci, F., Volpe, G., Micheli, L., & Palleschi, G. (2007). A review on novel developments and applications of immunosensors in food analysis. *Analytica Chimica Acta*, *605*(2), 111–129.

Richard, J. L. (2000). Mycotoxins—An overview. *Romer Labs' Guide to Mycotoxins*, *1*, 1–48.

Richard, J. L., Bray, G. A., & Ryan, D. H. (1991). Mycotoxins as immunomodulators in animal systems. *Mycotoxins, cancer and health. Pennington Centre Nutrition Series*, *1*, 196–220.

Richard, J. L., Meerdink, G. A. V. I. N., Maragos, C. M., Tumbleson, M. I. K. E., Bordson, G. A. R. Y., Rice, L. G., & Ross, P. F. (1996). Absence of detectable fumonisins in the milk of cows fed *Fusarium proliferatun* (Matsushima) Nirenberg culture material. *Mycopathologia*, *133*(2), 123–126.

Rotimi, O. A., Rotimi, S. O., Oluwafemi, F., Ademuyiwa, O., & Balogun, E. A. (2018). Oxidative stress in extrahepatic tissues of rats co-exposed to aflatoxin B1 and low protein diet. *Toxicological Research*, *34*(3), 211–220.

Rushing, B. R., & Selim, M. I. (2019). Aflatoxin B1: A review on metabolism, toxicity, occurrence in food, occupational exposure, and detoxification methods. *Food and Chemical Toxicology*, *124*, 81–100.

Sargent, A., & Sadik, O. A. (1999). Monitoring antibody–antigen reactions at conducting polymer-based immunosensors using impedance spectroscopy. *Electrochimica Acta*, *44*(26), 4667–4675.

Sava, V., Reunova, O., Velasquez, A., Harbison, R., & Sanchez-Ramos, J. (2006). Acute neurotoxic effects of the fungal metabolite ochratoxin-A. *Neurotoxicology*, *27*(1), 82–92.

Scarpari, M., Punelli, M., Scala, V., Zaccaria, M., Nobili, C., Ludovici, M., . . . Fanelli, C. (2014). Lipids in *Aspergillus flavus*-maize interaction. *Frontiers in Microbiology*, *5*, 74.

Schmidt-Heydt, M., Abdel-Hadi, A., Magan, N., & Geisen, R. (2009). Complex regulation of the aflatoxin biosynthesis gene cluster of *Aspergillus flavus* in relation to various combinations of water activity and temperature. *International Journal of Food Microbiology*, *135*(3), 231–237.

Shimizu, K., & Keller, N. P. (2001). Genetic involvement of a cAMP-dependent protein kinase in a G protein signaling pathway regulating morphological and chemical transitions in *Aspergillus nidulans*. *Genetics, 157*(2), 591–600.

Shotwell, O. L., Goulden, M. L., & Hesseltine, C. W. (1974). *Aflatoxin: Distribution in contaminated corn.*

Sirma, A. J., Ouko, E. O., Murithi, G., Mburugu, C., Mapenay, I., Ombui, J. N., . . . Korhonen, H. (2015). *Prevalence of aflatoxin contamination in cereals from Nandi County.*

Smith, J. F., Di Menna, M. E., & McGowan, L. T. (1990). Reproductive performance of Coopworth ewes following oral doses of zearalenone before and after mating. *Reproduction, 89*(1), 99–106.

Soriano, J. M., Rubini, A., Morales-Suarez-Varela, M., Merino-Torres, J. F., & Silvestre, D. (2020). Aflatoxins in organs and biological samples from children affected by kwashiorkor, marasmus and marasmic-kwashiorkor: A scoping review. *Toxicon, 185*, 174–183.

Sugri, I., Osiru, M., Larbi, A., Buah, S. S., Nutsugah, S. K., Asieku, Y., & Lamini, S. (2015). Aflatoxin management in Northern Ghana: Current prevalence and priority strategies in maize (*Zea mays* L). *Journal of Stored Products and Postharvest Research.*

Sun, C. A., Wu, D. M., Wang, L. Y., Chen, C. J., You, S. L., & Santella, R. M. (2002). Determinants of formation of aflatoxin-albumin adducts: A seven-township study in Taiwan. *British Journal of Cancer, 87*(9), 966–970.

Sun, L. H., Zhang, N. Y., Zhu, M. K., Zhao, L., Zhou, J. C., & Qi, D. S. (2015). Prevention of aflatoxin B1 hepatoxicity by dietary selenium is associated with inhibition of cytochrome P450 isozymes and up-regulation of 6 selenoprotein genes in chick liver. *Journal of Nutrition, 146*(4), 655–661.

Tesfamariam, K., De Boevre, M., Kolsteren, P., Belachew, T., Mesfin, A., De Saeger, S., & Lachat, C. (2020). Dietary mycotoxins exposure and child growth, immune system, morbidity, and mortality: A systematic literature review. *Critical Reviews in Food Science and Nutrition, 60*(19), 3321–3341.

Tjamos, S. E., Antoniou, P. P., Kazantzidou, A., Antonopoulos, D. F., Papageorgiou, I., & Tjamos, E. C. (2004). *Aspergillus niger* and *Aspergillus carbonarius* in Corinth raisin and wine-producing vineyards in Greece: Population composition, ochratoxin A production and chemical control. *Journal of Phytopathology, 152*(4), 250–255.

Trebak, F., Alaoui, A., Alexandre, D., El Ouezzani, S., Anouar, Y., Chartrel, N., & Magoul, R. (2015). Impact of aflatoxin B1 on hypothalamic neuropeptides regulating feeding behavior. *Neurotoxicology, 49*, 165–173.

Tubajika, K. M., & Damann Jr, K. E. (2002). Glufosinate-ammonium reduces growth and aflatoxin B1 production by *Aspergillus flavus*. *Journal of food protection, 65*(9), 1483–1487.

Turner, P. C. (2013). The molecular epidemiology of chronic aflatoxin driven impaired child growth. *Scientifica, 2013.*

Twyman, R. M. (2005). *Immunoassays, applications. Clinical.*

Udomkun, P., Wiredu, A. N., Nagle, M., Müller, J., Vanlauwe, B., & Bandyopadhyay, R. (2017). Innovative technologies to manage aflatoxins in foods and feeds and the profitability of application–A review. *Food control, 76*, 127–138.

Välimaa, A. L., Kivistö, A. T., Leskinen, P. I., & Karp, M. T. (2010). A novel biosensor for the detection of Zearalenone family mycotoxins in milk. *Journal of Microbiological Methods, 80*(1), 44–48.

Van Egmond, H. E. (1989). Aflatoxin M1: Occurrence, toxicity, regulation. *Mycotoxins in Dairy Products*, 11–55.

Var, I., & Kabak, B. (2009). Detection of aflatoxin M1 in milk and dairy products consumed in Adana, Turkey. *International Journal of Dairy Technology, 62*(1), 15–18.

Verma, R. J. (2004). Aflatoxin cause DNA damage. *International Journal of Human Genetics, 4*(4), 231–236.

Vries, H. D., Maxwell, S. M., & Hendrickse, R. G. (1989). Foetal and neonatal exposure to aflatoxins. *Acta Paediatrica, 78*(3), 373–378.

Wacoo, A. P., Wendiro, D., Vuzi, P. C., & Hawumba, J. F. (2014). Methods for detection of aflatoxins in agricultural food crops. *Journal of Applied Chemistry, 2014*(706291), 706291.

Waliyar, F., Osiru, M., Ntare, B. R., Kumar, K., Sudini, H., Traore, A., & Diarra, B. (2015). Post-harvest management of aflatoxin contamination in groundnut. *World Mycotoxin Journal, 8*(2), 245–252.

Wee, J., Day, D. M., & Linz, J. E. (2016). Effects of zinc chelators on aflatoxin production in *Aspergillus parasiticus*. *Toxins, 8*(6), 171.

Wild, C. P., & Turner, P. C. (2002). The toxicology of aflatoxins as a basis for public health decisions. *Mutagenesis, 17*(6), 471–481.

Williams, J. H., Phillips, T. D., Jolly, P. E., Stiles, J. K., Jolly, C. M., & Aggarwal, D. (2004). Human aflatoxicosis in developing countries: A review of toxicology, exposure, potential health consequences, and interventions. *American Journal of Clinical Nutrition*, *80*(5), 1106–1122.

Wilson, T. M., Ross, P. F., Rice, L. G., Osweiler, G. D., Nelson, H. A., Owens, D. L., . . . Pickrell, J. W. (1990). Fumonisin B1 levels associated with an epizootic of equine leukoencephalomalacia. *Journal of Veterinary Diagnostic Investigation*, *2*(3), 213–216.

Wogan, G. N. (1966). Chemical nature and biological effects of the aflatoxins. *Bacteriological Reviews*, *30*(2), 460–470.

Woloshuk, C. P., Foutz, K. R., Brewer, J. F., Bhatnagar, D., Cleveland, T. E., & Payne, G. (1994). Molecular characterization of aflR, a regulatory locus for aflatoxin biosynthesis. *Applied and Environmental Microbiology*, *60*(7), 2408–2414.

World Health Organization (WHO). (2018, February). *Aflatoxins*. Food Safety Digest, Department of Food Safety and Zoonoses. Ref. No: WHO. NHM/FOS/RAM/18.1.

Wu, Q., Jezkova, A., Yuan, Z., Pavlikova, L., Dohnal, V., & Kuca, K. (2009). Biological degradation of aflatoxins. *Drug Metabolism Reviews*, *41*(1), 1–7.

Yang, J., Bai, F., Zhang, K., Bai, S., Peng, X., Ding, X., . . . Zhao, L. (2012). Effects of feeding corn naturally contaminated with aflatoxin B1 and B2 on hepatic functions of broilers. *Poultry Science*, *91*(11), 2792–2801.

Yarru, L. P., Settivari, R. S., Antoniou, E., Ledoux, D. R., & Rottinghaus, G. E. (2009). Toxicological and gene expression analysis of the impact of aflatoxin B1 on hepatic function of male broiler chicks. *Poultry Science*, *88*(2), 360–371.

Yazdanpanah, H., Mohammadi, T., Abouhossain, G., & Cheraghali, A. M. (2005). Effect of roasting on degradation of aflatoxins in contaminated pistachio nuts. *Food and Chemical Toxicology*, *43*(7), 1135–1139.

Yiannikouris, A., & Jouany, J. P. (2002). Mycotoxins in feeds and their fate in animals: A review. *Animal Research*, *51*(2), 81–99.

Yoshinari, T., Noda, Y., Yoda, K., Sezaki, H., Nagasawa, H., & Sakuda, S. (2010). Inhibitory activity of blasticidin A, a strong aflatoxin production inhibitor, on protein synthesis of yeast: Selective inhibition of aflatoxin production by protein synthesis inhibitors. *Journal of Antibiotics*, *63*(6), 309–314.

Younis, Y. M., & Malik, K. M. (2003). TLC and HPLC assays of aflatoxin contamination in Sudanese peanuts and peanut products. *Kuwait Journal of Science and Engineering*, *30*(1), 79–93.

Yousef, M. I., Salem, M. H., Kamel, K. I., Hassan, G. A., & El-Nouty, F. D. (2003). Influence of ascorbic acid supplementation on the haematological and clinical biochemistry parameters of male rabbits exposed to aflatoxin B1. *Journal of Environmental Science and Health, Part B*, *38*(2), 193–209.

Yunus, A. W., Nasir, M. K., Aziz, T., & Böhm, J. (2009). Prevalence of poultry diseases in district Chakwal and their interaction with mycotoxicosis: 2. Effects of season and feed. *Journal of Animal and Plant Sciences*, *19*(1), 1–5.

Zaika, L. L., & Buchanan, R. L. (1987). Review of compounds affecting the biosynthesis or bioregulation of anatoxins. *Journal of Food Protection*, *50*(8), 691–708.

Zain, M. E. (2011). Impact of mycotoxins on humans and animals. *Journal of Saudi Chemical Society*, *15*(2), 129–144.

Zhang, J., Zheng, N., Liu, J., Li, F. D., Li, S. L., & Wang, J. Q. (2015). Aflatoxin B1 and aflatoxin M1 induced cytotoxicity and DNA damage in differentiated and undifferentiated Caco-2 cells. *Food and Chemical Toxicology*, *83*, 54–60.

Zhang, Y., Li, M., Liu, Y., Guan, E., & Bian, K. (2021). Degradation of aflatoxin B1 by water-assisted microwave irradiation: Kinetics, products, and pathways. *LWT*, *152*, 112310.

Zhou, L., He, X., He, D., Wang, K., & Qin, D. (2011). Biosensing technologies for *Mycobacterium tuberculosis* detection: Status and new developments. *Clinical and Developmental Immunology*, *2011*.

Zuberi, Z., Eeza, M. N., Matysik, J., Berry, J. P., & Alia, A. (2019). NMR-based metabolic profiles of intact zebrafish embryos exposed to aflatoxin B1 recapitulates hepatotoxicity and supports possible neurotoxicity. *Toxins*, *11*(5), 258.

4 An Insight into Infectious and Noninfectious Diseases of Sorghum Species

Hafiz Muhammad Usman Aslam, Nick T. Peters, Nabil Killiny, Hasan Riaz, Akhtar Hameed, Saba Aslam, and Qaiser Shakeel

4.1 INTRODUCTION

The grain sorghum, *Sorghum bicolor*, is cultivated either as an annual or a perennial grass belonging to the family Poaceae. Sorghum has an upright, robust stem with one or more tillers (subsequent shoots that grow out of the parent shoot) and lance-shaped, 30–135 cm (12–53 in.) long leaves positioned alternately on the stems. Racemes of spikelets are grouped in clusters at the plant's crown to form the inflorescence. Two florets are found in each of the spikelets, which are paired. The anthers of the flowering plant are yellow. Sorghum is an annual plant that can reach a height of 4 m before it is harvested (13 ft). Sorghum, often known as broomcorn, is said to have originated in Ethiopia, where it was likely domesticated from wild relatives (Cox *et al.*, 2018).

Pakistan occupies a total of 796,095 km^2 of land. Its coordinates in the northern hemisphere are 23–27 degrees North and 61–76 degrees East longitude. Punjab and Sindh are two provinces, while the Northwest Frontier Province and Baluchistan are the others (Shah *et al.*, 2020).

Sorghum is one of the most significant coarse-grain summer crops because it thrives in arid, open conditions and on marginal soils. Over the past 60 years, as new barrages have allowed winter irrigation of wheat to be a more economically viable proposition, sorghum output has declined as people have shifted their diets away from it in favor of wheat (Anwaar *et al.*, 2020). The value of sorghum in the poultry industry is projected to rise as more people realize it can also be utilized as a feed and fodder source.

Pakistan's primary sorghum-producing provinces are Punjab and Sindh, which account for 47% and 26% of the country's total acreage, respectively. About 60% of the entire area planted with this crop receives irrigation, while the remaining 40% relies only on natural rainfall (Iqbal *et al.*, 2015). Sorghum's decline in popularity as a food source has caused a corresponding shift in the land dedicated to the crop. Farmers switched from growing sorghum in the spring to growing cotton in the summer and wheat in the fall. Even though weather patterns have not shifted, the agricultural sector has seen a dramatic transformation due to the advent of perpetual irrigation.

Sorghum is known for its resilience and ability to grow well in semi-arid and drought-prone regions, making it a crucial crop in many parts of the world, particularly in Africa, Asia, and the Americas (Hassan *et al.*, 2018). Sorghum and millet are grown on an average of 1.5 million ha of land in Pakistan, yielding about 5.4 tonnes per acre. The output exceeds wheat (3.1 tonnes/ha) and rice (2.2 tonnes/ha). However, these crops' yields pale compared to millet and sorghum

DOI: 10.1201/b23394-4

(Mahmood *et al.*, 2017). Nevertheless, wheat is essentially a winter crop, whereas sorghum is a summer crop; therefore, demand for millet and sorghum is much lower and there is no direct rivalry. Sorghum's low yields are a result of its limited distribution to waterlogged locations, where it has no competition from rice.

4.2 USES

In Africa and Asia, sorghum is a staple diet because the grain may be popped, roasted, boiled, and eaten like rice (Aslam *et al.*, 2019). The flour milled from sorghum grain can be used to bake many kinds of bread. Animal feed and fodder made from sorghum grain are likewise in high demand. For example, syrups and molasses can be made from sweet sorghum types.

4.3 PROPAGATION

4.3.1 Basic Requirements

Sorghum is a warm season grass that thrives in various climates and soils, including those with high levels of water and nutrient deficiencies. Planting sorghum in deep, fertile, and well-draining loam soils with a pH of around 6.0 and 7.5 will produce the best results. Sorghum thrives in hot, humid, tropical, or semitropical climates but may survive in various settings (Hamid, 1980). Although the crop does better in hot, dry circumstances than corn or soybeans, frost will still kill it. Germination can only occur at soil temperatures over 18.3°C (65°F).

4.4 PROPAGATION

Plants of the genus *Sorghum* are grown from seeds. Soil temperatures must reach at least 18.3°C (65°F) before the seed may be planted. Seeds should be sown in a carefully prepared seed bed, but they are more commonly sown in furrows after plowing or broadcasting. Sorghum can be grown effectively with a wide variety of row spacings, with optimal row distances depending on the local climate and the amount of water that can be irrigated into the soil. When growing sorghum, it is best to leave 45–60 cm (17.7–23.6 in.) between rows and space plants 12–20 cm (4.7–7.9 in.) apart (Zarif *et al.*, 2013). It was determined that a density of about 120,000 plants per hectare was achieved with this spacing.

4.5 GENERAL CARE AND MAINTENANCE

Sorghum is primarily a rainfed crop; thus, it is sown after the start of the monsoon season in tropical and subtropical climates. If the average annual rainfall in a temperate region is not high enough (between 400 and 600 mm) to support the crop's needs for growth, then additional irrigation may be necessary. Farming for subsistence typically does not involve the use of fertilizers. Large-scale sorghum cultivation occurs in countries like the United States, where nitrogen fertilizer rates of up to 150 lb of nitrogen per acre are used, although these rates are modified based on the amount of nitrogen already present in the soil, the amount of moisture in the soil, and the expected yields (MacDonald, 2010). Sorghum fields should be weed-free since they lower crop yields and impair profitability. Whether the field is vast or small, weeds between the rows are often cleared by hand or by tilling. Herbicides are typically administered before planting since they are ineffective at killing weeds once they have sprouted (Teka, 2014).

4.6 DISEASES

Many diseases can impact grain sorghum, resulting in significant losses in yield and economic value. The widespread occurrence of some diseases may result in a substantial overall loss, even though individual cases may not be particularly worrisome. Some diseases are easy to spot because their symptoms are so obvious and cause a noticeable drop in production. Sorghum farmers need to be familiar with their region's most common plant diseases (Anitha *et al.*, 2020).

New pathotypes or races of plant pathogens can appear after only a brief amount of time due to mutation. Plant disease incidence needs to be routinely monitored in sorghum production due to the dynamic nature of plant diseases and annual fluctuations in environmental variables.

Hybrids exhibit a wide range of resistance in their host plants. A wide variety of microorganisms, including fungi, bacteria, and viruses, cause diseases in sorghum. Some diseases, for various reasons, do not manifest in a particular region. Diseases prevalent in the region's growth conditions should inform hybrid selection. Diseases currently difficult to control without highly resistant hybrids may be mitigated through crop rotation and other cultural measures.

4.7 SEEDLING BLIGHTS

Damping off diseases is a common name for seedling blights. Soil-borne pathogens, soil and weather climatic variables, seedling vigor, or some combination of these can all contribute to the spread of disease in young plants. Pioneer is very particular about the quality of their seeds, so they put each batch through rigorous testing for things like resistance to disease and optimal germination temperatures. As a result, growers may rest assured that they have access to premium seeds.

In some climates, soil-borne fungi can harm seeds and seedlings even when they are high quality. Soils that are too cold and damp can devastate sorghum germination and growth because they weaken the seed and make it easier for diseases to invade the young plant. Pioneer treats their seeds with fungicidal agents to prevent disease in seedlings.

The first sign of seedling blight is the inability to germinate the seed, leading to a rotten seed. It is also possible to notice necrotic tissue, seed germination, or a water-soaked appearance in the young roots or leaves (Ajeigbe *et al.*, 2020). Other times, seedlings develop but quickly wither and perish. A wide variety of fungi causes seedling blights. *Pythium*, *Fusarium*, *Aspergillus*, and *Rhizoctonia* are seedlings' most common fungal, bacterial, and viral agents of illness. The difficulties with the seedling disease can be caused by any of these pathogens or a combination. Seedling diseases can be kept to a minimum with high-quality seed, planting only when soils are above 65°F, and the application of suitable seed treatments.

4.8 LEAF BLIGHT

The fungus *Exserohilum turcicum* causes a foliar disease known as sorghum leaf blight. In 1996, the disease was most widespread throughout western Texas but still occurred intermittently in the sorghum-growing regions of the United States and northern Mexico. This disease thrives in moderate temperatures and high humidity in many regions where sorghum is cultivated (60–80°F). Heavy dew can help spread the infection, which can be slowed by dry conditions.

Infected seedlings will have tiny patches on their leaves that are either red or tan. These spots will get bigger as the disease develops, the leaves will droop and turn a purplish gray, and the seedling may die. Long, oval lesions that are either reddish purple or yellowish tan appear on the elder leaves of mature plants. The color and size of these lesions change as resistance does. Many lesions first appear on older leaves and eventually spread to younger ones (Anitha *et al.*, 2020).

Older plants often develop yellow or gray lesions in the middle with reddish edges. When the disease is present before the boot stage, yield losses of up to 50% are likely. It is possible to prevent sorghum leaf blight through crop rotation away from vulnerable plants and use fungicides applied topically to the plant's foliage, provided these products are approved for use on sorghum. The best way to control sorghum leaf blight is to plant resistant hybrids.

4.9 SOOTY STRIPE

The warm and humid climates of Asia are ideal for developing this disease. The pathogen can infect plants at any time during their development. Yellow rings often surround circular specks of reddish brown or tan color that appear on the leaves. The lesions grow longer or more elliptical, with centers that are a straw hue. These lesions will darken to a sooty black color if the warm, humid weather persists (Singh *et al.*, 2020).

The fungi *Ramulispora sorgi* and *Ramulispora andropogonis* are responsible for sooty stripe. These infections are typically found in leaf tissue above or below the soil. Johnsongrass, also used as an inoculum source, is infected with several viruses. The spores of these infections are carried on the wind and in the rain. Crop rotation and the disposal of diseased leaf waste are effective methods of eradicating the disease. In addition, using hybrids that are resistant to the disease will help keep it under control.

4.10 ZONATE LEAF SPOT

The fungus *Gloeocercospora sorghi* D. Bain & Edgerton is responsible for the Zonate Leaf Spot of Sorghum, commonly known as Copper Spot, on many other types of grass. Spreading rapidly on forage sorghums in India, it is now a problem in all sorghum-growing regions worldwide. It thrives in warm, humid climates and survives the winter on crop residue. Initially appearing as small, circular lesions around the plant's base, the disease eventually takes on the look of a target as it spreads upward. Leaf lesions near the leaf margins tend to be semicircular, and the hybrid response to disease is typically variable (Jiang *et al.*, 2018). In extreme circumstances, the lesions will join up and wipe off a lot of the plant's photosynthesis machinery. Splashing or blowing the colorless asexual conidia from these lesions allows the disease to spread rapidly from plant to plant within the same growing season. Tiny, dark sclerotia grow in the lesion's necrotic tissue and survive the winter.

Since this disease has only recently emerged, very little is known about it. Fungicides rarely provide a positive return on investment due to the costs of producing the crop. Crop rotation and deep tillage are effective management tools when possible.

4.11 SMUTS

The fungus *Sporisorium reilianum* lives in soil and is the culprit in the development of head smut. The pathogen may quickly adapt to new environments and overcome plants' defenses. Head smuts come in a few different varieties, including the Southern variety. Fungal spores overwinter in the soil, germinating at the first sign of warmer weather. The spores will enter the sorghum plant actively in the nodal region at the tip of the shoot. The plant's reproductive tissues will be progressively destroyed as the disease spreads (Little *et al.*, 2011). A black mass of spores will completely or partially replace the sorghum head. Some hybrids, when infected, become dwarfed and tiller heavily (with all tillers producing smutted heads). The spores, carried by the wind and the rain, act as inoculum in the soil the next year. Each diseased plant releases millions of spores, increasing the likelihood of infection the following year, and the head smut fungus's soil spores can survive for years. Neither crop rotation nor fungicides have effectively prevented or cured this disease. Only through genetic manipulation is head smut controllable. Pioneer has developed several hybrids that are immune to head smut, and the company maintains a rigorous screening procedure to check for the appearance of new strains of the disease in its hybrid stock.

Fungicide seed treatments are effective against both loose kernel smut and coated kernel smut. Widespread usage of highly effective seed treatments employed in sorghum means these pathogens should not be an economic problem.

4.12 DOWNY MILDEWS

4.12.1 Crazy Top

The mildew disease crazy top is caused by *Sclerophthora macrospora* and is commonly known by its name (Wrather and Sweets, 2022). Infected plant tissue or spores can live long in the soil. For the spores to become active and infect the sorghum plants, the soil must be saturated for at least 24 hours, preferably 48. Infected plants typically manifest in flooded areas at a young age. Hard, splashing rain or flood conditions are ideal for transporting these spores into the plant whorls. Upon release, crazy top spores can be carried by water over great distances without losing viability. Leaves of young plants affected by crazy top develop a spotted, blotchy appearance, much like they might after a viral infection (Anitha *et al.*, 2020). The disease causes the plant's leaves to thicken and twist as it spreads. Depending on the severity of the infection, infected plants will die or become infertile. A "crazy" appearance is expected from the plant. Fields that are poorly drained or frequently flooded are especially vulnerable to this disease. So far, no hereditary protection against the disease has been identified, so there is a possible threat to all sorghum hybrids. In flooded situations, where this disease is present, yield losses of up to 50% are possible. Agricultural land should be kept away from areas prone to flooding. The most effective method of management is to ensure proper drainage of fields.

4.13 DOWNY MILDEW

The fungal pathogen *Peronosclerospora sorgi* is responsible for sorghum downy mildew. Sorghum seedlings become infested with downy mildew when spores germinate and penetrate the plant's roots. Systemic infections spread rapidly throughout a plant's tissues and tissues, eventually affecting the entire plant. Young plants that have been infected may develop chlorosis and eventually perish.

Yellowing is often limited to the leaf's underside on the first affected leaf. There may be a white "downy" substance on the underside of the leaves if the weather is chilly. As the disease worsens, horizontal green and white tissue streaks appear on the growing leaves. Necrosis will set in on this white tissue, and the leaves will disintegrate (Sharma *et al.*, 2010). Typically, infection causes sterility in plants, and the disease can also spread on Johnsongrass.

The sorghum downy mildew has the potential to evolve and give rise to new races. Most sorghum hybrids have genetic resistance to the downy mildew of pathotype 1, whereas resistance to the downy mildew of pathotype 3 is far less common in the sorghum sector.

Sorghum downy mildew is indigenous to the Texas Gulf Coast, but fortunately, many Pioneer brand sorghum hybrids marketed there are resistant to both pathotypes. Systemic fungicides, such as mefanoxam or metalaxyl, are required to protect susceptible hybrids from this disease. Recently, pathotype variations of downy mildew resistant to metalaxyl and mefenoxam have been found in Texas's Upper Golf Coast, which are pathotypes 3a and 6, respectively. The first line of defense in treating this new fungicide-resistant variety of downy mildew is using hybrids that have significant genetic resilience to both pathotype 3 and pathotype 6 of the disease. Furthermore, eradicating shattercane and Johnsongrass, even during crop rotation, can aid in reducing disease levels in succeeding growing seasons (Werle *et al.*, 2016).

4.14 ROOT AND STALK DISEASES

The pathogen *Fusarium moniliforme* causes fusarium infections of the stalk, root, head, and seedling. Other *Fusarium* species may also harm root and stalk tissue. Diseases caused by *Fusarium* typically cause a 5–10% loss in crops, while losses of up to 100% have been reported in certain regions (Beacorn and Thiessen, 2021). Poor grain fill, weaker peduncles, or stuck plants are common causes of yield reductions.

The occurrence of the disease varies from year to year based on several different circumstances, including the type of soil, the weather, the fertility of the soil, the drainage system, the presence of insects, the management techniques employed, and the host plant's genetics. Roots, seeds, and stalks may develop elongated or circular patches ranging in color from light crimson to dark purple if *Fusarium* is present. As the growing season develops, the fungus can infect more parts of the plant, including the roots and stalks. Extensive amounts of crimson pith can be seen where insects chew through the wood. When the plant is cut in half vertically, the crimson pith may usually be seen in the lower two or three internodes. The symptoms of this disease include the premature demise of plants while the grains are still developing. Although the outer stalk tissue has remained green, the leaves look like they have been sunburned or destroyed by freezing temperatures, or peduncles and stalks can be used as lodging (Perumal *et al.*, 2020). *Fusarium* survives the winter in plant detritus and spreads to new plants by open wounds, insect damage, or mechanical injury. The disease seems to worsen when cool, wet ones follow hot, dry, or otherwise stressful growing circumstances.

The *Fusarium* disease can be contained in sorghum fields with careful crop management. When dealing with irrigation, giving adequate moisture from bloom to hard dough is important. Tension will decrease if weeds are effectively managed and get rid of any pest insects that could cause damage. Preserve a healthy reproductive rate while a lack of potassium and an abundance of nitrogen causes stunted growth. It is best to avoid hybrid plants with strong resistance to *Fusarium* and large populations of plants in general (Das, 2019).

4.15 CHARCOAL ROT

Charcoal rot is caused by the fungus *Macrophomina phaseolina* (Chattannavar and Bannur, 2020). Charcoal rot is a major problem in sorghum crops, but it also affects corn, soybeans, sunflowers, cotton, and another 400 or so plant varieties. Sorghum planted in the dry parts of Pakistan is particularly susceptible to charcoal rot (Iqbal *et al.*, 2014). Sorghum grown in dryland areas is more susceptible to an infection caused by charcoal rot. Charcoal rot can occur in grain sorghum whenever the plant is under duress after bloom and during the fill phase. Post-flowering stress can be exacerbated by factors such as high plant populations, hail damage, leaf diseases, mechanical damage, excessive nitrogen fertilizer, and insect feeding. When hot, dry conditions occur during grain fill after warm, moist growing conditions, this disease can do a lot of damage. Root infections cause water-soaked sores that eventually turn black. Lodging is inevitable and may have a widespread impact on entire fields. Once cut open, a stuck plant's pith will be dark and resemble charcoal. *Fusarium* rot control and charcoal rot control have comparable strategies. In dryland environments, it may not always be possible to eliminate stress after a bloom; therefore, the frequency and severity of charcoal rot can be significantly reduced by managing nitrogen rates, employing suitable plant populations, and limiting insect feeding (Akhtar *et al.*, 2011). The use of hybrid selection is crucial in the fight against this disease. Even though there is no gene for resistance to charcoal rot, several Pioneer hybrids are surprisingly resistant to the disease. To lessen the impact of charcoal rot, choose hybrids with high ratings for resistance to the disease, standability, and drought.

4.16 ANTHRACNOSE

Diseases produced by the fungus *Colletotrichum graminicola* (Ces.) are classified into three groups according to the part of the plant that is affected: leaf anthracnose, panicle (grain) anthracnose, and anthracnose stalk rot. The plant's lower leaves are the first to show signs of a lesion, which then spreads up the plant. Lesions tend to seem like ovals or lengthy streaks that have been soaked in water. The lesions' borders, which might be reddish orange, purple, or tan, become increasingly noticeable as the lesions get larger (Ali and Warren, 1992). Conidiospores are released from dark fruiting bodies called acervuli once the host tissue dies. Conidiospores, which range from white to rosy pink, are the inoculum that starts secondary and panicle infections.

Symptoms of panicle anthracnose appear when plants are reaching their full growth potential. It is common for the panicle to become infected when rain or irrigation water carries conidiospores from the leaf lesions and splashes them onto the panicle. The infected panicle is usually lighter than healthy, matures before normal, and shows different degrees of sterility.

Stalk rot caused by anthracnose occurs similarly to panicle anthracnose. While infections can arise at any moment during the growth season, mature plants are especially vulnerable. The conidiospores infect the stalks when splashed or blown over the plants. The initial sign is a discolored, water-soaked appearance at the stalk's base. These spots will turn red and mingle with the healthy stalk tissue.

The anthracnose fungus can survive the winter in the seed and any plant material left on the soil's surface (Conner *et al.*, 2019). The pathogen thrives in warm, damp environments and rapidly multiplies and spreads. Resistant hybrid planting, seed treatments, crop rotation, and tillage can partially manage anthracnose.

4.17 SORGHUM ERGOT

Sorghum ergot is a fungal disease historically restricted to just the continents of Africa and Asia. In 1996, sorghum ergot was discovered for the first time in the Caribbean and Mexico (Pazoutova *et al.*, 2000). Since then, cases of ergot have been reported in every region of North America where sorghum is cultivated. However, sorghum farmers in Australia and the Americas have been hit hard by a particularly virulent disease in recent years.

Disease spores spread easily through the air and potentially infect any sorghum field. Infection occurs in the florets' pollinating organs. Infectious disease is no longer a concern once pollination has taken place. Producing rich, sugary honeydew from diseased florets is the first visible sign of illness. Infectious spores bloom on the honeydew's surface, giving the substance a white, powdery appearance. Later, nonpathogenic fungi may colonize the sweet honeydew, turning it into a mass of brown or black spores (Velasquez-Valle *et al.*, 1998). In commercial growth or seed production, the intensity of infection can be exacerbated by anything that reduces pollen output or viability. Because male-sterile plants in hybrid seed production fields require pollen from other vegetation to fertilize, these areas will be particularly vulnerable. If the disease is prevalent, fields that do not "nick" tightly will be at extreme risk. Forage hybrids with a high infection risk tend to be male-sterile forage/hay hybrids rather than commercial grain sorghum. As a result, sorghum ergot spores in the environment will almost certainly infect these hybrids.

As a preventative measure, these hybrids should be harvested for hay before heading, and heavy grazing on heading plants should be kept to a minimum. Conditions leading to low pollen yield and viability can increase the likelihood of infection in commercial grain sorghum production. Late-planted fields that see pollination in the early fall or late summer, when temperatures are low, are a prime illustration of this phenomenon (Singh *et al*., 2020). Sorghum's ability to produce pollen is limited when temperatures are low. The disease will spread nationwide if the weather is very humid and winds blow toward the known infectious sites. Each major sorghum-producing region in the Plains states is at a moderate to high risk of suffering sorghum ergot. Currently, the risk level in the southeastern states is rated as low to moderate.

A variety of events can have a negative effect on the economy. Injuries to workers in areas where seeds are grown are the most common cause of delay. If there is a high risk of infection in seed-producing areas, a preventive spray program using a fungicide that has been licensed for use will need to be implemented. Second, commercial fields, especially those that bloom later in the season, risk losing grain yield. Fields with honeydew on their leaves and grain heads are difficult to harvest, perhaps the greatest severe risk to the commercial grain-producing industry. Honeydew is a sticky plant secretion that can cling to harvesting machinery, causing damage to grain augers, headers, and sieves.

The commercial sorghum hybrids of today have not been able to incorporate any sources of resistance against sorghum ergot. The use of biotechnology to introduce resistance genes is potentially effective in the long run. In the field of sorghum biotechnology, DuPont Pioneer is an industry leader (Kamanga *et al*., 2014). In contrast to other industrial crops like maize, soybeans, and cotton, however, little progress has been made in grain sorghum biotechnology. Whether through biotechnology or conventional plant breeding, marketable hybrid resistance is far off.

Successful chemical control strategies in sorghum seed production involve applying a triazole fungicide, such as propiconazole, three or four times at five- to seven-day intervals. Fungicide use in conventional grain or forage production is probably not economically viable. Good ergot control is likely to cost between US$30–50 per acre due to the needed rates and sequence of treatments. The crusty residue covering that frequently accompanies an ergot infection can harbor viable spores in seeds grown in an infected environment (Bandyopadhyay, 1992). Seed treatment fungicides can disinfect seeds even if they have some residue of ergot on the seed coat. The spores on ergot-coated seeds can be eliminated with topical treatments like captan and fludioxonil.

The ideal conditions for an ergot infection are low temperatures and high humidity (70°F). The majority of sorghum pollinates when temperatures are high, which is often in the spring and summer. This may make the impacts of the fungus in these expanding areas less severe. Commercial sorghum production is vulnerable to ergot infection in the High Plains due to the cold temperatures that can be present during pollination. Extremely late-planted crops are the most vulnerable because of the extra time needed for pollination when temperatures drop below 60°C.

4.18 BACTERIAL LEAF SPOT

Pseudomonas syringae pv. *syringae* is the bacterium responsible for the disease known as bacterial leaf spot (Stoliar and Kliuchevych, 2022). Lesions of a tiny cylinder shape appear in groups across the leaf, and the leaf may also fail to exit the whorl. As the lesions age, they may develop a tan center, making distinguishing from pesticide harm or another fungal infection difficult. The bacterium infects and causes disease during cool, damp seasons, either at the beginning of the growing season when seeds germinate or when the plants are fully grown. The pathogen can be carried great distances on the wind or in the water or even on the seed itself. Control actions against bacterial leaf spots are probably unnecessary in Arkansas because of the disease's low economic impact.

4.19 BACTERIAL LEAF STREAK

Xanthomonas campestris var. *holcicola* is the causative agent of bacterial leaf streak. Lesions manifest initially as purple patches and then spread out. They tend to stay in the leaf's veins, giving the impression of a "streak" across the leaf. The leaf turns tan, blights, and eventually dies when the disease can thrive in warm, moist conditions or when a cultivar is especially vulnerable to it (Lang *et al.*, 2017). The tips of leaves frequently become shredded or tattered. Wind and water can carry the bacterium from infected plants to healthy ones. The bacterium enters leaves via stomata, tiny openings in the leaves, or through cuts. The economic impact of bacterial leaf streak is probably low, so there is no need to take preventative actions against it.

4.20 BACTERIAL TOP AND STALK ROT

Erwinia chrysanthemi is the bacterium responsible for causing bacterial top and stalk rot. The disease is widespread across the state but only occasionally shows up in fields. Four or five dead leaves at the top of the whorl is the most obvious sign of the disease. The plant's inside is scarlet and water-soaked and carries a putrid odor when the stalk is pulled from the soil and split longitudinally (Kharayat and Singh, 2013). Control methods for bacterial top and stalk rot have not been documented. No grain sorghum should be planted in a field with a high disease concentration the following year.

4.21 VIRAL DISEASES

Sorghum is susceptible to infection from a wide variety of viruses. However, viruses are unique among plant diseases because they do not actively invade plant tissue. Insects like aphids and leafhoppers are often responsible for vectoring viruses to their plant hosts.

Maize dwarf mosaic virus (MDMV) and sugarcane mosaic virus (SMV) are the two most common viruses that infect sorghum (Klein and Smith, 2020). Aphids, which feed on infected plants, are responsible for spreading both diseases. Johnsongrass and other grass species are major transmitters of the MDMV. In addition to MDMV, sugarcane can serve as a vector for spreading SMV (Gordon and Thottappilly, 2013). Suppression can be achieved, to some extent, through the management of insect vectors and alternative hosts. The greatest way to manage these diseases is through genetic resistance.

4.22 NEMATODES

Several species of plant-parasitic nematodes use grain sorghum as a host; however, there is currently scant evidence that they constitute a global economic danger to the sorghum crop. Many nematode species, including the spiral nematode (*Helocotylenchus*), the lesion nematode (*Pratylenchus*), the stubby-root nematode (*Paratrichodorus*), the southern root-knot nematode (*Meloidogyne incognita*), and the stunt nematodes (*Tylenchorhynchus* and *Quinisulcius*) are commonly associated with grain sorghum (Bado *et al.*, 2011; De Brida *et al.*, 2018). Large numbers of lesion nematodes in the roots of grain sorghum raise the possibility that these pests are responsible for yield losses in specific environments. Grain sorghum is neither a host for the reniform nematode (*Rotylenchulus reniformis*) nor the soybean cyst nematode (*Heterodera glycines*)

(Kobayashi-Leonel *et al.*, 2017). However, grain sorghum is prone to *M. incognita*, even though hybrids differ in their ability to maintain root-knot populations. As a result, in areas where one or more of these nematodes are present and causing problems, growing grain sorghum in rotation with soybeans, corn, or cotton may or may not be beneficial.

4.23 CONCLUSION

Resistance and cultural control strategies are usually sufficient for managing sorghum diseases. The majority of the issues associated with sorghum diseases can be prevented by taking one or more of the following practical measures:

- It is recommended to use locally adapted, disease-resistant germplasm if it is available.
- Grow healthy plants from fungicide-treated (high-quality) seed.
- Grow plants in well-drained, fertile (neutral to alkaline) soils with a pH of 6.0 to 6.5.
- Sow seeds when the soil is at least 65°F (averaged daily).
- To avoid damage from water, it is important to drain fields properly.
- Eliminate grassy weeds (especially Johnsongrass and related *Sorghum* spp.) that could serve as reservoir hosts near production fields.
- Plant hybrids with improved stalk strength and other structural properties.
- Weed, insect, and plant population control can help reduce plant stress. Plant populations that are too dense in dryland production systems are often best avoided.
- To prevent the spread of soil-borne plant pathogens, sorghum should be rotated with noncereal crops regularly.

Sorghum disease can be managed somewhat through the aforementioned general practices, but difficulties persist. Knowledge of etiological agents and their interaction with the sorghum host and the environment is still lacking in many cases, including viral diseases, bacterial leaf diseases, nematode pathogens, and many minor fungal foliar diseases. It is possible to learn quite a bit about the life cycle of certain pathogens, such as anthracnose, ergot, and grain mold.

Most agricultural regions will experience significant global climate change in the coming century, with warmer and drier conditions expected. The frequency or severity of plant pathogen–caused epidemics may change due to these shifting environmental conditions. However, the directionality of these alterations may differ depending on the pathosystem and agricultural ecosystem, and this will be determined as more empirical evidence is collected. Due to the ever-changing climate brought on by global warming, germplasm must be constantly evaluated to pinpoint the origins of resistance to both major and minor diseases and their integration with abiotic stress tolerance.

REFERENCES

Ajeigbe, H., I. Angarawai, F. Akinseye, A. Inuwa, T. Abdulazeez and M. Vabi. 2020. Handbook on Improved Agronomic Practices for Sorghum Production in North East Nigeria. Manual. The U.S. Government's Global Hunger & Food Security Initiative, Nigeria.

Akhtar, K., G. Sarwar and H. Arshad. 2011. Temperature response, pathogenicity, seed infection and mutant evaluation against *Macrophomina phaseolina* causing charcoal rot disease of sesame. Archives of Phytopathology and Plant Protection. 44:320–330.

Ali, M. and H. Warren. 1992. Anthracnose of sorghum. Sorghum and millet diseases: A second world review. In: WAJ Milliano, RA Frederiksen and GD Bengston, eds. International Crop Research Institute for the Semi-Arid Tropics. Patancheru. India. 203–208.

Anitha, K., I. Das, P. Holajjer, N. Sivaraj, C.R. Reddy and S.B. Balijepalli. 2020. Sorghum Diseases: Diagnosis and Management. Sorghum in the 21st Century: Food–Fodder–Feed–Fuel for a Rapidly Changing World. Springer, Singapore. 565–619.

Anwaar, H.A., R. Perveen, M.Z. Mansha, M. Abid, Z.M. Sarwar, H.M. Aatif, U. Ud Din Umar, M. Sajid, H.M.U. Aslam and M.M. Alam. 2020. Assessment of grain yield indices in response to drought stress in wheat (*Triticum aestivum* L.). Saudi Journal of Biological Sciences. 27:1818–1823.

Aslam, H.M.U., M.L. Gleason, A. Abbas, Z. Ul Abdin and L. Amrao. 2019. Molecular characterization of *Magnaporthe oryzae* in Punjab, Pakistan and its *in vitro* suppression by fungicides and botanicals. International Journal of Agriculture and Biology. 22:1459–1466.

Bado, V., A. Sawadogo, B. Thio, A. Bationo, K. Traoré and M. Cescas. 2011. Nematode infestation and N-effect of legumes on soil and crop yelds in legume-sorghum rotations. Agricultural Sciences. 2:49.

Bandyopadhyay, R. 1992. Sorghum ergot. Sorghum and millets diseases: A second world review. In: WAJ de Milliano, RA Frederiksen and GD Bengston, eds. International Crops Research Institute for the Semi-Arid Tropics. Patancheru, India. 235–244.

Beacorn, J.A. and L. Thiessen. 2021. First report of *Fusarium lacertarum* causing *Fusarium* head blight on sorghum in North Carolina. Plant Disease. 105:699–699.

Chattannavar, S. and V.A. Bannur. 2020. Field evaluation of sorghum genotypes against charcoal rot caused by *Macrophomina phaseolina* (Tassi) Goid. Journal of Pharmacognosy and Phytochemistry. 9:1654–1655.

Conner, R.L., C.L. Gillard, K.B. Mcrae, S.-F. Hwang, Y.-Y. Chen, A. Hou, W.C. Penner and G.D. Turnbull. 2019. Survival of the bean anthracnose fungus (*Colletotrichum lindemuthianum*) on crop debris in Canada. Canadian Journal of Plant Pathology. 41:209–217.

Cox, S., P. Nabukalu, A.H. Paterson, W. Kong and S. Nakasagga. 2018. Development of perennial grain sorghum. Sustainability. 10:172.

Das, I. 2019. Advances in sorghum disease resistance. In: Breeding Sorghum for Diverse End Uses. Elsevier.

De Brida, A.L., B.M.D.C. E Castro, J.C. Zanuncio, J.E. Serrao and S.R.S. Wilcken. 2018. Oat, wheat and sorghum cultivars for the management of *Meloidogyne enterolobii*. Nematology. 20:169–173.

Gordon, D. and G. Thottappilly. 2013. Maize and sorghum. Virus and Virus-Like Diseases of Major Crops in Developing Countries. 295.

Hamid, S. 1980. Sorghum Diseases in Pakistan. Sorghum Diseases, a World Review: Proceedings of the International Workshop on Sorghum Diseases, Sponsored Jointly by Texas A&M University (USA) and ICRISAT. Patancheru, AP.340–341.

Hassan, M.U., M.U. Chattha, A. Mahmood and S.T. Sahi. 2018. Performance of sorghum cultivars for biomass quality and biomethane yield grown in semi-arid area of Pakistan. Environmental Science and Pollution Research. 25:12800–12807.

Iqbal, M.A., A. Bilal, M. Shah and A. Kashif. 2015. A study on forage sorghum (*Sorghum bicolor* L.) production in perspectives of white revolution in Punjab, Pakistan: Issues and future options. Agriculture and Environment Science. 15:640–647.

Iqbal, U., T. Mukhtar and S.M. Iqbal. 2014. *In vitro* and *in vivo* evaluation of antifungal activities of some antagonistic plants against charcoal rot causing fungus, *Macrophomina phaseolina*. Pakistan Journal of Agricultural Sciences. 51:689–694.

Jiang, Y., J. Xu, L. Hu, K.-J. Liu, X.-D. Xu, Z. Liu and W.-L. Meng. 2018. First report of sorghum zonate leaf spot caused by *Gloeocercospora sorghi* in China. Plant Disease. 102:1033–1033.

Kamanga, G.D., M.F. Wambugu, S. Obukosia, R. Gidado and I. Suleiman. 2014. Why communication and issues management (CIMS) must occupy a central role in GM projects: Case study of the Africa biofortified sorghum (ABS) project. In: Biotechnology in Africa: Emergence, Initiatives and Future. Science Policy Reports, Springer. 225–241.

Kharayat, B.S. and Y. Singh. 2013. Unusual occurrence of Erwinia stalk rot of sorghum in Tarai region of Uttarakhand. International Journal of Agricultural Science. 9:809–813.

Klein, P. and C.M. Smith. 2020. Host plant selection and virus transmission by *Rhopalosiphum maidis* are conditioned by potyvirus infection in *sorghum bicolor*. Arthropod-Plant Interactions. 14:811–823.

Kobayashi-Leonel, R., D. Mueller, C. Harbach, G. Tylka and L. Leandro. 2017. Susceptibility of cover crop plants to *Fusarium virguliforme*, causal agent of soybean sudden death syndrome, and *Heterodera glycines*, the soybean cyst nematode. Journal of Soil and Water Conservation. 72:575–583.

Lang, J., E. Ducharme, J. Ibarra Caballero, E. Luna, T. Hartman, M. Ortiz-Castro, K. Korus, J. Rascoe, T. Jackson-Ziems and K. Broders. 2017. Detection and characterization of *Xanthomonas vasicola* pv. *vasculorum* (Cobb 1894) comb. nov. causing bacterial leaf streak of corn in the United States. Phytopathology. 107:1312–1321.

Little, C.R., R. Perumal, T.T. Tesso, L.K. Prom, G.N. Odvody and C.W. Magill. 2011. Sorghum pathology and biotechnology-A fungal disease perspective: Part I. Grain mold, head smut, and ergot. European Journal of Plant Science and Biotechnology. 6:10–30.

Macdonald, J.M. 2010. Manure Use for Fertilizer and for Energy: Report to Congress. DIANE Publishing.

Mahmood, I., S. Hassan, A. Bashir, M. Qasim and N. Ahmad. 2017. Profitability analysis of carrot production in selected districts of Punjab, Pakistan: An empirical investigation. Journal of Applied Environmental and Biological Sciences. 7:188–193.

Pazoutova, S., R. Bandyopadhyay, D.E. Frederickson, P.G. Mantle and R.A. Frederiksen. 2000. Relations among sorghum ergot isolates from the Americas, Africa, India, and Australia. Plant Disease. 84:437–442.

Perumal, R., S.S. Tomar, A. Bandara, D. Maduraimuthu, T.T. Tesso, P.V. Prasad, H.D. Upadhyaya and C.R. Little. 2020. Variation in stalk rot resistance and physiological traits of sorghum genotypes in the field under high temperature. Journal of General Plant Pathology. 86:350–359.

Shah, A.A., Z. Gong, M. Ali, R. Sun, S.A.A. Naqvi and M. Arif. 2020. Looking through the Lens of schools: Children perception, knowledge, and preparedness of flood disaster risk management in Pakistan. International Journal of Disaster Risk Reduction. 50:101907.

Sharma, R., V. Rao, H. Upadhyaya, V.G. Reddy and R. Thakur. 2010. Resistance to grain mold and downy mildew in a mini-core collection of sorghum germplasm. Plant Disease. 94:439–444.

Singh, Y., D. Sharma and B.S. Kharayat. 2020. Major diseases of sorghum and their management. In: Diseases of Field Crops: Diagnosis and Management. Apple Academic Press.

Stoliar, S. and M. Kliuchevych. 2022. Sorghum diseases in Polissia of Ukraine. Sciences of Europe:3–6.

Teka, H.B. 2014. Advance research on Striga control: A review. African Journal of Plant Science. 8:492–506.

Velasquez-Valle, R., J. Narro-Sanchez, R. Mora-Nolasco and G. Odvody. 1998. Spread of ergot of sorghum (*Claviceps africana*) in central Mexico. Plant Disease. 82:447–447.

Werle, R., A.J. Jhala, M.K. Yerka, J. Anita Dille and J.L. Lindquist. 2016. Distribution of herbicide-resistant shattercane and johnsongrass populations in sorghum production areas of Nebraska and northern Kansas. Agronomy Journal. 108:321–328.

Wrather, A. and L. Sweets. 2022. Management of Grain Sorghum Diseases in Missouri. Agricultural MU Guide. University of Missouri Extension.

Zarif, M., B. Sadia, R.A. Kainth and I.A. Khan. 2013. Genotypes, explants and growth hormones influence the morphogenesis in Pakistani Sorghum (*Sorghum bicolour*): Preliminary field evaluation of *Sorghum Somaclones*. International Journal of Agriculture & Biology. 15.

5 Infectious and Noninfectious Diseases of the Alfalfa Crop

Aqleem Abbas, Mustansar Mubeen, Qaiser Shakeel, Yasir Iftikhar, Amjad Ali, and Azhar Hussain

5.1 INTRODUCTION

Alfalfa (*Medicago sativa* L.), often known as lucerne, is the most profitable forage crop extensively grown in warm temperate and cool subtropical regions of the world (Luo et al., 2014). Due to its large production, nutritional content, and flexibility, it is often called the "King of Forages." Additionally, it enhances global food security by reducing soil erosion and increasing soil fertility (Abbas et al., 2022). It is a productive, high-quality, long-lasting legume adapted to hay, silage, grazing, and for improving soil fertility (Bhattarai et al., 2020). The plant's crown is the hardened upper part of the root, which may produce several new shoots from the same root. Because of this physiological trait, alfalfa may be harvested again over several years (Chen et al., 2022). Three to four harvests yearly are typical for alfalfa, but in desert conditions, that number may increase to as many as twelve. Growing alfalfa requires deep, well-drained, neutral pH; nevertheless, it may be cultivated on clay loam soils, provided there are no subterranean obstructions (Chen et al., 2022). The demand for alfalfa feed has skyrocketed in the past 50 years due to the plant's importance in fostering the growth of grassland animal husbandry and ensuring the longevity of agricultural ecosystems (Li and Su, 2017). However, alfalfa production is hampered by various infectious organisms (biotic agents), including fungi, bacteria, nematodes, and viruses (Zhou et al., 2019; Cai et al., 2021; Zhao et al., 2021). These biotic agents cause various diseases, drastically affecting alfalfa's output and lowering the forage's nutritional value. The crown, branches, roots, and leaves are all vulnerable. Further abiotic agents (noninfectious) caused by environmental factors like froze or freeze injury, high acidity, high temperatures or moisture, mineral deficiencies, droughts, soil compaction, and soil saturation also hampered alfalfa production (Yinghua et al., 2019; Sharath et al., 2021). Alfalfa's life span, productivity, and quality have all suffered due to the increased prevalence of biotic and abiotic agents in recent years (Wang et al., 2016; Fang et al., 2019; Wang et al., 2020). Cultural, physical, biological, chemical, breeding, and transgenic approaches have all been employed to combat biotic and abiotic agents. However, these efforts have only partially succeeded (Pan et al., 2015; Zhang et al., 2021). This chapter summarizes the main biotic and abiotic factors that hampered alfalfa production worldwide.

5.2 BIOTIC FACTORS CAUSE INFECTIOUS DISEASES

Diseases caused by biotic factors are often called infectious diseases. Biotic factors that cause infectious diseases of alfalfa are discussed in the following sections.

5.2.1 Fungi with Major Importance

Various fungal and oomycetes pathogens belong to the genera *Phytophthora*, *Pythium*, *Pseudopeziza*, *Peronospora*, *Microdochium*, *Rhizoctonia*, *Plectosphaerella*, *Macrophomina*,

DOI: 10.1201/b23394-5

Fusarium, *Verticillium*, *Stagonospora*, *Uromyces*, *Colletotrichum*, *Stemphylium*, *Phoma*, *Marasmius*, *Myrothecium*, *Paraphoma*, *Alternaria*, *Cylindrocladium*, *Aphanomyces*, *Sclerotinia*, *Bipolaris*, and *Chaetomium* are fungal pathogens of alfalfa.

5.2.1.1 *Fusarium* spp.

Fusarium spp. are typical soil-borne pathogens that mainly infect roots. *Fusarium* spp., including *F. oxysporum*, *F. avenaceum*, *F. solani*, *F. acuminatum*, *F. semitectum*, *F. moniliforme*, *F. culmorum*, *F. tricinctum*, *F. sambucinum*, *F. fusarioides*, *F. equisti*, *F. culmorum*, *F. poae*, *F. chlamydosporum*, *F. sporotriviides*, *F. nivale*, *F. graminearum*, *F. lateritum*, *F. campocceras*, and *F. proliferatum*, have been reported from infected alfalfa (Hwang and Flores, 1987; Huang et al., 2013). Among these species, *F. oxysporum*, *F. solani*, *F. moniliforme*, and *F. avenaceum* are highly pathogenic to alfalfa (Abbas et al., 2022). On plant wastes, seeds, and soils, *chlamydospores*, mycelial fragments, and spores (microconidia and macroconidia) endure the winter (Azam et al., 2018). Macroconidia have several nuclei, whereas microconidia have none. Most *Fusarium* spp. have fusiform or sickle-shaped macroconidia and spindle- or dumbbell-shaped microconidia. Since they are more resilient than other latent forms, the *chlamydospores* are the main form of *Fusarium* spp. that persist in the soil for decades (Figure 5.1). Mycelium and chlamydospores are two examples of main infection sources. Upon favorable environmental circumstances, chlamydospores germinate in reaction to the organic chemicals in alfalfa's root exudates, including sugars and amino acids. The infection's hyphae, which are produced by chlamydospores and mycelium, penetrate rootlets, taproots, and stem bases either directly or by way of wounds. The hyphae of the infection penetrate the dermis and go to the root cortex, tearing the skin. Consequently, necrotic patches ranging from brown to black develop around the roots. The hyphae then spread to the xylem and other vascular tissues, preventing water flow and causing the leaves to wilt and become yellow. The root stele eventually rots away, leaving a hollow collar and interior. No new shoots are being produced at the base, and many lateral roots will have rotted away. Leaves and stems may become a rosy color after that, and the tissues inside them may bleach. Internally, roots often exhibit dark reddish-brown streaks. In later phases, the whole inner stem may become a different color. High soil temperatures encourage the spread of *Fusarium* wilt, a once-major disease that has since become uncommon because of the introduction of *Fusarium*-resistant varieties (Li, 2003). The disease progresses to a point where the root becomes hollow, and in extreme cases, the roots rot. As time passes, the plants weaken, so they may be readily dug out and discarded. Alfalfa is susceptible to secondary infection from micro and macroconidia, which may live on the surface of infected plants and spread to healthy ones nearby. Therefore, chlamydospores are very important to the development and spread of root rot. The prevalence and severity of chlamydia depend on the quantity and persistence of chlamydospores. The damage caused by *Fusarium* spp. is especially severe for alfalfa plants that are already vulnerable due to exposure to various abiotic (nonliving) stressors. An alfalfa stand may easily last 10 years or more with proper care. Once *Fusarium* spp. has infected alfalfa, it seldom produces a good harvest the following year. The root rot that occurs depends on climate and time of year (Li, 2003). Furthermore, the primary inoculum, application of fungicides, cultivation measures, soil characteristics like soil temperature and moisture contents, environmental conditions, root rot disease produced by alfalfa's host circumstances, incidence, and prevalence all play a part regarding *Fusarium* spp. (Zaccardelli et al., 2006). When alfalfa is present in an environment conducive to fungal growth, the fungus infects the plant through wounds in the stem or rootlets, causing rapid root rot. However, it might take a few months to a year for the tap roots or collar to decay after infection. According to the findings of previous studies, *Fusarium* spp. take anything from a few months to a year to grow in the taproot or collar (Berg et al., 2017). Recent research suggests that alfalfa seeds are similarly susceptible to infection from *Fusarium* spp. Examples include the work of Kong et al. (2018), who

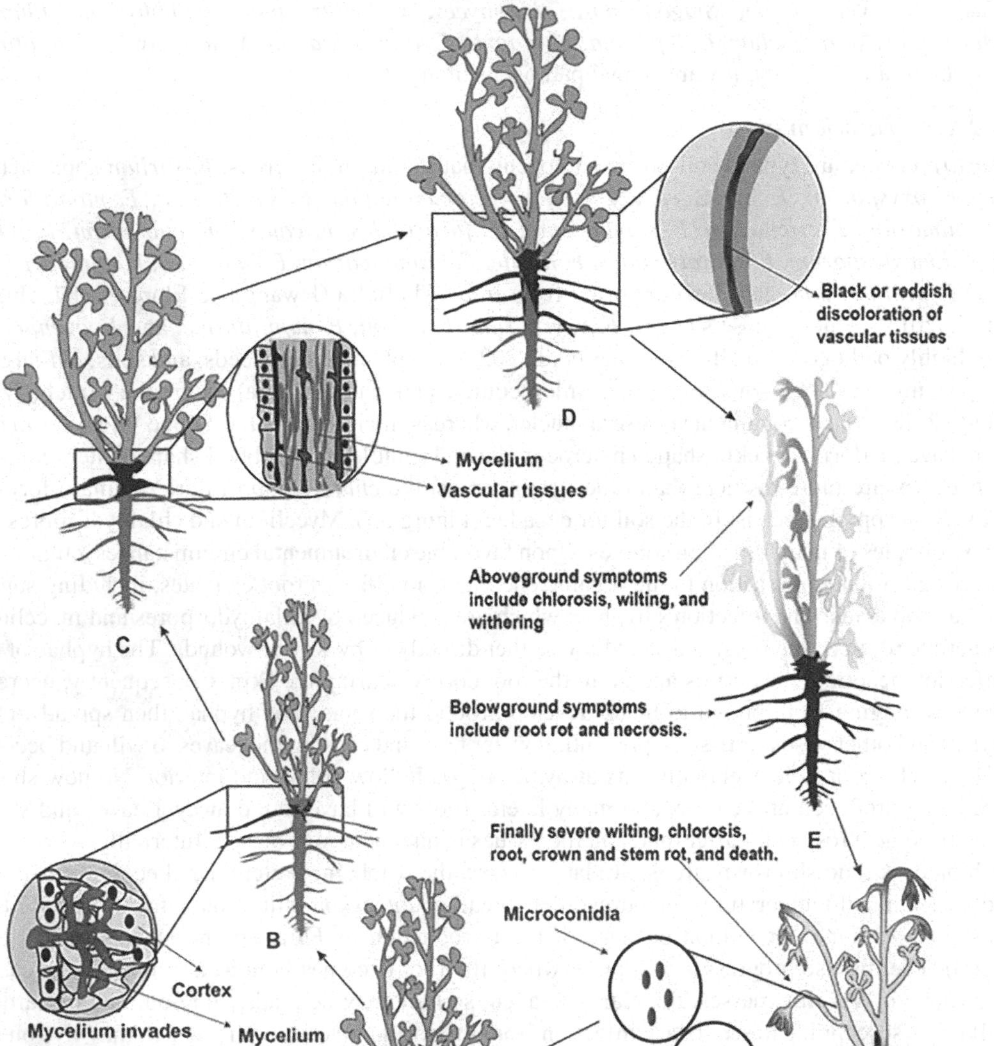

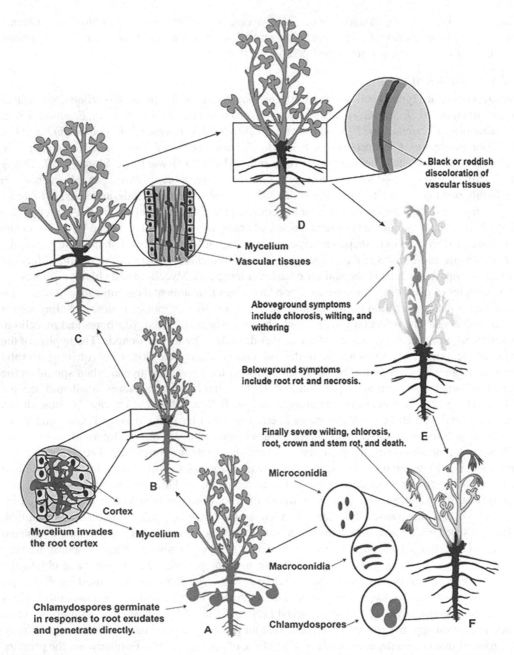

FIGURE 5.1 *Fusarium* spp. alleged disease cycle on alfalfa is depicted schematically. (A) The alfalfa plant releases root exudates, which cause *Fusarium* spp. spores (chlamydospores) to germinate and form infection hypha in order to enter the root epidermis at the root tip. (B) The hypha multiplies within the cortex of the root and enters the xylem vessels. (C) It overgrows and obstructs the vessels in the body, resulting in brown discoloration. (D) The young leaves have wilted as a result of the symptoms first developing at the stem's base and then moving upward. (E) Mature leaves are the principal site of partial or total chlorosis. (F) The alfalfa plants eventually die after going from wilt to severe root and crown rot. Dead alfalfa plant tissues produce fungal spores such microconidia, macroconidia, and chlamydospores, which are disseminated in the soil.

used alfalfa root samples to extract 150 *Fusarium* strains and then used those strains to induce pathogenicity in alfalfa seedlings (Kong et al., 2018). All of the identified *Fusarium* strains were harmful to germinating alfalfa seeds; however, their relative virulence varied. The ease with which inoculum may spread to unaffected areas is a reason for concern.

5.2.1.2 *Rhizoctonia solani*

Rhizoctonia solani (syn. *Thanatephorus cucumeris*) is the most researched and significant species of the genus *Rhizoctonia.* Kuhn initially discovered it on rotting potato tubers in 1858 (Vincelli and Herr, 1992). The 200 host plants, including cereals, vegetables, agricultural trees, horticultural trees, forest trees, weeds, ornamentals, and fodder crops are all harmed by this pervasive soil-borne necrotroph. Important characteristics of *R. solani* include septate hyphae, multinucleate cells in young hyphae, the brown coloration of mature hyphae, right-angled hyphal branching, constriction at the point of branching, and dolipore septa that allow unrestricted cell-to-cell movement of cytoplasm, mitochondria, and nuclei (Oladzad et al., 2019). Only one species of fungus, *T. cucumeris*, has been shown to have evidence of clamp connections, rhizomorphs, conidia, sexual states, and hyphal pigmentations different than brownish adult hyphae (Stalpers and Andersen, 1996). *R. solani* isolates differ substantially in terms of host range, shape of the colony and culture, molecular and biochemical markers, pathogenicity, virulence, nutritional requirements, and other characteristics. As a result of all the different cultures existing side by side, a system of categorization based on hyphal anastomosis emerged. There are now 14 anastomosis groups (AGs) recognized for *R. solani* (AG1–AG13), plus an AG-bridging isolate (AG-B1) (Ajayi-Oyetunde and Bradley, 2018). So far, no binucleate *Rhizoctonia* has been identified in diseased alfalfa; however, anastomosis groups AG1, AG2, AG4, and AG5 are all responsible for the disease. While isolates AG1, AG2, and AG5 have been observed to produce little harm to alfalfa, strain AG4 is extremely pathogenic (Balali and Kowsari, 2004). Alfalfa may be affected by various diseases, with symptoms ranging from stunting, wilting, chlorosis, and stem cankers to root canker and crown rot and pre- and post-emerging damping-off (Figure 5.2). The most typical sign of *R. solani* is a slowing down of growth before and after emergence. Furthermore, the fungus causes root and crown rot and stem rot when soil temperatures and moisture levels are optimal for its growth. Reddish-brown lesions (cankers) emerge on the stems and roots of the surviving seedlings (Williamson-Benavides and Dhingra, 2021). The pathogen seldom produces vegetative or asexual spores (like conidia). It is unknown if sexual spores, like basidiospores, serve as a source of inoculum for alfalfa root rot. The pathogen is a facultative parasite, which may successfully compete with other saprophytes that originate from the soil. This means that sclerotia formation is essential for the plant's continued existence in the soil. These propagules originate in the undifferentiated hyphae or monilioid cells and may grow without external nutrients. Sclerotia germinate and produce mycelia when circumstances are right. In response to root exudates, these mycelia are drawn to alfalfa roots, where they develop into hyphae, which then grow alongside the root's epidermal cells to produce appressoria, which the fungus uses to insert itself into the plant's tissues through infection pegs. Therefore, sclerotia are often regarded as the main inoculum of alfalfa root rot disease. According to Ajayi-Oyetunde and Bradley (2018), this pathogen is necrotrophic, using extracellular enzymes or toxins to destroy its host before colonization. *R. solani* has been linked to root discoloration and necrosis in alfalfa, although it is unknown whether these symptoms are caused by *R. solani* itself or by any of the toxins or enzymes it secretes (Ajayi-Oyetunde and Bradley, 2018). The pathogen may also persist in the environment by producing mycelium in dead plant matter. Sclerotia may continually colonize the roots of alfalfa plants if circumstances are favorable, at which point the sclerotia germinate, produce hyphae, penetrate the root cortex, and begin expanding internally and externally. Therefore, the roots develop wet, longitudinal ulcers.

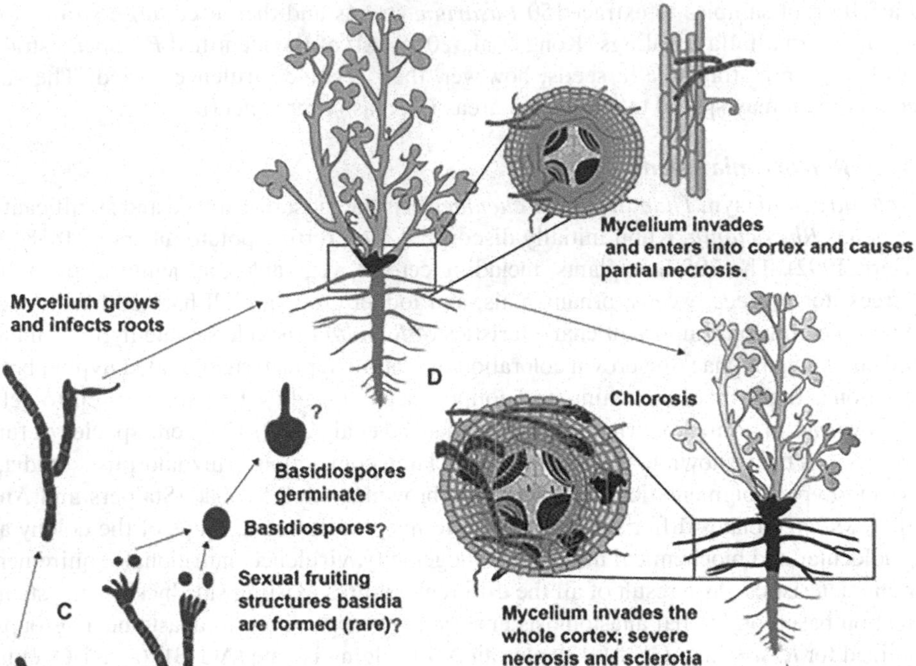

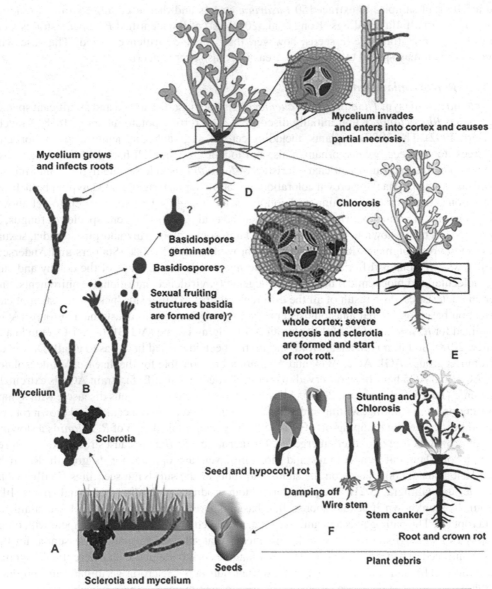

FIGURE 5.2 *Rhizoctonia solani*'s disease cycle in alfalfa is depicted schematically. (A) The mycelium of the fungus overwinters in plant debris and seeds, as well as in the soil as sclerotia and mycelium. Young hyphae (B, C) germinate and grow in a favorable environment; sexual fruiting structures like basidia and basidiospores are uncommon. (D) The mycelium invades intracellular and intercellular gaps in roots close to the soil line. (E) The cortex's continued proliferation of mycelium leads to necrosis, sclerotia in and on infected tissues, tissue disintegration, and acute root rot. (F) The fungus affects seeds and seedlings and produces damping-off, which results in chlorosis, blights, stunting, and ultimately death.

Later in the infection process, the roots begin to rot and disintegrate. The last stage is the development of sclerotia, which may re-germinate and produce additional infections if a suitable host is present. They hibernate until the host becomes available again. Sclerotia can survive for years in extreme environments, including extreme heat, cold, dehydration, and exposure to chemicals and radiation. Previous studies have shown that *R. solani* sclerotia may serve as the major inoculum and stay viable in the soil for 8–10 years without a host. Irrigation water, rain, and floods may all carry the mycelia and sclerotia to new alfalfa fields. Environmental variables such as temperature and humidity greatly influence root rot disease. The optimal conditions for the spread of disease are temperatures of 28–32°C and relative humidity of 80–95%. This way, root rot disease tends to flourish during the rainy months (Garibaldi et al., 2021).

5.2.1.3 *Phoma* and *Paraphoma* spp.

They are widespread pathogens and are often found in soil. The fungus *Phoma* spp., including *P. medicaginis* (syn. *Ascochyta medicaginicola*) and *P. sclerotioides* (syn. *Plenodomus meliloti*), is doing havoc on alfalfa all over the globe. For instance, *P. sclerotioides*, a native of Canada's prairies, is responsible for winterkill in alfalfa fields since it causes infected plants to break dormancy faster than healthy ones at higher rates elevations (Hollingsworth et al., 2003). Two varieties of *P. medicaginis*, var. *macrospora* and var. *medicaginis*, are also responsible for the spring black stem and leaf spot of alfalfa. They are widespread in North America, Europe, and Africa, where they reduce crop yields, degrade fodder quality, and negatively impact cattle (Fan et al., 2018). The conidia produced by *P. medicaginis* var. *macrospora* are bigger, and the strain's virulence on alfalfa is higher than that of *P. medicaginis* var. *medicaginis* (Li et al., 2021). Root and crown rot caused by *Paraphoma* spp., a group of soil-borne diseases, affects many plant species across the globe (Rahman et al., 2018). *Paraphoma* was formerly considered a subgenus of *Phoma*, but it was separated from its genus due to its ability to create setose, pseudo-parenchymatous pycnidia (de Gruyter et al., 2009). Most scientists agree that infected plant debris, such as stubble, is the primary vector for many diseases.

Moreover, the pathogen relies heavily on chlamydospores, quiescent mycelia, and pseudosclerotia for life (Chen et al., 2015). Despite the importance of these places and contaminated plant leftovers or stubble, life may also be ensured by using weeds as a substitute for alfalfa when the latter is unavailable. The next most significant are the asexual, dark, spherical, or flask-shaped pycnidia, which release tiny conidia. Alfalfa's principal inoculum comes from infected or contaminated seeds. Through the process of heterothallism, infected seeds not only facilitate the wide dissemination of the pathogen but also lead to the creation of virulent pathotypes. Wounds are a common route for the spread of infection. For *P. medicaginis* var. *medicaginis* to produce crown and root rot in alfalfa, for instance, it must first get access to the plant through a wound. Infected plants will show symptoms such as leaf blight, necrotic patches, chlorotic halos, spring black stems, and root and crown rot with brown lesions (Figure 5.3) (Cao et al., 2020). These problems reduce alfalfa's productivity and performance. The pathogen overwinters as pycnidia and pseudothecia in infected crop residues and field stubbles. During the same growing season, secondary infections are most often spread by the conidia produced by pycnidia (Akamatsu et al., 2008).

Following the development of germ tubes, the conidium germinates on the main infection site. In most instances, the germ tubes pierce the cuticle of leaves and stem. Stomata are delicate natural holes that are the most frequent way for fungi to enter a host; however, in *P. tracheiphila* and *P. medicaginis* var. *medicaginis*, this has never been shown to happen (occasionally reported). Additionally, a significant portion of *Phoma* spp. also enters via wounds. After successful penetration, the host's plasmalemma and chloroplast are damaged, which causes the cellular integrity to disintegrate. The virus then spreads intercellularly across the plant tissues, colonizing

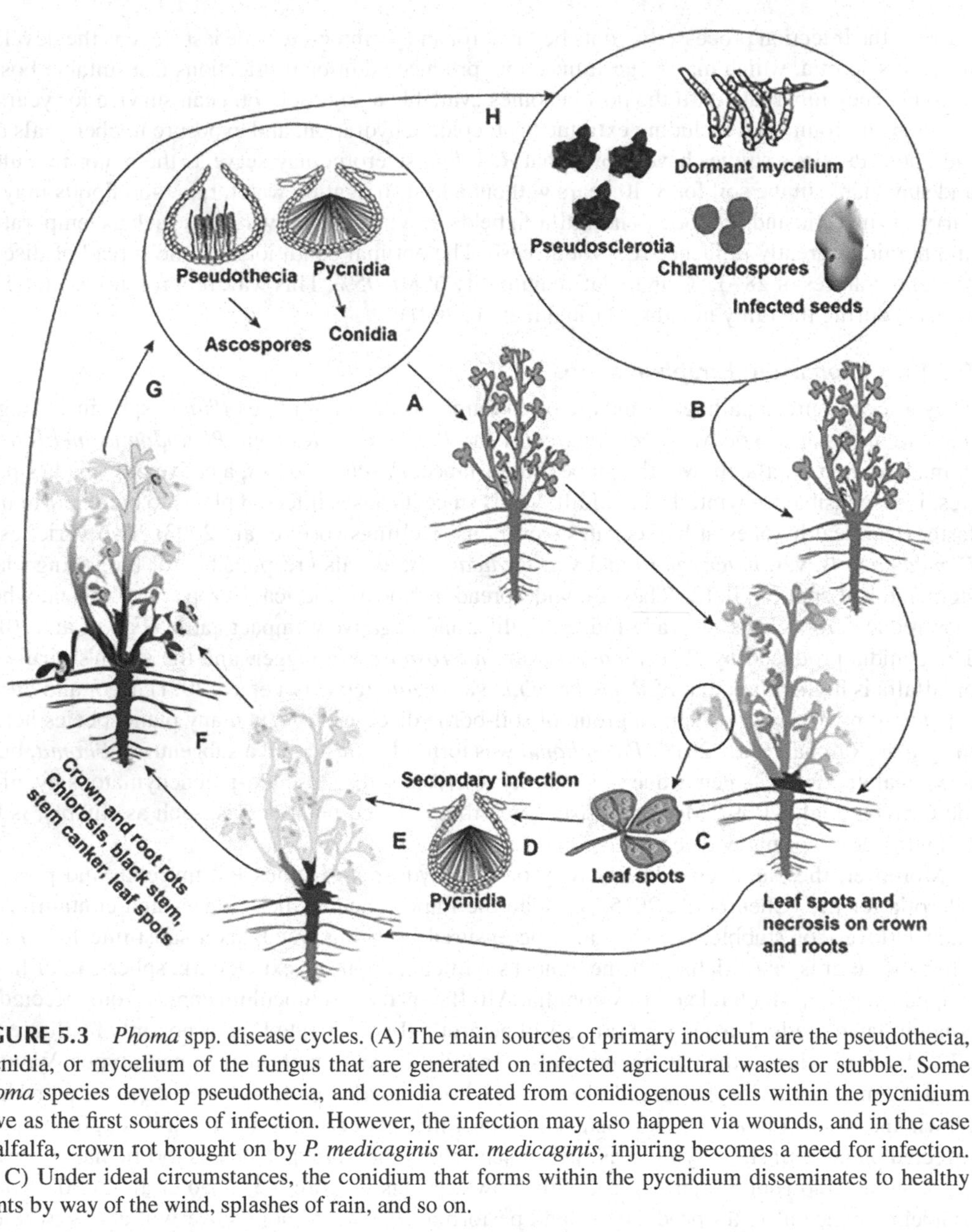

FIGURE 5.3 *Phoma* spp. disease cycles. (A) The main sources of primary inoculum are the pseudothecia, pycnidia, or mycelium of the fungus that are generated on infected agricultural wastes or stubble. Some *Phoma* species develop pseudothecia, and conidia created from conidiogenous cells within the pycnidium serve as the first sources of infection. However, the infection may also happen via wounds, and in the case of alfalfa, crown rot brought on by *P. medicaginis* var. *medicaginis*, injuring becomes a need for infection. (B, C) Under ideal circumstances, the conidium that forms within the pycnidium disseminates to healthy plants by way of the wind, splashes of rain, and so on.

them before necrosis sets in. As the infection spreads basipetal, the whole alfalfa plant becomes infected, causing it to die. The pathogen generates pycnidia, pseudothecia, or extra dermal mycelium after successfully invading plant tissues. Pycnidia and pseudothecia, which may be dispersed by wind or rain splash, generate a lot of conidia and ascospores (in teleomorphs of *Phoma*). The source of secondary inoculum during the same growing season is the recurring crop of conidia generated on infected plants. After harvest, the agricultural stubble left in the field may also distribute the mycelial pieces and conidia. Chlamydospores and conidiomata are the best structures for long-term survival in damaged crop stubble. A conidioma is a collection of conidiophores that only grow in coelomycetous fungus and create asexually by producing conidia. Chlamydospores are thick-walled resting spores with cytoplasm containing lipid reserves that allow the pathogen to survive for a year. Also, pseudo sclerotia are created, which survive, stay contagious for 14 months, and may reappear the next season to spread infection. New infections from these infected structures might happen in freshly planted crops the following season.

5.2.1.4 *Verticillium* spp.

Alfalfa output loss may be substantial because of a vascular disease called *Verticillium* wilt (Atkinson, 1981; Huang, 2003). Rapid disease spread in fields may reduce crop yields by as much as half within 3 years. The disease's effects on alfalfa crops are similar to those of rust: decreased yields and a shorter harvest window (Peaden et al., 1985). Swedish physicians initially documented this disease in 1918, and it quickly gained attention as a major health concern throughout the temperate Northern European regions (Hedlund, 1923; Sheppard, 1979). *Verticillium* wilt first affected the alfalfa industry in the United States and Canada; this disease was often found in colder places but was also reported in warm, dry regions (Sheppard, 1979; Howell and Erwin, 1995). Before the publication of Xu et al. (2016), the *Verticillium* wilt of alfalfa had only been documented in Japan and Iran (Xu et al., 2016). *Phytophthora* root rot, *Fusarium* wilt, Anthracnose, and *Gibellulopsis* spp. (*Verticillium nigrescens*) can all cause wilting in alfalfa, but only *Verticillium* wilt causes the characteristic V-shaped lesions on the leaf tips of infected plants, with green stems and discolored vascular tissue (Hu et al., 2011). All 25 Chinese alfalfa cultivars examined by Huang et al. (1999) were shown to be vulnerable to *Verticillium* wilt of alfalfa, a disease that is considered a Class A quarantined disease in China (Huang et al., 1999). Regulatory and phytosanitary measures were implemented to stop the spread of *Verticillium* wilt from contaminated seeds and hay into disease-free areas. *Verticillium* wilt might spread from field to field if the causative agent was introduced from diseased seeds or hay to a previously disease-free zone. Pathogens may infect plants via wounds in their stems and roots, and their conidia can be carried by the wind, water, and agricultural equipment like balers, mowers, and tractor wheels (Howard, 1985). In addition, insect species, both pests and beneficial ones like bees and wasps, have been identified as efficient vectors for acquiring and transmitting the virus from infected alfalfa locations to surrounding or distant wilt-free fields in acceptable conditions (Huang, 2003). Commercial alfalfa plants in Liuxin village, Minle County, western Gansu, China showed characteristic signs of *Verticillium* wilt in 2014, with an average incidence of 45.3%, and were proven to be infected with *Verticillium alfalfae* despite regulatory precautions (Xu et al., 2016). It was previously believed that *Verticillium albo-atrum* caused vascular wilt in alfalfa and that the strains obtained from affected plants were mostly host specific, meaning that they could only infect alfalfa. Thanks to multigene phylogenetic systematics research, the connection between *V. albo-atrum* and *Verticillium nonalfalfae*, two species with comparable morphological traits, has been elucidated (Inderbitzin et al., 2011). *V. alfalfae* is the species name after it was first discovered in alfalfa (Xu et al., 2019).

There are two ways alfalfa seed might harbor the fungus *Verticillium albo-atrum*: inside and externally (Xu et al., 2019). Other sources of sustenance for the fungus include alfalfa hay and animal dung. Alfalfa roots may be exposed to it directly or through wounds. Infected stems may cover a large area while swathing an alfalfa field. Trucked-in sheep herds graze pastures throughout the winter and have tested positive for the fungus. *Verticillium* wilt of alfalfa shows up as a variety of symptoms, including infected plants occurring frequently or sporadically in the fields. Infected plants show stunting and wilting symptoms. The infected stem remains green with severe chlorotic and wilt leaves with blotches. V-shaped tips of chlorotic leaves as the disease progresses. Finally, chlorotic leaves twist loosely or tightly along the midribs and remain open. Planting resistant cultivars is the most effective method of prevention. Aware measures should be taken to avoid importing contaminated seed or plant materials into regions where the disease does not exist (Xu et al., 2019).

5.2.1.5 *Pseudopeziza medicaginis*

One of the most common and damaging diseases affecting alfalfa is the common leaf spot (CLS), which is caused by the fungus *Pseudopeziza medicaginis* (Lib.) Sacc (Gui et al., 2016). Reduced yields and poor pasture quality are two major consequences of CLS. Due to CLS injury, dry matter production in alfalfa was reduced by nearly 40%. Additionally, infected plants' crude protein content and digestibility were reported to decrease by 16% and 14%, respectively. Also, ovulation and pregnancy in female cattle are negatively impacted by the estrogenic activity stimulated by infected plants. Tiny brown or black patches on the upper surface of leaves are one sign of alfalfa (Nutter Jr et al., 2002). In most cases, the edges of spots will be serrated or otherwise irregular. Infected leaves become yellow and fall off as the illness develops. Apothecia are round, brown fruiting structures produced by the fungus inside the spots during chilly, damp weather (Morgan and Parbery, 1980). Spores are released into the air by the apothecia in a forceful manner during times of chilly, damp weather. The infection of alfalfa begins when spores from the airborne pathogen fall on the leaves (Li et al., 2018). Undecomposed leaves and leaf litter on the soil's surface are the fungus' main source of sustenance. Harvesting infected alfalfa immediately is important since the illness becomes more severe as the plant matures. The defoliation caused by the disease does not kill the plants but affects plant vitality, hay quality, and production. As a result of CLS, irrigated crops may lose even more leaves after harvesting than they did previously. As a result of a lack of knowledge regarding disease-resistant cultivars, most farmers can only accept this problem as a fact of life (Li et al., 2018).

5.2.1.6 *Stagonospora* spp.

Among the *Stagonospora* spp., *Stagonospora meliloti* is the most severe pathogen of alfalfa (Summers and Gilchrist, 1991). This fungus causes spots on leaves and stems and is most prevalent in the spring when the weather is chilly and damp. Look for a light tan core and a brown, diffuse border to identify a spot. In the middle of the lesion, tiny black dots called fruiting bodies (pycnidia) develop (Musial et al., 2007). Multiple infections on a leaf usually result in defoliation once pycnidia appear. The spores in pycnidia are carried by rain or irrigation water and can infect the plant at any stage of its life cycle, including the leaves, stems, and crown (Musial et al., 2007).

5.2.1.7 *Stemphylium botryosum*

The irregularly shaped lesion with a tan core and a black border that is indicative of the fungus *Stemphylium botryosum* caused leaf spot disease is known as a leaf spot (Borges et al., 1976). After establishing a boundary, the area will not grow any larger. In the middle of the lesion, spores begin to develop. Infection and transmission are encouraged by the combination of cool and damp conditions (Cowling et al., 1981). First and second cuts are the most susceptible to disease. *Stemphylium* leaf spot is less dangerous than other leaf spot infections because defoliation only occurs under a high disease load. When temperatures are high, some types of this fungus thrive. Large areas of leaves will get covered in spots that lack a distinct boundary and continue to spread. No effective methods of prevention are currently available. Early pruning may reduce the onset of the sickness and prevent significant leaf loss in years where the disease is particularly severe. Although seed companies occasionally test for and report resistant cultivars, this is uncommon (Samac et al., 2014). In addition to dead stems and seeds, the fungus may persist in undecomposed leaf tissue on the soil's surface. The conidia and ascospores of this fungus are carried by the wind, splashed about by rain, and blown onto new plants. During the summer and autumn, thick stands are particularly susceptible to the fungus *Stemphylium* leaf spot, which flourishes during warm, rainy conditions. Small, round, or oval patches of white or tan color with a brown border appeared first and eventually grew 5 or 8 mm in diameter as the disease progressed. Massive lesions often join together to form larger ones. Leaves, petioles, stems, peduncles, and seed pods develop tiny oval dark brown marks. Dead leaves stuck to the main stem at ground level. The depressed areas often grow in size and zone out over time. Frequently, a subtle yellow 'halo' surrounds them, emphasizing their gradations of brown from light to dark. Yellowing and early drop of infected leaves are frequent disease symptoms (Samac et al., 2014). Managing this disease effectively requires timely harvesting and crop rotations that do not include forage legumes for at least 2 consecutive years. The two most common methods of managing *Stemphylium* leaf spots in alfalfa are chemical management, such as using chlorothalonil, cupric hydroxide, and mancozeb, and host resistance (Samac et al., 2014). *Pseudomonas* strain 679–2 is particularly efficient against the *Stemphylium* leaf spot (Samac et al., 2014).

5.2.1.8 Alfalfa Rust

Alfalfa suffers significant yield loss due to leaf rust produced by *Uromyces striatus*, a fungus species in the Pucciniaceae family (Adhikari and Missaoui, 2019). Spores of the rust fungus are readily obtained by scrubbing infected alfalfa leaves, petioles, and occasionally stems, which are a distinctive reddish brown (Webb and Nutter Jr, 1997). The spores survive the winter, and when conditions are right, they move to alfalfa fields, creating epidemics (Webb and Nutter Jr, 1997). The observable effects of severe rust in alfalfa include leaf shriveling, early defoliation, degradation of biomass quality, and decreased yield. Disease impact on freshly planted seedlings causes plant mortality and stand weakening, while rust also affects the persistence of existing alfalfa. Alfalfa rust infections are susceptible to environmental elements, including temperature, leaf wetness, humidity, and light. Initial pathogen invasion efficiency, latent time, and rates at which pustules form on alfalfa leaves are all influenced by temperature and leaf moisture (Webb et al., 1996). At high humidity and temperatures between 21°C and 29°C, *U. striatus* grows rapidly (Webb and Nutter Jr, 1997). The frequency of rust infections on alfalfa may vary depending on the host cell's qualities and growing conditions. Some studies have shown a weak to moderate correlation between the amount of acid detergent lignin (ADL) in alfalfa leaves and their rust resistance. Agronomic measures or resistant cultivars may prevent the spread of alfalfa rust. There have been calls for reducing the pathogen burden in the field by focusing on cypress spurge (*Euphorbia cyparissias*) and leafy spurge (*Euphorbia esula*) (Adhikari and Missaoui, 2019). Other successful methods of controlling alfalfa rust include commercial fungicides and timely trimming. Additionally, glyphosate application was proposed as an additional strategy for managing *U. striatus* in glyphosate-resistant alfalfa. In any case, the greatest long-lasting method of protecting alfalfa against rust is to plant rust-resistant varieties.

5.2.1.9 Anthracnose

One of the most significant diseases affecting alfalfa globally is anthracnose, caused by the fungus *Colletotrichum* and reduces production and fodder quality (Porto et al., 1988; Yang et al., 2008). Anthracnose may spread to other plant sections, although it mainly targets the stems and the crowns. Reduced production and quality may occur during the stem disease phase (Elgin et al., 1981), while the crown phase can reduce the number of stems per plant and increase plant mortality, both of which can contribute to a drop in alfalfa stands (Jones et al., 1978; Porto et al., 1988). Because of this, yields might drop by as much as 25% in locations with regular or heavy rainfall or irrigated areas (Juliatti et al., 2017). Most cases of anthracnose in alfalfa may be traced back to a fungus called *Colletotrichum trifolii*. Although the disease manifests in many ways (leaves, stems, crowns), crown rot is the most devastating. A dry, bluish-black V-shaped decay characterizes the crown rot. The root decay causes a change in hue from tan to brick red as the rot progresses. The whitewashing of dead stems is a common occurrence around such crowns. The abrupt demise of the stem causes the leaves to stay linked to it even after the plant has died. Small, irregularly shaped blackened patches on stems evolve into huge, oval, or diamond-shaped, straw-colored lesions with black margins, which are diagnostic of anthracnose. Lesions eventually generate tiny fruiting bodies (acervuli) home to spores of a crimson hue. In severe cases, lesions may consolidate as they get larger, resulting in the girdling and eventual death of afflicted stems. Dead, white shoots are all over the field in the summer and autumn. Like a shepherd's crook, curling over is a common symptom of the afflicted stem. Fungal spores may live in alfalfa waste and crowns for a long time. The peak of the disease occurs in the late summer and early autumn when the temperature and humidity are at their highest. Spores on stem lesions serve as an inoculum source throughout the growing season. Splashing water from rain and irrigation washes spores onto the developing stems and petioles. The threshing process risks spore transmission if contaminated seeds are dispersed. The adoption of anthracnose-resistant cultivars is crucial in its prevention. Alfalfa may still be harvested from infected fields before significant losses occur. To eliminate inoculum in the field, rotate out clover and alfalfa for 2 years and plant something else.

5.2.1.9.1 *Sclerotinia*

Under moist, chilly circumstances typical of foggy Central Valley winters, *Sclerotinia* stem and crown rot caused by *Sclerotinia trifoliorum* (Eriks.) or *S. sclerotiorum* ([Lib.] de Bary) may do significant harm to alfalfa. It is also unusual in the desert or during dry winters. White cottony fungal threads (mycelia) and hard black formations (sclerotia) near the base of crowns and stems, or within stems, are signs of this disease. Round or irregular in form (0.1–0.3 in. [2–8 mm] in diameter), sclerotia when opened, a white inside may be seen. *Sclerotinia* stem and root rot of alfalfa are caused by two different fungal species with similar life cycles. The fungus spends the summer as sclerotia, which may be either partly buried or on the soil's surface. Small (0.25 in. [5 mm] in diameter or smaller), orange, yellow, or tan mushroom-like formations (apothecia) emerge on the soil surface when sclerotia sprout in late autumn or early winter when temperatures decrease, and soils get moist from rain or irrigation. Wind releases millions of spores from apothecia and spreads to nearby plants and fields. Spores that land on moist, cool alfalfa tissue will germinate and infect the plant if temperatures are between 10°C and 20°C. White, cottony mycelial growth will appear on stems and around the crown if the right circumstances persist for disease development. Commercial nondormant variants do not yet have any effective genetic resistance. The easiest way to prepare established fields for winter is to mow or graze off as much vegetation as possible in the fall. Together, this and effective weed management open the canopy, letting sunlight and wind reach the plant bases and lowering the humidity and moisture the fungus needs to develop and sustain infections. Most sclerotia in the soil will not germinate if plow the land is plowed deeply before planting. However, new infections might occur from alfalfa fields or weed hosts nearby. The outcomes of fungicide studies are encouraging, but the emergence of resistance to these chemicals is a major setback.

5.2.2 Fungi with Minor Importance

Bipolaris spp. are also responsible for causing root rots of various plants worldwide among the *Bipolaris* spp. *B. sorokiniana* causes diseases in several forage crops, including alfalfa (Al-Sadi, 2021; Jiang et al., 2021). The teleomorph for this fungus is *Cochliobolus sativus*. However, the sexual reproduction of *C. sativus* has rarely been reported. By comparison, *B. sorokiniana* reproduces asexually by producing conidia on conidiophores (Sultana et al., 2018). The fungus is reported to infect seeds, and when in the next season, the infected seeds are grown, which may result in root rot disease of alfalfa (Al-Sadi and Deadman, 2010). In May 2017 in Harbin, Heilongjiang Province, disease symptoms caused by *Bipolaris* spp. were observed on alfalfa plants, including stunting, foliar chlorosis, and wilting with brownish or brownish-reddish water-soaked root lesions and root rot. The disease covered 40% of the area in two commercial alfalfa fields. Isolates were collected from the infected plants and, based on morphological and molecular strategies, identified as *B. sorokiniana*. Under greenhouse conditions, the alfalfa plants were inoculated with *B. sorokiniana*, and plants were maintained at 23 ± 3°C and 65% relative humidity in a greenhouse. After 20 days, root symptoms were visible (Li et al., 2019).

The other potential fungi that cause root rot of alfalfa are *Plectosphaerella* spp. The members of this genus are causing considerable losses to vegetables and trees worldwide (Carlucci et al., 2012). *P. cucumerina* was reported on alfalfa fields in Huanxi Country, Gansu Province, China, in 2015. A year later, in 2016, brown root rot disease was observed on the alfalfa collected from several fields in Tongliao City, Inner Mongolia Autonomous Region of China. The incidence was about 50–70% in the 2-year-old alfalfa fields. Infected alfalfa plants' roots were decayed. About 44 isolates were recovered; six resembled the genus *Plectosphaerella*. The isolates identified *P. cucumerina* based on morphological and molecular characteristics (Zhao et al., 2021). The pathogenicity was confirmed by root dipping in the conidial suspensions of the isolates. In the greenhouse temperature ranges, 25–28°C under a 12-h photoperiod was maintained. After 25 days, the roots of the alfalfa showed a brownish lesion. Other fungi responsible for causing roots to rot in the alfalfa are *Microdochium* spp. In 2015, crown and root rot was detected in the alfalfa fields of Huanxian County in Gansu Province. About 11 fungal species were recovered from the alfalfa roots. These isolates resembled the genus *Microdochium*. Molecular work identified the isolates as *M. tabacinum* (Wen et al., 2015).

Other fungi which cause root rot are *Macrophomina* spp. (Marquez et al., 2021). These species are generally soil-borne and found all over the world. They affect more than 100 families and at least 500 different plant species. They result in seedling blights, stem and root rot, and charcoal rot. In high temperatures (30–35°C) and poor soil moisture (below 60%) (Marquez et al., 2021). It exists in the soil as solid masses of toughened fungal mycelium called microsclerotia. They are spherical, oval, or oblong and are initially light brown before turning darker (brown to brown) and then black as they mature. They also produce pycnidia, which are hardly ever detected under natural conditions (Lakhran et al., 2018). From 2017 to 2019, surveys on disease in the major alfalfa-growing region in Northwest China were conducted. Root and crown rot disease of alfalfa was commonly found, with incidence ranging from 30% to 80%. Infected plants were stunted and wilted, and red-brown to dark brown discoloration was observed in the cortex/vascular tissues in the crown area. Moreover, lateral roots and stem tissues also showed similar symptoms. In severe cases, roots and crowns were decayed, and shoots of alfalfa plants were wilted. Two hundred fifty symptomatic plants were randomly collected in five alfalfa fields in Gansu Province. About 21 isolates were primarily identified as *Macrophomina phaseolina* (Wang et al., 2020). *M. phaseolina* has also been reported to cause root and crown rot on alfalfa in United States and many legume crops worldwide (Pandey et al., 2020).

5.2.3 Oomycetes

Oomycetes, also known as "water molds," are a group of several hundred organisms that resemble fungi in their growth habits and nutritional strategies (Harper et al., 2005). However, they form a distinct phylogenetic lineage and are evolutionarily distant from true fungi. They are more closely related to green plants and algae. They are members of the Stramenopiles kingdom and contain the most dangerous plant pathogen. They bring about root rot, downy mildew, foliar blights, damping-off, and seedling blights.

Its cell walls include cellulose, have tubular mitochondrial cristae, and are vegetative diploid, in contrast to real fungi (Van West et al., 2003). They have the capacity to endure in both watery and arid conditions. The majority of terrestrial oomycetes are parasites of alfalfa, and they contain a number of significant diseases such *Aphanomyces* spp., *Pythium* spp., and *Phytophthora* spp. that cause root rot. They do so by producing oospores, which are hard, durable structures. When alfalfa plant roots exude chemical signals, oospores respond by germinating. They produce a germ tube and directly infect the alfalfa roots by producing infection hyphae or proliferating as sporangia. Within the sporangia, zoospores with heterokont flagella (one tinsel and one whiplash) are produced. Zoospores are expelled from sporangia and use their flagella to swim through water-filled soil pores. A zoospore loses its flagella and encysts as it touches the alfalfa root surface and then begins to germinate by creating a germ tube. The epidermis is penetrated by hyphae from the germ tube, which populate the roots. The hyphae within the roots separate into antheridia and oogonia, and oospores are produced. Without alfalfa, oospores can persist for several years in the soil (Figure 5.4).

5.2.3.1 *Aphanomyces* spp.

Root rot of legume crops caused by *Aphanomyces* spp. was first described by Jones and Drechsler in 1927. Among *Aphanomyces* spp. that cause root rot, *A. cochlioides* and *A. euteiches* have become a severe problem for many plants in the world (Karppinen et al., 2020). *A. cochlioides* cause severe root rot and damping-off in sugar beet, spinach, and other plants of the families, that is, *Chenopodiaceae* and *Amaranthaceae*. Epidemics of *A. cochlioides* root rot are common in the sugar beet–growing regions of the world. *A. euteiches* causes seedling damping-off and root rot disease in alfalfa; as a result, severe yield losses have been observed. For example, 10–80 % yield losses occur due to *A. euteiches* (Gaulin et al., 2007; Richard et al., 1991). Both *A. euteiches* and *A. cochlioides* are strictly soil-borne pathogens that may thrive in soil for up to 10 years. Therefore, no adequate control strategy is available. The only way to control these pathogens is to not cultivate plants in infected fields for 10 years. Against these pathogens, so far, no completely resistant cultivars have been created (Jacquet and Bonhomme, 2019). *A. euteiches* is also causing significant yield losses in alfalfa-growing regions of the world. It limits water and nutrient passage from the roots to aboveground parts of the plant; consequently, plants become stunted, and yield losses occur (Cao et al., 2020). In China, about 235 fungal pathogen isolates were isolated from the legume crops in Gansu province in 1992. *A. euteiches* was identified as the main pathogen in the root rot complex (Wu et al., 2018). Similarly, in 1996, various fungal pathogens, including *A. euteiches*, were isolated as a component of legume root rots from the central dry region of Qinghai Province (Zhimin, 1996). *A. euteiches* is mainly present in China's wet and poorly drained alfalfa fields. So far, no report regarding *A. cochlioides* infecting alfalfa is available from China (Wu et al., 2018). Infected seedlings develop yellow cotyledons followed by chlorosis of other leaflets. Roots and stems initially appear gray and water-soaked, then turn light to dark brown, and finally rotted. Alfalfa seedlings become stunted, and leaves become yellowish eventually; the death of alfalfa seedlings occurs with full stand loss in wet and poorly drained soils. Even if seedlings survive, the damage to seedling roots reduces forage yields and winter survival. *Aphanomyces euteiches* also infect the adult plant*s* in the wet periods, which causes loss of feeder roots and root nodules, affecting nitrogen fixation, yield, and stand life (Karppinen et al., 2020).

5.2.3.2 *Phytophthora* spp.

There are more than 100 species in the *Phytophthora* genus, the majority of which are virulent plant diseases that severely damage agricultural, horticultural, and pasture crops (Cai et al., 2021). The phrase "plant destroyer," *Phytophthora*, was used in the 19th century after they decimated Irish potato farms, resulting in the Great Irish Famine (Yang et al., 2017). *Phytophthora* spp.,

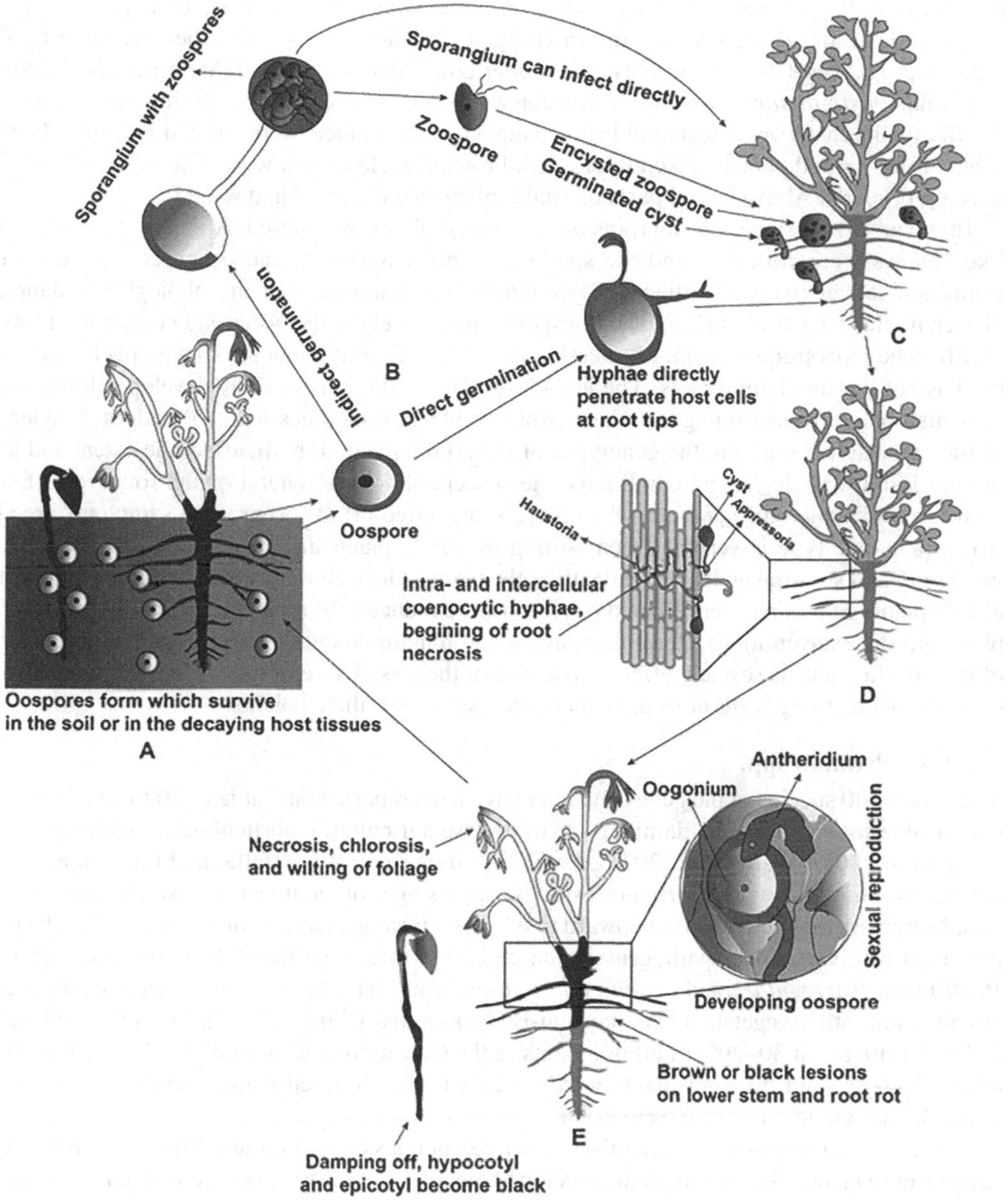

FIGURE 5.4 Disease cycle of oomycetes on alfalfa. (A) Oomycetes survive in the soil or the decaying host tissues in the oospores form. (B) Oospores germinate directly, produce hyphae, and penetrate the host at root tips or indirectly by producing sporangium with zoospores and these zoospores infect host plants. (C) Intra- and intercellular invasion of coenocytic hyphae takes place, and root necrosis begins. (D, E) Brown or black lesions on the lower stems form and root rot; symptoms on young alfalfa seedlings include damping-off, blackish hypocotyl, and epicotyl. Finally, oospores are formed by sexual reproduction, and the disease cycle is completed.

P. infestans, *P. citrophthora*, *P. cactorum*, *P. cinnamomic*, *P. fragariae*, *P. sojae*, *P. capsici*, *P. nicotianae*, and *P. medicaginis* are in charge of producing root rot in the agriculturally significant crops, vegetables, fruit trees, and fodder crops across the world (Steinmetz et al., 2020). In China, *Phytophthora cactorum* has become a severe problem in the alfalfa-growing region. Alfalfa plants that were 2 years old in Jinchang, Gansu Province, were stunted in 2018. The taproots of the stunted plants showed red to dark brown discolorations, while lateral roots could not develop properly. Aboveground parts of alfalfa plants were stunted and wilted.

In the advanced stages, the taproots of the stunted plants were rotted. About 32 *Phytophthora*-like isolates were collected, and the sporangia, chlamydospores, and oospores were carefully examined. Isolates were identified as *Phytophthora cactorum* based on morphological and molecular characteristics (Cai et al., 2021). Zoospores move freely in the water and contact the roots of alfalfa. The most frequent point of infection is the tips of rootlets and the spongy phellem cells at the base of the fine lateral roots. The lesions on the alfalfa roots initially develop yellowish and brownish before transforming into dark brown or black, sometimes with halo edges. The length of the moist soil conditions, the genotypes of the alfalfa plant, or both, affect the extent and kind of root lesions. Under severe conditions, the susceptible plants' lateral or tap roots rot off, and aboveground symptoms appeared when these roots rotted off. Aboveground, symptoms are also two types. One type is general green wilting of alfalfa plants and then death, and the other is yellowing and stunting and then the death of the plants. Once aboveground symptoms appear, the alfalfa plants cannot recover. When the soil moisture reduces, the rot stops advancing further. The plants produce adventitious lateral roots again and remain productive. However, the productivity of the alfalfa plants is severely affected because of the loss of several roots. They persist in soil as oospores with strong walls or as mycelia in the tissues of afflicted plants.

5.2.3.3 *Pythium* spp.

More than 200 species of the genus *Pythium* have been reported, and at least 10 to 15 of them are responsible for root rots and damping-off in diverse agricultural, horticultural, and forage crops (Berg et al., 2017; Wang et al., 2020). *Pythium* causes root rot in alfalfa, and the symptoms are similar to those of *Phytophthora*; however, the early stages of the infection result in necrotic root tips. Moreover, the main roots rot upward to the stem, turning black (Zhang et al., 2021). The most prevalent root rot-causing pathogens in the globe are three significant *Pythium* spp., including *P. ultimum*, *P. irregulare*, and *P. aphanidermatum*, which have been found in fields, sands, ponds, streams, and other vegetation (Williamson-Benavides and Dhingra, 2021). From the summer of 2017 to 2019, about 30–80% of alfalfa plants in the Gansu Province rotted. To identify the causal agent, 3–4-year-old diseased plants were collected. The diseased plants were poorly growing, stunted, and wilted, and some were dried.

Moreover, on the taproots, irregular brown necrotic lesions were seen. The lateral roots were furthermore brown-discolored, underdeveloped, necrotic, and rotten. We recognized the morphological traits of sporangia, oogonia, and antheridia. The isolates' morphological traits and genetic identification indicated they were *Pythium coloratum* (Zhang et al., 2021). Root rot on alfalfa caused by *Pythium* spp. has been reported in other countries (Berg et al., 2017). This was the first report showing that *Pythium* spp. causes alfalfa's root rot in China.

5.2.3.4 *Peronospora trifoliorum*

Peronospora trifoliorum (de Bary), the causal agent of downy mildew, thrives under low temperatures and high humidity (Obert et al., 2000). Only during prolonged rainy seasons does this fungus produce spores; only then can those spores germinate and cause illness. Most diseases are kept at bay when temperatures are above 65°F (18°C) and low humidity (Yu et al., 2022). Infected leaves' top surfaces become pale green and seem mottled yellow. On the bottom of the affected area, bluish-gray mycelia patches, spores (sporangia), and branching, sporangiophores under a hand lens can be seen. Morning, when the canopy's humidity is highest, is when you are most likely to come upon one of them. Infected buds and leaves may sometimes spread throughout the plant, resulting in a systemic infection that causes the leaves to warp and become yellow. The risk of infection is highest in spring-planted fields because the illness thrives in the moist, warm conditions that prevailed then. Dropping infected leaves may reduce both production and quality. The fungus relies on mycelium in systemically infected crown buds, shoots, and resistant resting spores (oospores) in detritus to survive the hot summer. Plant mortality is uncommon due to this illness, although stand longevity is typically unaffected. To this day, the pathogen may be found in the soil, where it has found a suitable habitat among the fallen leaves that serve as the principal source of infection. Air currents carry fungal spores that have settled on the undersides of the leaves to new locations. Lucerne, white, and red clover are all hosts to this fungus. It is most dangerous in the spring when the weather is rainy and humid (20–30°C temperatures and 85% humidity). Fungus propagates in contaminated crown buds and crown shoots. Upper leaves develop spots that range in color from pale green to yellow. The tops of the shoots typically wither, and the leaves roll or twist. On the underside of the leaves, you can frequently see the mycelium and fruiting structure of a fungus, which looks like a cottony gray growth. In the advanced stages of the illness, plant leaves dry up and fall off, exposing the naked stems and branches underneath (Yu et al., 2022). To delay the onset of disease, it is helpful to use cultural measures such as thorough summer plowing, crop rotation, and removing and burning residual plant waste. By minimizing the amount of diseased material that may serve as a source of survival and lowering the humidity inside the stand, keeping pasture short or cutting for hay can assist in suppressing disease growth. The disease may be controlled by using disease-resistant cultivars and cymoxanil at a rate of 0.1% through spraying. The cultivation of disease-resistant cultivars is the most effective method of disease management. Downy mildew resistance testing on alfalfa revealed that varieties from diverse origins varied greatly in this regard (Yu et al., 2022).

5.2.4 Bacteria

5.2.4.1 *Clavibacter michiganensis*

The *Clavibacter michiganensis* subspp. *Insidiosus* is responsible for the widespread but increasingly uncommon occurrence of bacterial wilt worldwide (Samac et al., 1998). Few alfalfa types are issued now without at least some degree of resistance to bacterial wilt since this was the first disease for which resistant cultivars were produced. Stands are often destroyed after 3 to 4 years; therefore, it is possible that this illness is not taken seriously since symptoms do not manifest until the second or third year of a stand. Symptoms aboveground include reduced growth and yellowing of the leaves (Víchová and Kozová, 2004). The leaflets may be speckled and have a little upward cup or curl. Affected plants may have soft, flimsy stems.

The disease shows itself most clearly in the new growth after harvest. Inside an infected taproot, the cross section shows a yellowish-brown tint. Occasionally, you may see brown spaces within the bark tissue. Infected plants often die off and cannot be saved. Plants usually perish after 5–8 months of presenting symptoms. Plants with the disease are more likely to die during the cold months. The bacteria overwinter in plant debris in the soil and gain access to plants via cuts in the roots, crown, or recently mowed stems. This bacterium may live dormant in dry plant material for up to 10 years, and it can travel great distances in seed and dried hay. However, as contaminated plant material decays, the organism's population in the soil rapidly decreases. The bacteria are quickly disseminated by running water, tilling, and harvesting machinery. Fields with poor drainage have the highest disease prevalence, and enormous areas may get affected during prolonged rainy spells.

In most cases, this disease is kept in check by resistant cultivars. If bacterial wilt is found in a field, it is essential to mow other fields before returning to that one. Infested sections of a field should be mown last and never while the grass is damp (Víchová and Kozová, 2004).

5.2.4.2 *Agrobacterium tumefaciens*

The pathogenic bacteria *Agrobacterium tumefaciens* (Smith & Townsend) causes crown gall, which affects several plant species (Du et al., 1994). It has been discovered in the low desert environment of the Imperial Valley, where it seldom appears in alfalfa. Galls develop on crown branches at or near the soil line as a defense mechanism against infection. The germs get into the body via fewer than 24-h-old cuts. Since this disease has seldom happened, we do not know the extent of the possible yield loss (Du et al., 1994).

5.2.4.3 *Pseudomonas syringae*

Alfalfa is susceptible to a wide range of diseases, but bacterial stem blight caused by *Pseudomonas syringae* is a major concern in alfalfa-growing regions of the world (Kataria et al., 2022). The disease has also been reported in western Iran, the United States, Australia, and Europe (Harighi, 2007). This illness significantly reduces crop quality and quantity (Lipps et al., 2019). The ubiquitous epiphyte *P. syringae* pv. *syringae* is a major pathogen in many types of cultivated plants. These are gram-negative rod-shaped bacteria having an entirely sequenced genome. *P. syringae* is a useful model for researching pathogen-host interactions because of its Type III secretion system (T3SS), ice nucleation activity, poisonous compounds, cell wall–disintegrating enzymes, and exopolysaccharides, among other virulence characteristics (Lipps et al., 2019). The toxin produced by *P. syringae* strains called syringomycin is thought to contribute to the pathogen's virulence. Most strains can act as ice nuclei at subfreezing temperatures, damaging plants. This disease has two stages of infection: blight (localized tissue death in the leaves) and vascular wilt (affecting the entire plant). At frost injury sites, the bacterium penetrates the host stem and causes water-soaked lesions and the development of slender, aged stems with a blackened appearance. Yield losses of up to 50% or more may occur if the disease spreads rapidly (Kataria et al., 2022). Ice-nucleation protein (INP) surrounds *P. syringae's* outer membrane and initiates ice formation, connecting bacterial stem blight and frost. The pathogenicity of this protein is just one reason it is of such great interest to scientists; there may be other uses for it, such as in the freezing of food or the creation of snow (Harighi, 2007).

5.2.4.4 Alfalfa Dwarf

Dwarfing of alfalfa was initially identified as a disease in that plant in Southern California in the 1920s (Sisterson et al., 2010). Small, bluish-green leaves and thin stems characterize infected plants, which are often stunted. However, the taproot tissue is unnaturally yellow when cut through, with black streaks of dead tissue. Yellowing in freshly infected plants often occurs in a circular pattern beneath the bark. The pathogen was xylem-limited, bacterial pathogen *Xylella fastidiosa*. Xylem-feeding insects transmit this pathogen. Compared to bacterial wilt, there are no infected pockets under the bark. Infected plants wilt and die over time. In alfalfa, the dwarf is not considered a marketable disease. Alfalfa may have significantly influenced the spread of Pierce's disease. Pierce's disease has been found at higher rates in grapes close to alfalfa fields with diseased plants. Pierce's disease was transmitted mainly through xylem-feeding insects, such as numerous species of sharpshooter leafhoppers until the arrival of the glassy-winged sharpshooter (*Homalodisca coagulata*; Lopes et al., 2010). These pinpoint shooters are predominantly grass eaters and may be discovered in the grassy weeds that plague alfalfa. It is thought that the alfalfa in these fields gets infected almost by mistake when the insects randomly poke the plant while foraging for other hosts. Although the glassy-winged sharpshooter has been seen feeding on alfalfa, this vector prefers other plants when alternative options are available. This suggests that the glassy-winged sharpshooter is unlikely to have a significant role in spreading alfalfa dwarf or Pierce's disease in the genus *Medicago* (Daugherty et al., 2010).

5.2.5 Viruses

There are several viruses that cause significant economic damage to alfalfa crops. The most prevalent viruses found in alfalfa are alfalfa mosaic virus (AMV, genus *Alfamovirus*), bean leafroll virus (BLRV, genus *Luteovirus*), pea streak virus (PeSV, genus *Carlavirus*), red clover vein mosaic virus (RCVMV, genus Carlavirus), alfalfa enation virus (AEV, genus Nucleorhabdovirus) and lucerne *transient* streak virus (LTSV, genus *Sobemovirus*; Nemchinov et al., 2022). Alfalfa mosaic virus (AMV, genus *Alfamovirus*), bean leafroll virus (BLRV, genus *Luteovirus*), pea streak virus (PeSV, genus *Carlavirus*), and red clover vein mosaic virus (RCVMV, genus *Carlavirus*), are single-stranded (ss), positive-sense (+) RNA viruses are aphid-transmitted. Alfalfa enation virus (AEV, genus *Nucleorhabdovirus*) and lucerne transient streak virus (LTSV, genus *Sobemovirus*), are transmitted by aphids and beetles, respectively. In addition, a group of persistent viruses were identified in alfalfa, including *Medicago sativa* alphapartitiviruses 1 and 2 (MsAPV1 and MsAPV2), *Medicago sativa* deltapartitivirus 1 (MsDPV1), and *Medicago sativa* amalgavirus 1 (MsAV1). Symptoms include mosaic, yellow mottling, or streaking on leaves, which may sometimes be noticed but can sometimes be hidden, such that leaves seem normal at other times. While some viruses, like AMV, BLRV, and persistent viruses, are widespread in alfalfa crops grown all over the world, many other viruses identified in alfalfa show a more restricted, localized distribution typical of a single country or even a smaller geographic region. Recently, novel viruses such as the flavi-like virus with an unusual genome organization were discovered, dubbed Snake River alfalfa virus (SRAV), which has a positive-sense (+) RNA genome (Dahan et al., 2022). In conclusion, alfalfa may be a host for viruses crucial to other crops' growth.

5.3 ABIOTIC FACTORS CAUSE NONINFECTIOUS DISEASES

Abiotic factors also cause diseases, and such diseases are called non-infectious diseases. The pathogen is not involved in these diseases and, therefore, not transmitted from the diseased to healthy plants. Diagnosing these diseases is often made easy by the characteristic's symptoms known to be produced by the lack or excess of the particular factor. Sometimes symptoms of noninfectious diseases are not too different from those of infectious diseases. These diseases can be managed by ensuring the alfalfa plant is not exposed to extreme abiotic factors and stresses. Stresses such as drought, high temperatures, pollutants, salinity, waterlogging, salinity, frost, and winterkill may all kill alfalfa. Alfalfa losses from exposure to one or more of these stressors are expected to grow due to the forecasted global environment changes. Every stage of plant development, from germination through seed set, is vulnerable to disruption under stressful circumstances. To deal with environmental stresses, plants employ various physiological and biochemical mechanisms, including changes to the cell wall, the restoration of intracellular ion balance, the synthesis of osmotic material, the activation of antioxidant or stress response signal pathways, and alterations to or variations in the genetic material. Sometimes, the signs of nutritional deficits, herbicide burns, and insect feeding are mistaken for disease symptoms. Insects, herbicide damage, nutritional shortage (such as low phosphorus levels), disease, general stunting, yellowing, and leaf deformation may all be causes. Sometimes suppressing the activity of a gene or two might boost resistance to stress. Transgenic methods have proven more effective in silencing certain genes than traditional breeding. The public's perception of transgenics and their legality are additional obstacles.

5.3.1 Unfavorable Temperature and Soil Moisture

Temperature causes alfalfa injury and also helps in the progression of infectious diseases. Furthermore, alfalfa's developmental stages differ in their ability to withstand extreme temperatures. Mature plants can resist cold winters (low temperatures) more than young alfalfa seedlings. However, cold winters or low temperatures cause far more damage to alfalfa than high temperatures. The cold winters are often not a problem for alfalfa. However, infrequent frost may cause the upper leaves of plants to turn brown if exposed. Tall, uncut alfalfa is more likely to be infected than shorter forms of alfalfa because it has a greater surface area to absorb the fungus. If a cold period (below −3°C) hits a freshly planted alfalfa plant while just a few unifoliolate (single) leaves have formed, the plant may be damaged. Discolored, weakened, and ultimately dead plants are the outcome of injury. However, high temperatures cause alfalfa injury faster than low temperatures. For example, sun scald injuries on the sun-exposed sides of alfalfa plants. Mainly when alfalfa is produced in heavy soils, scald is a risk. It is an abiotic disorder brought on by prolonged exposure to high soil temperatures and waterlogged soil. Waterlogged soil is one of the soil moisture disturbances and is responsible for poor root health of alfalfa and low production because of lack of oxygen.

Moreover, a combination of low oxygen and high soil or air temperatures causes roots to collapse in alfalfa. The first condition reduces the amount of oxygen available to the roots, whereas the other enhances the oxygen requirements of the plants. A scalding effect may occur when the soil remains wet for an extended time after irrigation or rainfall. This is often the case only in the hot desert valleys. As little as 3 or 4 days after watering, affected plants may perish. Plant symptoms include wilting and a change in color (to a white or tan tone) despite the soil being moist. Roots may decay and emit a foul stench when dug out. Infected roots will have discolored, necrotic xylem, the tissue responsible for transporting water throughout the plant.

Alfalfa is vulnerable to flooding harm when air temperatures exceed 45°C. Poor management is a critical trigger on heavy soils, and the lack of appropriate soil aeration at high temperatures is likely the primary cause. Water conservation is the principal method of control. A field's susceptibility to scald increases significantly after it has been mowed, compared to when it is closer to harvest time. The risk of scald may be mitigated by limiting the irrigation duration to a short time frame (less than 4 hours) or by doing the irrigation at night during the extreme daytime heat. However, due to factors including heavy clay soil, poor drainage, the slope of the field, and the duration of the irrigation flow, certain soils stay saturated for extended periods even after irrigation has ended. When temperatures get over 109°F, it is best to hold off watering the garden (43°C). It is advisable to wait at least 3–6 days after mowing before watering newly cut plants so that enough new growth has occurred to prevent whole plants from becoming submerged. Soil saturation is necessary for both scald and *Phytophthora* root rot, which leads to confusion. Soil surface temperatures below 100°F are safe for alfalfa; above 100°F will cause scalding. In addition, to scald, prolonged droughts cause less alfalfa growth; small, scorched leaves; short twigs; defoliation, wilting; and, finally, the death of alfalfa plants.

5.3.2 Air Pollution

Negative effects on agricultural output and quality are a direct result of crop harm brought on by air pollution, such as slowed photosynthesis and premature aging. Common pollutants causing serious damage to alfalfa plants include primary pollutants such as sulfur dioxide, fluorides, nitrogen oxides, chlorine, ethylene, and particulate matter and secondary pollutants such as ozone and peroxyacetyl nitrate (PAN). The primary pollutants originate in the form toxic to alfalfa plants, whereas the secondary pollutants originate in the atmosphere usually photochemically

TABLE 5.1
Air Pollution Injury to Alfalfa Plants

Pollutant	Symptoms	References
Sulfur dioxide	Increased susceptibility to diseases and older and middle leaf injury	(Thomas and Hill, 1935)
Nitrogen dioxide	Reduction in alfalfa growth	(Hou et al., 1977)
Fluorides	Necrosis at leaf margins and leaf tip chlorosis	(Hindawi, 1968)
Chlorines	Older and middle leaves become susceptible to diseases	(Hindawi, 1968)
Ethylene	Dwarf plants, premature senescence, less flower and fruit production, sepal and petals withering, necrosis, flower drops	(Zhang et al., 2013)
Particulate matter	Premature leaf fall, leaf burning, poor growth, and chlorosis	(Hindawi, 1968)
Ozone	Stippling, mottling, chlorosis of leaves, shite to tan or brown or black spots on leaves	(Oshima et al., 1976)
	Premature defoliation and stunting, affected chloroplasts	
Peroxyacyle nitrate	The underside of the leaf's silvery or glazed look, metallic shine, similar to that of silver or copper	(Thompson et al., 1976)

by reactions among pollutants. Primary pollutants such as sulfur dioxides and nitrogen oxides convert into sulfuric and nitric acids by atmospheric moisture and are washed down with air. This rain, known as acid rain, is phytotoxic to alfalfa plants. Leaves show pits, spots, curling, and reduction in biomass. The only air pollutants of significance are the photochemical oxidants (ozone and PAN), although others can also damage alfalfa. Sunlight catalyzes the formation of both via the interaction of oxygen, nitrogen oxides, and organic molecules.

Vehicle tailpipes are the most significant contributor to these atmospheric precursor chemicals, although industrial operations and other combustion sources are also substantial. Due to high ozone levels, there is a bleached stippling on the top leaf surfaces and isolated necrotic patches dispersed between the veins of wounded leaves. Symptoms often manifest on older, mature leaves. Leaves may turn brown and fall if this happens. Ozone levels tend to be high when there are few clouds, plenty of sunshine, and slow-moving winds. As gas is absorbed via the stomates (pores in the leaves), nearby mesophyll cells collapse, causing PAN damage to begin. Air pockets are formed between the lower epidermis and the palisade cells due to the collapsing tissues. The underside of the leaf's silvery or glazed look is believed to be due to light refraction via these air pockets. Symptoms on alfalfa leaves are similar to those reported for ozone damage, albeit the lesions may be more severe. Infected leaves often have a metallic shine similar to silver or copper. Injuries caused by PAN are often confined to certain metropolitan locations with heavy traffic and steep terrain, both of which serve to concentrate the airborne contaminants that cause the condition (Table 5.1).

5.3.3 Mineral Toxicities

It is very uncommon for soil to have dangerous concentrations of either necessary or nonessential components that may kill alfalfa plants. Excessive amounts of an element may cause varying degrees of damage to plants, depending on how severely the plant cells are affected. It may cause symptoms of a deficit in another element by interfering with its absorption or action. As a result, plants exposed to high salt levels develop a calcium deficit. The addition of sodium salts has also made the soil more alkaline. However, alfalfa has a high tolerance for alkali harm. Like other plants, Alfalfa needs various minerals for healthy development. The disease may be brought on by a lack of one or more of these factors. Nitrogen, phosphorus, potassium, calcium, sulfur, and magnesium are all examples of significant elements because plants need them in relatively high concentrations; by comparison, iron, zinc, manganese, copper, boron, and molybdenum are all examples of trace elements because they are needed only in extremely minute amounts. The effects of nutrient deficiencies and excesses in alfalfa symptoms are discussed in Table 5.2.

For almost a century, alfalfa has been developed actively to increase its tolerance to various stress agents and other important agronomic features. Much research using genomic, proteomic, and metabonomic analysis has been conducted on alfalfa with an emphasis on abiotic stress. Some of the most critical genes in stress tolerance were separated, and their roles were determined. All these investigations showed that, like other plants, alfalfa has a very sophisticated regulatory

TABLE 5.2
Symptoms of Alfalfa Due to Excesses and Deficiencies of Mineral Nutrients

Element	Deficiency	Excess	References
Nitrogen	Poor nodule development on the roots, chlorotic and etiolated plants mixed with normal tall green plants, lower leaves turn yellow or brown	Succulent plants prone to lodging and attack by pests and diseases, crop maturity is delayed	(He et al., 2020)
Phosphorus	Stunted plants; similar to the effects of water stress; poor growth leaves; brown, purple, or bronze spots on lower leaves; flowering delayed	Early crop maturity, other elements such as iron and zinc deficiencies, reduced yield	(He et al., 2020)
Potassium	Dieback, chlorosis with browning of tips, scorching of leaf edges, and brown spots near the edges of older leaves	Excess of potassium causes a reduction of uptake of other elements such as magnesium	(Collins et al., 1986)
Magnesium	Mottling and chlorosis in the older leaves followed by younger leaves, and finally, leaves become reddish; sometimes necrotic spots appear on the tips and edges of leaves turn upward; defoliations	Excess of magnesium cause a deficiency of potassium and calcium	(Miller and Sirois, 1983)
Calcium	In the lower leaves, tips hook back and edges curl upward; leaves finally become distorted with brown spots and scorched	Excess cause deficiency of other elements such as K, Fe, B, Mn, Zn, and Cu	(Miller and Sirois, 1983)
Sulfur	Reduced growth and yellow leaves	Enhance soil acidity	(Collins et al., 1986)
Manganese	Green spots on chlorotic leaves	Crinkling of leaves	(McCauley et al., 2009)
Iron	Upper leaves become severely chlorotic; central veins remain green	Toxic to plants in acidic soil	(Carrillo-Castañeda et al., 2002)
Zinc	Poor growth of internodes and leaves resulting in a rosette; leaves show interveinal chlorosis; white spots on leaves	Disturb auxin level resulting in abnormal growth	(Grewal and Williams, 1999)
Boron	Stem and leaves distorted, cracks on the surface of root and stem	Necrosis at margins of older leaves	(Chen et al., 2020)
Copper	Burning of leaf margins, chlorosis, resetting	Phytotoxic	(Printz et al., 2016)
Molybdenum	Stunted plants with yellow leaves	Excess causes abnormal levels of sugar and ascorbic acid	(Giddens and Perkins, 1960)

system that controls its response to abiotic stress, including changes in cellular architecture and composition, the propagation of signals via molecules, and genetic variability.

5.4 CONCLUSION

In conclusion, alfalfa is grown all over the globe as a feed for many species of animals due to its high nutritional value. Its cultivation likely predates historical records, making it one of the earliest crops cultivated by humans. Because it is a legume, alfalfa is a great way to increase the amount of biological nitrogen in the soil while providing a useful by-product. While the average lifetime of a perennial crop is about 5 years, alfalfa fields may continue to produce for much longer in other parts of the globe. Long periods of persistence provide plenty of time for many organisms to settle down and create a complex community structure. An alfalfa field offers temporal stability unusual among field crops, despite system disturbances produced by regular harvests, periodic pesticide treatments, and other abiotic stresses. This predictability allows alfalfa to host many biotic agents that may rival or surpass that of riparian habitats. Although most biotic agents have little effect on alfalfa as a crop, a select few, such as *Fusarium*, *Rhizoctonia*, *Sclerotinia*, and oomycetes, do significant harm. Significant production and quality decreases and, in many cases, a shortened productive life of the stand may be attributed to biotic and abiotic agents, as described earlier.

REFERENCES

Abbas, A., Mubeen, M., Sohail, M.A., Solanki, M.K., Hussain, B., Nosheen, S., et al. (2022). Root rot a silent alfalfa killer in China: Distribution, fungal, and oomycete pathogens, impact of climatic factors and its management. *Front. Microbiol.* 13, 961794.

Adhikari, L., and Missaoui, A.M. (2019). Quantitative trait loci mapping of leaf rust resistance in tetraploid alfalfa. *Physiol. Mol. Plant Pathol.* 106, 238–245. https://doi.org/10.1016/j.pmpp.2019.02.006.

Ajayi-Oyetunde, O., and Bradley, C. (2018). *Rhizoctonia solani*: Taxonomy, population biology and management of *Rhizoctonia* seedling disease of soybean. *Plant Pathol. J.* 67(1), 3–17.

Akamatsu, H.O., Chilvers, M.I., and Peever, T.L. (2008). First report of spring black stem and leaf spot of alfalfa in Washington state caused by *Phoma medicaginis*. *Plant Dis.* 92(5), 833–833. doi: 10.1094/PDIS-92-5-0833A.

Al-Sadi, A.M. (2021). *Bipolaris sorokiniana*-induced black point, common root rot, and spot blotch diseases of wheat: A review. *Front. Cell. Infect. Microbiol.* 11, 118.

Al-Sadi, A.M., and Deadman, M.L. (2010). Influence of seed-borne *Cochliobolus sativus* (Anamorph *Bipolaris sorokiniana*) on crown rot and root rot of barley and wheat. *J. Phytopathol.* 158(10), 683–690.

Atkinson, T.G. (1981). *Verticillium* wilt of alfalfa: Challenge and opportunity. *Can. J. Plant Pathol.* 3(4), 266–272.

Azam, Badrhadad, F., Nazarian, F., and Ahmad, I. (2018). Fusion of a chitin-binding domain to an antibacterial peptide to enhance resistance to *Fusarium solani* in tobacco (*Nicotiana tabacum*). *Biotech.* 8, 331.

Balali, G.R., and Kowsari, M. (2004). Pectic zymogram variation and pathogenicity of *Rhizoctonia solani* AG-4 to bean (*Phaseolus vulgaris*) isolates in Isfahn, Iran. *Mycopathologia.* 158(3), 377.

Berg, L.E., Miller, S.S., Dornbusch, M.R., and Samac, D.A. (2017). Seed rot and damping-off of alfalfa in Minnesota caused by *Pythium* and *Fusarium* species. *Plant Dis.* 101(11), 1860–1867.

Bhattarai, S., Biswas, D., Fu, Y.-B., and Biligetu, B. (2020). Morphological, physiological, and genetic responses to salt stress in alfalfa: A review. *Agronomy* 10(4), 577.

Borges, O.L., Stanford, E.H., and Webster, R.K. (1976). The host-pathogen interaction of alfalfa and *Stemphylium botryosum*. *Phytopathology* 66(6), 749–753.

Cai, W., Tian, H., Liu, J., Fang, X., and Nan, Z. (2021). *Phytophthora cactorum* as a pathogen Associated with Root Rot on Alfalfa (*Medicago sativa*) in China. *Plant Dis.* 105(1), 231.

Cao, S., Liang, Q.W., Nzabanita, C., and Li, Y.Z. (2020). *Paraphoma* root rot of alfalfa (*Medicago sativa*) in Inner Mongolia, China. *Plant Pathol.* 69(2), 231–239. https://doi.org/10.1111/ppa.13131.

Carlucci, A., Raimondo, M., Santos, J., and Phillips, A. (2012). *Plectosphaerella* species associated with root and collar rots of horticultural crops in southern Italy. *Persoonia* 28, 34.

Carrillo-Castañeda, G., Muños, J.J., Peralta-Videa, J.R., Gomez, E., Tiemannb, K.J., Duarte-Gardea, M., et al. (2002). Alfalfa growth promotion by bacteria grown under iron limiting conditions. *Adv. Environ. Res.* 6(3), 391–399.

Chen, L., Beiyuan, J., Hu, W., Zhang, Z., Duan, C., Cui, Q., et al. (2022). Phytoremediation of potentially toxic elements (PTEs) contaminated soils using alfalfa (*Medicago sativa* L.): A comprehensive review. *Chemosphere*, 133577.

Chen, L., Xia, F., Wang, M., Wang, W., and Mao, P. (2020). Metabolomic analyses of alfalfa (*Medicago sativa* L. cv.'Aohan') reproductive organs under boron deficiency and surplus conditions. *Ecotoxicol. Environ. Saf.* 202, 111011.

Chen, Q., Jiang, J.R., Zhang, G.Z., Cai, L., and Crous, P.W. (2015). Resolving the *Phoma* enigma. *Stud. Mycol.* 82, 137–217. https://doi.org/10.1016/j.simyco.2015.10.003.

Collins, M., Lang, D.J., and Kelling, K.A. (1986). Effects of phosphorus, potassium, and sulfur on alfalfa nitrogen-fixation under field conditions 2. *Agron. J.* 78(6), 959–963.

Cowling, W.A., Gilchrist, D.G., and Graham, J.H. (1981). Biotypes of *Stemphylium botryosum* on alfalfa in North America. *Phytopathology* 71, 679–684.

Dahan, J., Wolf, Y.I., Orellana, G.E., Wenninger, E.J., Koonin, E.V., and Karasev, A.V. (2022). A novel flavi-like virus in alfalfa (*Medicago sativa* L.) crops along the Snake River Valley. *Viruses* 14(6). doi: 10.3390/v14061320.

Daugherty, M.P., Lopes, J.R.S., and Almeida, R.P.P. (2010). Strain-specific alfalfa water stress induced by *Xylella fastidiosa*. *Eur. J. Plant Pathol.* 127(3), 333–340.

de Gruyter, J., Aveskamp, M.M., Woudenberg, J.H.C., Verkley, G.J.M., Groenewald, J.Z., and Crous, P.W. (2009). Molecular phylogeny of *Phoma* and allied anamorph genera: Towards a reclassification of the *Phoma* complex. *Mycol. Res.* 113(4), 508–519. https://doi.org/10.1016/j.mycres.2009.01.002.

Du, S., Erickson, L., and Bowley, S. (1994). Effect of plant genotype on the transformation of cultivated alfalfa (*Medicago sativa*) by *Agrobacterium tumefaciens*. *Plant Cell Rep.* 13(6), 330–334.

Elgin Jr, J.H., Barnes, D.K., Busbice, T.H., Buss, G.R., Clark, N.A., Cleveland, R.W., et al. (1981). Anthracnose resistance increases alfalfa yields. *Crop Sci.* 21(3), 457–460.

Fan, Q., Creamer, R., and Li, Y. (2018). Time-course metabolic profiling in alfalfa leaves under *Phoma medicaginis* infection. *PLOS One* 13(10), e0206641.

Fang, X.L., Zhang, C.X., and Nan, Z.B. (2019). Research advances in *Fusarium* root rot of alfalfa (*Medicago sativa*). *Acta Prataculturae Sinica* 28(12), 169.

Garibaldi, A., Tabone, G., Luongo, I., and Gullino, M. (2021). First report of *Rhizoctonia solani* AG 2–1 causing crown rot on *Alcea rosea* in Italy. *Plant Dis.* 105(3), 707–707.

Gaulin, E., Jacquet, C., Bottin, A., and Dumas, B. (2007). Root rot disease of legumes caused by *Aphanomyces euteiches*. *Mol. Plant Pathol.* 8(5), 539–548.

Giddens, J., and Perkins, H.F. (1960). Influence of molybdenum on growth and composition of alfalfa and distribution of molybdenum in a Cecil-Lloyd soil. *Soil Sci. Soc. Am. J.* 24(6), 496–497.

Grewal, H.S., and Williams, R. (1999). Alfalfa genotypes differ in their ability to tolerate zinc deficiency. *Plant Soil.* 214(1), 39–48.

Gui, Z., Gao, J., Xin, N., Wang, Y., Pi, Y., Liu, H., et al. (2016). Association of polygalacturonase-inhibiting protein gene 2 (MsPGIP2) to common leaf spot resistance in alfalfa. *Eur. J. Plant Pathol.* 144(2), 245–256. doi: 10.1007/s10658-015-0720-x.

Harighi, B. (2007). Occurrence of alfalfa bacterial stem blight disease in Kurdistan Province, Iran. *J. Phytopathol.* 155(10), 593–595.

Harper, J.T., Waanders, E., and Keeling, P.J. (2005). On the monophyly of chromalveolates using a six-protein phylogeny of eukaryotes. *Int. J. Syst. Evol. Microbiol.* 55(1), 487–496.

He, H., Wu, M., Guo, L., Fan, C., Zhang, Z., Su, R., et al. (2020). Release of tartrate as a major carboxylate by alfalfa (*Medicago sativa* L.) under phosphorus deficiency and the effect of soil nitrogen supply. *Plant Soil.* 449(1), 169–178.

Hedlund, T. (1923). Om Nagrasjukdomar och skador pa vara lantbruksvaxter. *Sver. Allm. Jordbrukstidskr.* 5, 166–168.

Hindawi, I.J. (1968). Injury by sulfur dioxide, hydrogen fluoride, and chlorine as observed and reflected on vegetation in the field. *J. Air Pollut. Control Assoc.* 18(5), 307–312.

Hollingsworth, C.R., Gray, F.A., Koch, D.W., Groose, R.W., and Heald, T.E. (2003). Distribution of *Phoma sclerotioides* and incidence of brown root rot of alfalfa in Wyoming, USA. *Can. J. Plant Pathol.* 25(2), 215–217. doi: 10.1080/07060660309507071.

Hou, L.-Y., Clyde Hill, A., and Soleimani, A. (1977). Influence of CO2 on the effects of SO_2 and NO_2 on alfalfa. *Environ. Pollut.* 12(1), 7–16. https://doi.org/10.1016/0013-9327(77)90003-9.

Howard, R.J. (1985). Local and long-distance spread of *Verticillium* species causing wilt of alfalfa. *Can. J. Plant Pathol.* 7(2), 199–202.

Howell, A.B., and Erwin, D.C. (1995). Characterization and persistence of *Verticillium albo-atrum* isolated from alfalfa growing in high temperature regions of southern California. *Plant Pathol.* 44(4), 734–748.

Hu, X.P., Wang, M.X., Hu, D.F., and Yang, J.R. (2011). First report of wilt on alfalfa in China caused by *Verticillium nigrescens*. *Plant Dis.* 95(12), 1591–1591.

Huang, H.C. (2003). *Verticillium* wilt of alfalfa: Epidemiology and control strategies. *Can. J. Plant Pathol.* 25(4), 328–338.

Huang, H.C., Acharya, S.N., Hou, T.J., Erickson, R.S., Dalton, R.E., and Mueller, C.A. (1999). Susceptibility of Chinese alfalfa cultivars to *Verticillium* wilt. *Plant Pathol.* 8(2), 6–8.

Huang, N., Sun, X.-B., Wang, T.-M., and Lu, X.-S. (2013). Evaluation of resistance to *Fusarium* Wilt (*Fusarium oxyssorum*) and preliminary screening of check varieties of resistant evaluation from alfalfa (*Medicago sativa*) cultivar. *J. Chinese J. Grassl.* 1.

Hwang, S.E., and Flores, G. (1987). Effects of *Cylindrocladium gracile*, *Fusarium roseum* and *Plenodomus melilotion* crown and root rot, forage yield. *Can. Plant Dis. Surv.* 67(2), 31.

Inderbitzin, P., Bostock, R.M., Davis, R.M., Usami, T., Platt, H.W., and Subbarao, K.V. (2011). Phylogenetics and taxonomy of the fungal vascular wilt pathogen *Verticillium*, with the descriptions of five new species. *PLOS One*. 6(12), e28341.

Jacquet, C., and Bonhomme, M. (2019). Deciphering resistance mechanisms to the root rot disease of legumes caused by *Aphanomyces euteiches* with *Medicago truncatula* genetic and genomic resources. In: *F Bruijin, ed. The Model Legume Medicago Truncatula.* Hoboken, NJ, USA: John Wiley & Sons, 307–316.

Jiang, D., Xu, C., Han, W., Harris-Shultz, K., Ji, P., Li, Y., et al. (2021). Identification of fungal pathogens and analysis of genetic diversity of *Fusarium tricinctum* causing root rots of alfalfa in north-east China. *Plant Pathol.* 70(4), 804–814. https://doi.org/10.1111/ppa.13333.

Jones, E.R., Carroll, R.B., Swain, R.H., and Bell, K.W. (1978). Role of anthracnose in stand thinning of alfalfa in Delaware. *J. Agron.* 70(2), 351–353.

Juliatti, F.C., Polloni, L.C., de Morais, T.P., Zacarias, N.R.S., Silva, E.A., and Juliatti, B.C.M. (2017). Sensitivity of *Phakopsora pachyrhizi* populations to dithiocarbamate, chloronitrile, triazole, strobilurin, and carboxamide fungicides. *J. Biosci.* 33(4), 933–943.

Karppinen, E.M., Payment, J., Chatterton, S., Bainard, J.D., Hubbard, M., Gan, Y., and Bainard, L.D. (2020). Distribution and abundance of *Aphanomyces euteiches* in agricultural soils: Effect of land use type, soil properties, and crop management practices. *Appl. Soil Ecol.* 150, 103470.

Kataria, R., Duhan, N., and Kaundal, R. (2022). Computational systems biology of alfalfa—bacterial blight host-pathogen interactions: Uncovering the complex molecular networks for developing durable disease resistant crop. 12. doi: 10.3389/fpls.2021.807354.

Kong, Q., Liu, R., Liu, D., Liu, Z., and Wang, H. (2018). Biological characteristics of *Fusarium* causing alfalfa root rot in Hebei Province. *J. China Agric. Univ.* 23(8), 59–76.

Lakhran, L., Ahir, R., Choudhary, M., and Choudhary, S. (2018). Isolation, purification, identification and pathogenicity of *Macrophomina phaseolina* (Tassi) goid caused dry root rot of chickpea. *Journal of Pharmacognosy and Phytochemistry* 7(3), 3314–3317.

Li, M. (2003). Comparative pathogenicity of isolates of *Fusarium* spp. and cultivars resistance in alfalfa. *J. Grassland of China* 25(1), 39–43.

Li, Y., Duan, T., Nan, Z., and Li, Y. (2021). Arbuscular mycorrhizal fungus alleviates alfalfa leaf spots caused by *Phoma medicaginis* revealed by RNA-seq analysis. *Appl. Microbiol.* 130(2), 547–560.

Li, Y., Huang, H., Wang, Y., and Yuan, Q. (2018). Transcriptome characterization and differential expression analysis of disease-responsive genes in alfalfa leaves infected by *Pseudopeziza medicaginis*. *Euphytica*. 214(7), 126. doi: 10.1007/s10681-018-2204-5.

Li, Y., Meng, L., Gong, L., Zhao, T., Ji, P., Zhang, Q., et al. (2019). Occurrence of *Bipolaris* root rot caused by *Bipolaris sorokiniana* on Alfalfa in China. *Plant Dis.* 103(10), 2691–2691.

Li, Y., and Su, D. (2017). Alfalfa water use and yield under different sprinkler irrigation regimes in north arid regions of China. *Sustainability* 9(8), 1380.

Lipps, S.M., Lenz, P., and Samac, D.A. (2019). First report of bacterial stem blight of alfalfa caused by *Pseudomonas viridiflava* in California and Utah. *Plant Dis.* 103(12), 3274–3274.

Lopes, J.R.S., Daugherty, M.P., and Almeida, R.P.P. (2010). Strain origin drives virulence and persistence of Xylella fastidiosa in alfalfa. *Plant Pathol.* 59(5), 963–971.

Luo, D., Liu, W., Wang, Y., Zhang, J., and Liu, Z. (2014). Development of a rapid one-step PCR protocol to distinguish between alfalfa (*Medicago sativa*) and sweet clover (*Melilotus* spp.) seeds. *Seed Sci. Technol.* 42(2), 237–246.

Marquez, N., Giachero, M.L., Declerck, S., and Ducasse, D.A. (2021). *Macrophomina phaseolina*: General characteristics of pathogenicity and methods of control. *Front. Plant Sci.* 12(666). doi: 10.3389/fpls.2021.634397.

McCauley, A., Jones, C., and Jacobsen, J. (2009). Plant nutrient functions and deficiency and toxicity symptoms. *Nutrient Management Module*. 9, 1–16.

Miller, R.W., and Sirois, J.C. (1983). Calcium and magnesium effects on symbiotic nitrogen fixation in the alfalfa (*M. sativa*)*–Rhizobium meliloti system. Physiol. Plant*. 58(4), 464–470.

Morgan, W.C., and Parbery, D.G. (1980). Depressed fodder quality and increased oestrogenic activity of lucerne infected with *Pseudopeziza medicaginis. Aust. J. Agric. Res.* 31(6), 1103–1110.

Musial, J.M., Mackie, J.M., Armour, D.J., Phan, H.T.T., Ellwood, S.E., Aitken, K.S., et al. (2007). Identification of QTL for resistance and susceptibility to *Stagonospora meliloti* in autotetraploid lucerne. *Theor. Appl. Genet.* 114(8), 1427–1435.

Nemchinov, L.G., Irish, B.M., Grinstead, S., Shao, J., and Vieira, P. (2022). Diversity of the virome associated with alfalfa (*Medicago sativa* L.) in the U.S. Pacific Northwest. *Sci. Rep.* 12(1), 8726. doi: 10.1038/s41598-022-12802-4.

Nutter Jr, F.W., Guan, J., Gotlieb, A.R., Rhodes, L.H., Grau, C.R., and Sulc, R.M. (2002). Quantifying alfalfa yield losses caused by foliar diseases in Iowa, Ohio, Wisconsin, and Vermont. *Plant Dis.* 86(3), 269–277.

Obert, D.E., Skinner, D.Z., and Stuteville, D.L. (2000). Association of AFLP markers with downy mildew resistance in autotetraploid alfalfa. *Mol. Breed.* 6(3), 287–294.

Oladzad, A., Zitnick-Anderson, K., Jain, S., Simons, K., Osorno, J.M., McClean, P.E., et al. (2019). Genotypes and genomic regions associated with *Rhizoctonia solani* resistance in common bean. *Front. Plant Sci.* 10, 956.

Oshima, R.J., Poe, M.P., Braegelmann, P.K., Baldwin, D.W., and Way, V.v. (1976). Ozone dosage-crop loss function for alfalfa: A standardized method for assessing crop losses from air pollutants. *J. Air Pollut. Control Assoc.* 26(9), 861–865.

Pan, L., Zhang, L., Yang, C., Yuan, Q., Wang, Y., and Miao, L. (2015). Identification and biological characteristics of *Fusarium sporotrichioide* isolated from *Medicago sativa* root. *Acta Prataculturae Sinica* 24(10), 88–98.

Pandey, A.K., Burlakoti, R.R., Rathore, A., and Nair, R.M. (2020). Morphological and molecular characterization of *Macrophomina phaseolina* isolated from three legume crops and evaluation of mungbean genotypes for resistance to dry root rot. *Crop Prot.* 127, 104962.

Peaden, R.N., Gilbert, R.G., and Christen, A.A. (1985). Control of *Verticillium albo-atrum* on alfalfa. *Canadian Journal of Plant Pathology* 7(2), 511–514.

Porto, M.D.M., Grau, C.R., de Zoeten, G.A., and Gaard G. (1988). Histopathology of *Colletotrichum trifolii* on alfalfa. *Phytopathology* 78, 345–349.

Printz, B., Guerriero, G., Sergeant, K., Audinot, J.-N., Guignard, C., Renaut, J., et al. (2016). Combining-omics to unravel the impact of copper nutrition on alfalfa (*Medicago sativa*) stem metabolism. *Plant Cell Physiol.* 57(2), 407–422.

Rahman, M.T., Bhuiyan, M.K.A., Karim, M.A., and Rubayet, M.T. (2018). Screening of soybean resistance genotypes against *Fusarium oxysporum*, *Macrophomina phaseolina*, *Rhizoctonia solani* and *Sclerotium rolfsii*. *Res. Agric. Vet. Sci.* 2(3), 139–156.

Richard, C., Beghdadi, A., and Martin, J.G. (1991). *Aphanomyces euteiches*, a novel root pathogen to alfalfa in Québec. *Plant Dis.* 75(3).

Samac, D.A., Nix, R.J., and Oleson, A.E. (1998). Transmission frequency of *Clavibacter michiganensis* subsp. *insidiosus* to alfalfa seed and identification of the bacterium by PCR. *Plant Dis.* 82(12), 1362–1367.

Samac, D.A., Willbur, J., Behnken, L., Brietenbach, F., Blonde, G., Halfman, B., et al. (2014). First report of *Stemphylium globuliferum* causing *Stemphylium* leaf spot on alfalfa (*Medicago sativa*) in the United States. *Plant Dis.* 98(7), 993–993.

Sharath, C., Tarafdar, A., Mahesha, H., and Sharma, M. (2021). Temperature and soil moisture stress modulates the host defense response in chickpea during dry root rot incidence. *Front. Plant Sci.* 12, 932.

Sheppard, J.W. (1979). *Verticillium* wilt, a potentially dangerous disease of alfalfa in Canada. *Can. Plant Dis. Surv.* 59, 60.

Sisterson, M.S., Thammiraju, S.R., Lynn-Patterson, K., Groves, R.L., and Daane, K.M. (2010). Epidemiology of diseases caused by *Xylella fastidiosa* in California: Evaluation of alfalfa as a source of vectors and inocula. *Plant Dis.* 94(7), 827–834.

Stalpers, J.A., and Andersen, T.F. (1996). A synopsis of the taxonomy of teleomorphs connected with *Rhizoctonia* SL. In *Rhizoctonia Species: Taxonomy, Molecular Biology, Ecology, Pathology and Disease Control*, eds. B. Sneh, S. Jabaji-Hare, S. Neate and G. Dijst. Dordrecht: Springer Netherlands, 49–63.

Steinmetz, O.J., Huset, D.E., Rouse, D.I., Raasch, J.A., Gutiérrez, L., and Riday, H. (2020). Synthetic cultivar parent number impacts on genetic drift and disease resistance in alfalfa. *Crop Sci.* 60(5), 2304–2316.

Sultana, S., Adhikary, S., Rahman, S.M., and Islam, M. (2018). Sexuality and compatibility of *Bipolaris sorokiniana* and segregation pattern in teleomorph (*Cochliobolus sativus*): Geographic origin and segregation ratio. *Indian Phytopathol.* 71(3), 365–375.

Summers, C.G., and Gilchrist, D.G. (1991). Temporal changes in forage alfalfa associated with insect and disease stress. *J. Econ. Entomol.* 84(4), 1353–1363.

Thomas, M.D., and Hill, G.R. (1935). Absorption of sulphur dioxide by alfalfa and its relation to leaf injury. *Plant Physiol.* 10(2), 291–307. doi: 10.1104/pp.10.2.291.

Thompson, C.R., Kats, G., Pippen, E.L., and Isom, W.H. (1976). Effect of photochemical air pollution on two varieties of alfalfa. *Environ. Sci. Technol.* 10(13), 1237–1241.

Van West, P., Appiah, A.A., and Gow, N.A. (2003). Advances in research on oomycete root pathogens. *Physiol. Mol. Plant Pathol.* 62(2), 99–113.

Víchová, J., and Kozová, Z. (2004). The virulence of *Clavibacter michiganensis* subsp. insidiosus strains and tests of alfalfa varieties for resistance to the wilt pathogen. *J. Plant Prot. Res.* 44(2), 147–154.

Vincelli, P.C., and Herr, L.J. (1992). Two diseases of alfalfa caused by *Rhizoctonia solani* AG-1 and AG-4. *Plant Dis.* 76(12), 1283B.

Wang, Y., Yuan, Q.H., Miao, L.H., Zhang, L., and Pan, L.Q. (2016). The major types and epidemic trends of alfalfa diseases in Northeast and North China. *Acta Prataculturae Sinica.* 25(3), 52–59.

Wang, Z., Tian, H., Zhang, C., Fang, X., and Nan, Z. (2020). Occurrence of *Macrophomina phaseolina* causing root and crown rot on alfalfa (*Medicago sativa*) in China. *J. Plant Disease.* 104(9), 2521.

Webb, D.H., and Nutter Jr, F.W. (1997). Effects of leaf wetness duration and temperature on infection efficiency, latent period, and rate of pustule appearance of rust in alfalfa. *Phytopathol.* 87(9), 946–950.

Webb, D.H., Nutter Jr, F.W., and Buxton, D.R. (1996). Effect of acid detergent lignin concentration in alfalfa leaves on three components of resistance to alfalfa rust. *Plant Dis.* 80(10), 1184–1188.

Wen, Z., Duan, T., Christensen, M.J., and Nan, Z. (2015). *Microdochium tabacinum*, confirmed as a pathogen of alfalfa in Gansu Province, China. *Plant Dis.* 99(1), 87–92.

Williamson-Benavides, B.A., and Dhingra, A. (2021). Understanding root rot disease in agricultural crops. *Horticulturae.* 7(2), 33. doi: 10.3390/horticulturae7020033.

Wu, L., Chang, K.F., Conner, R.L., Strelkov, S., Fredua-Agyeman, R., Hwang, S.F., and Feindel, D. (2018). *Aphanomyces euteiches*: A threat to Canadian field pea production. *Engineering* 4(4), 542–551.

Xu, S., Christensen, M.J., Creamer, R., and Li, Y.Z. (2019). Identification, characterization, pathogenicity, and distribution of verticillium alfalfae in alfalfa plants in China. 103(7), 1565–1576. doi: 10.1094/pdis-07-18-1272-re.

Xu, S., Li, Y.Z., and Nan, Z.B. (2016). First report of *Verticillium* wilt of alfalfa caused by *Verticillium alfalfae* in China. *Plant Dis.* 100(1), 220–220.

Yang, S., Gao, M., Xu, C., Gao, J., Deshpande, S., Lin, S., et al. (2008). Alfalfa benefits from *Medicago truncatula*: The RCT1 gene from *M. truncatula* confers broad-spectrum resistance to anthracnose in alfalfa. *PNAS* 105(34), 12164–12169.

Yang, X., Tyler, B.M., and Hong, C. (2017). An expanded phylogeny for the genus *Phytophthora*. *IMA Fungus* 8, 355–384.

Yinghua, L., Chenghao, J., Meihua, X., and Xiangping, L. (2019). Research advances of root rot in *Medicago sativa* L. *J. Plant Dis. Pest.* 10(1), 15–27.

Yu, L., Lin, K., Xu, L., Cui, J., Zhang, Y., Zhang, Q., et al. (2022). Downy mildew (*Peronospora aestivalis*) infection of alfalfa alters the feeding behavior of the aphid (*Therioaphis trifolii*) and the chemical characteristics of alfalfa. *J. Pest Sci.*, 1–13.

Zaccardelli, M., Balmas, V., Altomare, C., Corazza, L., and Scotti, C. (2006). Characterization of Italian isolates of *Fusarium semitectum* from alfalfa (*Medicago sativa* L.) by AFLP analysis, morphology, pathogenicity and toxin production. *J Phytopathol.* 154(7–8), 454–460.

Zhang, C., Yu, S., Tian, H., Wang, Z., Yu, B., Ma, L., et al. (2021). Varieties with a high level of resistance provide an opportunity to manage root rot caused by *Rhizoctonia solani* in alfalfa. *Eur. J. Plant Pathol.* 160, 983–989 doi: 10.1007/s10658-021-02287-8.

Zhang, Y., Liu, J., Zhou, Y., Gong, T., Wang, J., and Ge, Y. (2013). Enhanced phytoremediation of mixed heavy metal (mercury)–organic pollutants (trichloroethylene) with transgenic alfalfa co-expressing glutathione S-transferase and human P450 2E1. *J. Hazard. Mater.* 260, 1100–1107.

Zhao, Y.Q., Shi, K., Yu, X., and Zhang, L.J. (2021). First report of alfalfa root rot caused by *Plectosphaerella cucumerina* in Inner Mongolia autonomous region of China. *Plant Dis.*, 1–4. doi: 10.1094/PDIS-03-21-0515-PDN.

Zhimin, D.I.A.O. (1996). Study on the pathogen and pathogenicity of root disease of peas in Qinghai. *J. Microbiol.* 1, 31–34.

Zhou, W., Guo, Z., Niu, J., Cao, X., Zhao, T., Jiang, W., et al. (2019). Effect of methyl jasmonate on resistance of alfalfa root rot caused by *Fusarium oxysporum*. *J. Acta Phytopathologica Sinica.* 49(3), 379–390.

6 Physiological Effects of Thermal, Mineral, and Water Stresses on Summer Fodders

Rabia Tahir Bajwa, Qaiser Shakeel, Muhammad Shahid Rizwan, Sajjad Ali, Muhammad Anjum Aqueel, Mustansar Mubeen, Yasir Iftikhar, and Ifrah Rashid

6.1 INTRODUCTION

Fodders are defined as plants and their parts used in feeding domestic livestock. When plants are subjected to adverse environmental conditions, they cannot reach their full genetic capacity regarding growth, development, reproduction, and overall crop production (Singh and Chahal, 2020). As a result of abiotic stress, farmers' incomes, and food security are threatened, as are the lives of animals and nations' economies. Raising livestock on fodder is an important step in sustainable agriculture. Fodder crops are important because they provide food for both wild and domestic animals and help preserve soil fertility and prevent crop damage. It also serves as a shelter for various flora and fauna. Fodder is essential to the biological system, including plant–soil and plant-animal interaction. The rising population in the next few decades will set a strain on arable land used to grow food and fiber. Therefore, we may be forced to cultivate fodder crops in areas with low-fertility and poorly managed soils (Singh and Chahal, 2020). Irrigation uses a disproportionately large amount of water, and if current trends continue, this could drastically reduce supply (Sanderson et al., 1997; Singh and Chahal, 2020). It is crucial for annual or perennial fodder crops and native plants to endure harsh abiotic conditions. When improving the long-term viability of fodder productivity, particularly in harsh environmental conditions, it is more important to employ a survival scheme instead of a growth scheme (Singh and Chahal, 2020). Thermal, water, and mineral stress are the major abiotic factors that significantly affect fodder crop production. This chapter highlights the physiological effects and mechanisms involved in the physiological variations and stress tolerance in summer fodders in response to these major abiotic stresses.

6.2 THERMAL STRESS

Temperatures more or less than the thermal threshold level required for effective biochemical, morphological, and physiological development are thermal stress (Asseng et al., 2015). Plant growth is stunted by higher temperatures and radioactivity, typically in tropical regions. Low-temperature-tolerant (or "psychophilic") plants are categorized separately from those that thrive in a more moderate range of temperatures (called "mesophilic") (Eggen et al., 2019). The severity of the negative effects of heat stress is conditional on factors including how long the stress lasts, how advanced a stage it is in, and how intense it is. The number of spikelets, as well as the florets number on each plant in cereal crops, and the seeds in summer fodder crops like pearl millet, sorghum, maize, cowpeas, Sudan grass, and guar are all negatively impacted by increased

DOI: 10.1201/b23394-6

thermal stress (Prasad et al., 2015). Starch, oil, and protein production all fall, lowering the overall quality. Frost stress is estimated to affect about 15% of farmable land (Jordan et al., 2012).

6.3 HIGH TEMPERATURE

High-temperature stress is defined as exposure to temperatures above the physiological ideal level that inhibits normal growth and development. As one of the most important abiotic stressors, high temperatures reduce crop yields. There will likely be an increase in the frequency of unforeseen short or lengthy events of high-temperature stress in the future (IPCC, 2013; IPCC, 2014), which can lead to a drastic drop in yield.

At the end of the 21st century, CO_2 levels and other greenhouse gas rises (Frank et al., 2015) are expected to raise the air temperature by 3.7–4.8°C worldwide (IPCC, 2014). Studies have shown that local and global agricultural production will be threatened by more frequent high-temperature events and higher average temperatures (IPCC, 2014). The air temperature in arid and semiarid regions is frequently more than 32°C, which is the optimum temperature for sorghum crop growth and yield. If temperatures rise above ideal, food grain crops like sorghum, maize (Prasad et al., 2019), and other summer fodder crops may experience less yield and low quality. Ultimately, fodder production in semiarid regions will be severely disrupted by the anticipated increases in temperature. Extreme temperatures negatively impact maize plants in many ways, including disruption of the photosynthetic activities, injury to plant membranes, decreased nutrient uptake, and inhibition of several enzyme functions. Overall, maize performance is diminished due to retarded growth and poor photosynthetic rates.

6.4 PHYSIOLOGICAL EFFECTS OF HIGH-TEMPERATURE STRESS

6.4.1 Membrane Damage

Since thermal stress primarily damages membranes, the plant cell membrane is its most vulnerable part. As a result of proteins degenerations and an increase in unsaturated fatty acids, membranes become more fluid in plants exposed to thermal stress, causing a change from a rigid gel to a more malleable crystalline liquid structure. A reduction of membrane stability because of structural alterations in protein constituents, improving the membrane heat stability, and leakage of organic and inorganic matter from the plant cells can be used to indicate plant damage due to thermal stress. For this reason, the electrolyte leakage value has been used to assess membrane thermostability under thermal stress (Nadeem et al., 2018). This value is an indicator of membrane damage and signifies stress-induced changes. Cell membrane stability can be gauged, and heat-stress tolerance can be screened by monitoring the rate at which ions of plant cells leak out due to increased permeability (ElBasyoni et al., 2017). Membrane damage caused by thermal stress has been documented in various agricultural species. Membrane injury and lipid peroxidation caused by heat stress have been observed in various agronomic and fodder crops. The plasma membranes become more permeable, and electrolyte leakage increases when exposed to heat stress, reducing the membrane's ability to retain solutes and water. Wheat cellular membranes were recently studied to see how they hold up in thermal and water stress. Transgenic and physiological research and mutation analysis have elucidated the role of membrane integrity in temperature tolerance. For instance, a fatty acid unsaturation-deficient soybean mutant showed remarkable resistance to heat stress. In addition, silencing a3-desaturase gene in tobacco increased lipid saturation, which also made the exposed plants relatively tolerant to thermal stress, and two mutants of the model plant (*Arabidopsis thaliana*) lacking unsaturation of fatty acid (fad5 and fad6) displayed high thylakoid membrane stability to thermal stress (Nadeem et al., 2018). When grain filling occurred in extreme temperatures, wheat lines with greater membrane thermal stability typically produced more grain than lines with lesser thermostability (Gupta et al., 2013). The role of HIT1 in membrane trafficking for thermal adaptation of the cell membrane in thermal stress-tolerant plants is being studied. Overexpression of the *PpEXP1* gene in transgenic tobacco resulted in less damage to cell structures, less escape of electrolytes, and lower concentrations of lipid peroxidation in cell membranes than in wild-type plants. Genotypes tolerant to thermal stress are more dynamic due to severe field stress being isolated by the research team. The thermostability membrane has been used to test heat stress tolerance in crops worldwide (Wang et al., 2011; Xu et al., 2014).

6.4.2 Photosynthesis and Respiration

As climate change continues, reports indicate that extreme weather events, such as those involving prolonged sunshine and high temperatures stress, have increased dramatically. It is well established that photosynthesis is essential for life on Earth but that abiotic stresses, including high-temperature stress, hinder it. Most of the detrimental impacts of thermal stress on growth, development, and yield can be traced back to the fact that it interferes with the photosynthetic process, one of the most temperature-sensitive features of plant systems (Wang et al., 2016). Under the co-stress of high light and temperature, comparative water content, total chlorophyll content, and photosystem II activity decreased (Chen et al., 2017). Several abiotic stresses have been identified as detrimental to plant photosynthesis at the photosystem II reaction center (Zivcak et al., 2014). Compared to photosystem I, photosystem II appears more sensitive to high thermal stress or light (Mishra et al., 2017). Since the enzyme chlorophyllase aids in the transition of chlorophyll into phytol and chlorophyllide, a drop in chlorophyll contents may be one of the major causes of lower photosynthesis when exposed to thermal stress.

Photosynthesis and photosynthesis-related structures are modified at each level of acclimation to different climate conditions (Brestic et al., 2014). Important photo-defense mechanisms in thylakoids include the gradient of protons and non-photochemical slake in photosystems I and II, respectively (Brestic et al., 2015). Chlorophyll concentrations and thylakoid protein's photochemical reaction are known to drop in plants when they are subjected to environmental stress (Su et al., 2014; Wang et al., 2015; Chen, Wang et al., 2016). Temperature is said to have a major impact on crop plants' ability to produce food through photosynthesis and the pathways through which this occurs (C3 or C4 plants). While C4 plants that are adapted to a hot environment, cultivated in the summer have shown active photosynthetic activity within the range of 7°C to 40°C, the optimal temperature range for cold season C3 crops is between 0°C and 30°C (Mishra et al., 2017). At present, CO_2 and light saturation levels, the photosynthetic response activity in C3 plant species to temperature is assessed through the bioavailability of phosphate compounds at relatively low temperatures during the photophosphorylation process. It relies on the Rubisco activity to fix atmospheric carbon at the optimum temperature range.

The mechanisms by which the photosynthetic activity is impacted at higher temperatures in the optimum temperature level in C4 plant species grown in hot climates are unclear. Photosynthetic activity is limited at low temperatures due to the availability of Rubisco. When temperatures rise above the optimum range, there is a dramatic drop in photosynthetic rates. Light harvesting in photosystem II is reduced due to cyclic electron flow, Rubisco restrictions, and thylakoid membrane instabilities, all contributing to a slower photosynthesis rate. Photoinhibition of photosystem II occurs at seven lower thermal stress ranges during photosynthesis. Researchers have found that moderate thermal stress causes little to no damage to photosystem II. Thermal stress slows down photosystem II repair by influencing reactive oxygen species (ROS) production throughout the thylakoid membrane. It has been discovered that the double bonds of fatty acids in the thylakoid membrane are largely responsible for its stability under thermal stress, which occurs between 32°C and 45°C. The overproduction of ROS under heat stress causes a reduction in fatty acid double bonds and a rise in membrane electron leakage, both of which contribute to the denaturation of proteins found in the thylakoid membrane. As a result of the increased turnover of the proteins complex, the amount of assimilates available for crop growth and development decreases under heat stress. Maintenance respiration rates in maize increase by more than 80% between 18°C and 33°C. Since the respiration rate increases under high temperatures far more than the photosynthesis rate eventually reduces, this could be a suitable indicator for stimulating plant response to heat stress (Nadeem et al., 2018).

6.4.3 Water Relations

Because of its far-reaching impact on water relations, thermal stress drastically reduces crop yields. Rapid water loss from the plant surface is associated with thermal stress, which can lead to dehydration and death. Enhancing transpiration and water movement is crucial for plant survival in hot conditions (Hasanuzzaman et al., 2013). Rapid soil water loss due to thermal stress affects soil temperature and transpiration, affecting the relationship between plants and water (Sita et al., 2017). Obstructed photosynthesis, increased respiration, decreased osmotic potential of leaves, and a lower sugar concentration level are all direct and indirect effects of thermal stress that cause osmotic adjustments in plants. Snap beans (*Phaseolus vulgaris* L.) under thermal stress lost more water during the day due to increased transpiration than during the night, leading to stress. Chickpeas with a high tolerance to thermal stress have a high stomatal conductance, which increases their transpirational heat dissipation as long as soil water is present (Kaushal et al., 2013). Despite having plenty of water in the soil, sugarcane experienced an instant reduction of water contents in leaf tissues when exposed to high temperatures. Reduced stomatal conductance under severe thermal stress conditions exacerbates leaf injury in tobacco (Nadeem et al., 2018).

6.4.4 Metabolic Disturbance and Imbalance of Primary and Secondary Metabolism

Metabolites are typically classified as either primary or secondary. Secondary metabolites focus on how an organism communicates with its surroundings, while primary metabolites fortify cells. Directly and crucially, primary metabolism contributes to plant growth, development, and reproduction by generating cofactors for secondary metabolite biosynthesis. Precursors for the biosynthesis of secondary metabolites are also produced (Medeiros et al., 2021). The chemical and functional variety of secondary metabolites. Plants use thousands of metabolites, or chemical compounds, as intermediaries in their interactions with their surroundings (Erb and Kliebenstein, 2020; Medeiros et al., 2021). When plants are under stress, they adjust their metabolic output to deal with the situation, but the mechanisms, causes, and controls for this procedure are poorly understood. Long-term treatments of salt, thermal, or water stress had the greatest impact on leaf metabolites compared to any other maize organ. The thermal stress metabolome status also highlighted metabolites from the raffinose and galactinol pathways and certain amino acids (tryptophan, threonine, histidine). It was found that 2.549 thousand genes were upregulated due to thermal stress, such as galactinol synthase (Zm00001d028931), stachyose synthase (Zm00001d039685), and a presumed inositol transporter (Zm00001d018803); however, the downregulation of 2.587 thousand genes was found. However, the pattern shown by the pathway of raffinose genes, in which the impacts of thermal and mineral stresses were interrelated, stands in stark contrast to the responses from water and thermal stressor interactions (Joshi et al., 2021).

By denaturing the enzymes involved in carbohydrate catabolism, heat stress causes an increase in sucrose and starch storage (Xalxo et al., 2020). Genes and potential role in carbohydrate metabolism showed variable expression patterns in heat-stressed *Arabidopsis*. Increased sucrose, maltose, and monosaccharides specific to the cell wall are also observed in response to heat stress. When plants were subjected to two abiotic stresses, like drought and heat, metabolic profiling revealed an increase in the accumulation of sugars like glucose, fructose, sucrose, trehalose, and maltose, which are involved in regulating turgor pressure of plant cells and stabilizing plasma membranes and proteins. However, prolonged heat stress induces loss of all carbohydrate reservoirs and deprivation in plants, so they cannot even digest starch molecules for energy when conditions are unfavorable (Rodziewicz et al., 2014; Sengupta et al., 2015; Kumar et al., 2021). In maize kernels, temperature fluctuations greatly impact the synthesis of starch, affecting the dry matter weight of grains. Mannose, fructose, sucrose, and proline are just a few examples of osmolytes that are produced in high enough quantities in response to heat stress to be crucial to thermal stress tolerance. The amount of sucrose in the kernels and the level of enzyme activity are the two most important factors in determining the rate and length of grain filling (Slama et al., 2015; Sharma et al., 2019; Alam et al., 2021). A study of maize metabolome perturbation found that the "tipping point" is delayed by more than a day under drought stress and that this delay is furthered under the co-effects of water and thermal stress (Bechtold et al., 2016). Water stress disrupts the osmotic balance in plant cells, while thermal stress induces mechanical shifts (Haswell and Verslues, 2015). Because of this, adaptations to water and high thermal stresses are likely at the root of abiotic stress-induced variations in metabolic responses (Khan et al., 2015).

Osmolytes have a major contribution in preserving the structure of plant membranes, defending cell organelles, scavenging ROS, preserving antioxidants, and keeping redox equilibrium stable (Sharma et al., 2019; Hasanuzzaman et al., 2020). Energy, plant nutrition, structural compounds, signaling molecules, and the development of seedlings are all greatly aided by osmolytes like sucrose, fructose, and mannose (Osuna et al., 2015). Heat-stressed maize (*Zea mays* L.) seedlings showed rapid glycan degradation and increased metabolism of mannose and fructose. Thermal stress in 21-day-old maize seedlings also upregulated the gene expression involved in the biosynthesis of mannose, fructose, and sucrose (Lieu et al., 2021; Stavridou et al., 2021). Thermal stress also wreaked havoc on the structures of mitochondria and nuclear membranes, especially in the heat-susceptible cultivars (Török et al., 2014; Li and Howell, 2021). Light energy utilization is also diminished due to mitochondrial membrane disruption, necessitating a rise in carbohydrate consumption to maintain adequate adenosine triphosphate (ATP) levels (Li and Howell, 2021). Furthermore, the nuclear genome is involved in encoding several proteins of chloroplast; therefore, damage to the nuclear envelope can impede the induction of photo-protection pathways, thereby exacerbating the damage to the photosynthetic activity and delay in the plant repair and structural impairment caused by photoinhibition (Kumar and Kaushik, 2021). One possible explanation for the greater photosystem II severity under high thermal stress is that the less clustered photosystem II units are more reactive to light.

6.4.5 Imbalance of Hormones

Some of the most important plant growth regulators in the face of abiotic stress are the phytohormones abscisic acid, gibberellic acid, indole acetic acid or auxin, cytokinin, ethylene, jasmonic acid, brassinosteroids, salicylic acid, and strigolactone (Sharma et al., 2019). Maize kernels develop incorrectly when exposed to high temperatures because of an overabundance of abscisic acid hormone and subsequent downregulation of the transcription factor cytokines (Niu et al., 2021). Benzyladenine increases heat tolerance in maize by keeping abscisic acid (ABA) and cytokinin (CTK) levels in check in young plants. Similar benefits are seen when Ca^{2+} ions solution and abscisic acid are applied to maize seedlings, increasing their heat resistance and improving their antioxidant enzyme activity and lipid peroxidation (Hossain et al., 2015; Yang et al., 2021).

Similar to how salicylic acid, gibberellic acid, and H_2S increase antioxidant activity in maize by stimulating the synthesis of trehalose, betaine, and proline (Li, 2015; Li et al., 2015; Zhou et al., 2018), these compounds also have a similar effect on tobacco. Tolerance to abiotic stress is increased after ZmbZIP4 overexpression causes lengthy primary roots, a profusion of lateral roots, and increased abscisic acid synthesis (Ma et al., 2018).

6.5 PHYSIOLOGICAL TRAITS RELATED TO HIGH THERMAL-STRESS TOLERANCE

For optimal photosynthesis and respiration, membrane stability must be maintained even under high thermal stress. Plant cell membranes are the main target of abiotic stresses, particularly high thermal stress, so mechanisms that help them remain stable under these conditions are considered adaptive. Reduced production of ROS, improved antioxidant rate, and antioxidant enzyme activities are linked to increased cell membrane stability (Djanaguiraman, Boyle et al., 2018). The photosystem I and photosystem II reaction centers of chloroplast, peroxisomes, and mitochondria are the primary sites of ROS production. Reduced membrane stability is due to ROS-induced lipid peroxidation. Furthermore, stomatal and non-stomatal limitations increase ROS formation because ATP and nicotinamide adenine dinucleotide hydrogen phosphate ($NADPH_2$) are underutilized in reactions without light (Calvin-Benson cycle) (Song et al., 2014). Since chlorophyll's minimum and maximum yield (Fo/Fm) proportion, fluorescence can be interpreted as thylakoid membrane destruction (Alemu, 2020). It was recommended that chlorophyll fluorescence analyses might be a useful nondestructive instrument for measuring thermal stress tolerance. Most studies have high genetic diversity and heritability for such physiological traits. Also, as a form of adaptation to thermal stress, the unsaturation rate of plastidic glycerolipids in the leaf decreased (Prasad et al., 2011). Identifying lipid molecular markers for sorghum thermal stress amelioration may require further research.

The effect of thermal stress has been observed on plant respiration is less well understood. Carbon starvation occurs when a plant's respiration rate rises faster than its capacity to absorb carbon dioxide through photosynthesis. This happens when the air temperature rises above a critical threshold. Respiration reaction to thermal stress varies with the maturity of the crop. From 0 to 35–40°C, the respiration rate rises exponentially, peaks at 40–45°C, and then declines with further temperature rises. The increased respiratory rate indicates a higher assimilate consumption for either maintenance or growth. Breathing rates increased at night in response to high day and nighttime temperatures (Djanaguiraman et al., 2013). Ultimately, boosting respiration efficiency can improve plant growth, yield, or tolerance to thermal stress, as well as the effective usage partitioning of carbon (Prasad et al., 2019). Higher respiration rates and lower yield were observed in the thermal-vulnerable rice genotype compared to the thermal-tolerant genotype. The diurnal variation in respiration rate, the genotypic diversity responding to thermal stress, and the quantitative trait loci (QTLs) for traits associated with respiration need more investigation in summer fodder crops. Sorghum grain yield is calculated by multiplying the plant spike/panicle production by its grain production per spike/panicle and finally by the weight of each grain. Reproductive success, measured by the number of seeds produced (or a seed set rate of >80%), is directly proportional to the fertility of the florets (viability of gametes). If increased seed production under thermal stress is not balanced by a corresponding decrease in individual seed size, selecting an optimized seed set and quantity will be beneficial. Reduced pollen development, pollen or stigma validity, and growth of pollen germination tube due to high temperature throughout gametogenesis or anthesis, infertility, and abortion of embryo at early stages are all factors that contribute to a lower seed set percentage. Degradation of the fibrous tissue layer and modified carbohydrate metabolism are linked to pollen viability loss under high temperatures (Prasad et al., 2021).

The premature degradation of tapetal cells under high temperatures produces sterile pollen under water deficit conditions and high thermal stress in wheat. The tapetal cells produce the required carbohydrate and nutrients for the developing pollen grains. Carbohydrates were found to be reduced in the sorghum anther wall, and pollen and pollen germination was also inhibited. How much a single grain weighs depends on how quickly and for how long it is filled. Since more nutrients will be available per grain, fewer seeds can be planted to get the same harvest (Prasad et al., 2021). Grain sorghum seed size did not expand when subjected to high thermal stress during early flowering, which reduced seed numbers. Gene variants in sorghum that affected ovary development also affected grain weight (Djanaguiraman et al., 2013; Prasad et al., 2015). In a study conducted by Singh et al. (2015), thermal stress's impact on seed-set percentage was found to be significantly more than the impact particularly on seed weight, suggesting that an increase in seed weight did not reimburse for a decrease in seed set in thermal-sensitive cultivars. To maximize grain yields under thermal stress, it is important to cultivate sorghum varieties with a high rate of seed filling and a long period of seed filling during the plant's physiological maturity. Prasad et al. (2019) report a few other possible traits related to tolerance to sorghum's thermal stress, including early morning inflorescence and dejection in canopy temperature. Plants that can flower early in the morning will be protected from the harmful effects of thermal stress. There was a wide range in anthesis time between rice genotypes across species and cultivars, suggesting that this area also needs investigation in sorghum. Another trait that is efficient in preventing high canopy or plant tissue temperatures is the plants' potential to keep their canopies cool under high thermal stress. Several sorghum genotypes were found to have higher grain yields, and these were the ones that had a relatively cooler canopy (escape) and a higher canopy temperature (tolerance) (Prasad et al., 2021).

6.6 PHYSIOLOGICAL MECHANISMS ASSOCIATED WITH HIGH-TEMPERATURE TOLERANCE

Researchers have found that sorghum yield decreases when the temperature rises above about 32°C. There has been a report of an overview of decisive temperatures for sorghum at various growth stages (Prasad et al., 2019). Therefore, the crucial phase must correlate with the optimum air temperatures in each environment for sorghum to reach its full yield potential. Sorghum, on the whole, does best in temperatures ranging from about 15°C to about 40°C. Reduced photosynthesis due to high temperature is linked to alterations in the structure and function of cell organelles. The destruction of thylakoid membranes, the chloroplast-protein complexes, and the chloroplast envelope are all part of the structural abnormalities (Chen et al., 2020).

Inactivation or hindrance of the enzymes in the Calvin-Benson cycle and modifications of plant structure in grana are two functional consequences of high temperature on photosynthesis. Plants under high-temperature stress have a lower concentration of pigments, an electron transport system, lower photosystem II quantum yield, activities of enzymes associated with photosynthesis, and decreased gas exchange. The Rubisco enzyme does not constrain photosynthesis at high temperatures. However, the photosynthetic rate was reduced because of a decrease in Rubisco activase activity in the presence of high temperatures. Chlorophyll and the photosynthetic apparatus may be affected by high temperatures because of the development of ROS like superoxide radical (O_2), hydroxyl radical (OH^{-1}), and hydrogen peroxide (H_2O_2) (Djanaguiraman, Perumal et al., 2018; Djanaguiraman, Narayanan et al., 2020). Increased temperatures during flower initiation and gametogenesis stage reduced grain yield per panicle without affecting specific grain weight (Prasad et al., 2015). Toughness under heat and humidity stress strongly correlates with membrane thermostability in many plant species. Both plastidic and extra-plastidic glycerolipid unsaturation levels were found to decrease at high temperatures. Figure 6.1 displays a schematic overview of high temperatures across the entire plant.

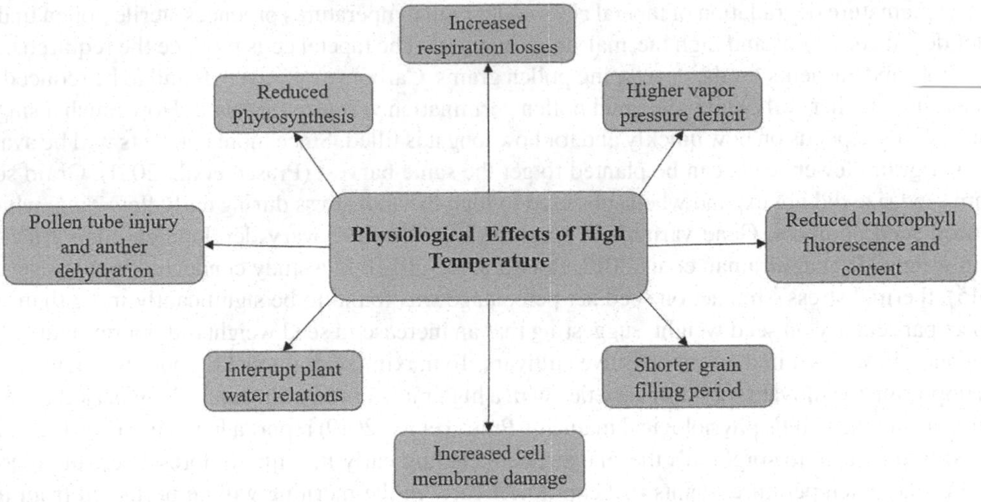

FIGURE 6.1 Physiological processes of a plant system in response to high-temperature stress.

6.7 LOW-TEMPERATURE STRESS

The general progression of crop development, from the earliest germination phase to the later reproductive phase, can be disrupted by low-temperature stress, characterized by the exposure of plants at a temperature less than 10°C for an adequate period (Farooq et al., 2009). Photosystem II potential, chlorophyll concentrations, and leaf area are all negatively impacted by cold stress. Delaying the onset of anthesis in maize by as much as a week can be achieved through a single series of low-temperature stress (i.e., less than 10°C for a week) during the plant growth stages (Hayashi, 2016).

6.7.1 Physiological Effects of Low-Temperature Stress

It has been found that temperatures between 4°C and 10°C can inhibit chlorophyll biosynthesis and cause an extreme suppression in photosystem II activity, while temperatures between 8°C and 10°C prolong seedling emergence and reduce the root-to-shoot proportion and total chlorophyll content during the early growing period in forage corn (Bano et al., 2015; Riva-Roveda et al., 2016). Reduced metabolite transport caused by low-temperature stress negatively affects chloroplast and thylakoid complexes, metabolic activities, and the Calvin-Benson cycle (Sun et al., 2017). Under chilling stress (12–14°C), the cell wall properties in a cold-susceptible corn hybrid were investigated, and the pectin level and pectin methyl esterase activity decreased. The photosynthetic mechanism, plasma membranes, and enzymatic activity all exhibit evidence of physiological and biochemical disturbances at low-temperature stress (Bilska-Kos et al., 2017). Chilling injury in early maize seedlings increases malondialdehyde (MDA) level and plasma membrane permeability while decreasing water, chlorophyll, and proline content. As a result of reduced metabolite transport, low-temperature stress also causes deficiencies in macronutrients (N, P, K, Ca, Mg) in shoots and roots (Liu et al., 2016). However, maize plants produce defense-signaling compounds (such as abscisic acid ABA and nitric oxide) when subjected to chilling temperatures of about 7–10°C (Esim and Atici, 2016). ROS are overproduced in response to low thermal stress, causing injury to biomolecules, cellular components, and

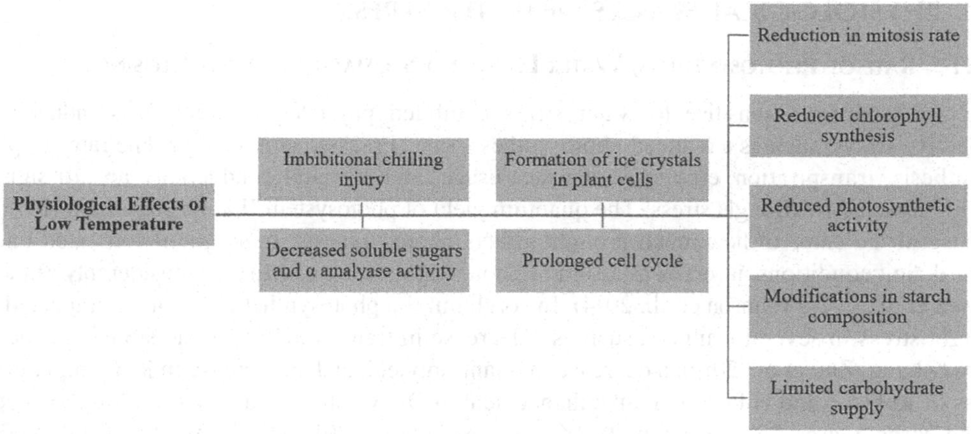

FIGURE 6.2 Physiological responses of plants under low-temperature stress.

membranes (Hussain et al., 2016). Plants increase antioxidant enzyme production like superoxide dismutase (SOD), peroxidase (POD), and proline during times of stress (Hussain, Khaliq et al., 2019; Hussain, Shengnan et al., 2019).

Starch composition in grains can be altered by low-temperature stress during grain filling, resulting in decreased amylose content, decreased water solubility, decreased swelling capacity of starch, and increased gelatinization temperatures (Lu et al., 2014). Reduced phosphoenolpyruvate carboxylase (PEPC), sucrose phosphate synthase (SPS), sucrose synthase, and the photosynthetic apparatus are observed at less than 15°C temperature during the late reproductive phase. Grain quality is typically diminished, with low-quality components and a rough physical texture of maize grain (Yan et al., 2015). Low-temperature stress decreases yield by slowing germination, growth, and photosynthesis. Consequently, temperatures below the optimum level can cause a serious reduction in yield if they occur during crucial reproductive phases because plants allocate >50% of photosynthates for grain production during this stage until physiological maturity (Panison et al., 2016). Low-temperature stress significantly reduces the height of the plant and the maize crop's total production (Liu et al., 2016). Cold-stressed plants have a slower rate of mitosis and a longer cell cycle, which slows down leaf development. Figure 6.2 provides a schematic summary of the several impacts and mechanisms of low-temperature stress.

6.8 WATER STRESS

About 28% of the Earth's land is unsuitable for farming (Izaurralde et al., 2011). There is poor plant physiology, growth, and development due to the increasing drought of the changing climate scenario. According to estimates, nearly 17% of tropical annual yield is lost during harvesting (Siebers et al., 2017). Because of the water stress, plants lose more water through transpiration and have less water available at their roots (Waqas et al., 2017). It recommends the water balance toward the downside, which negatively affects development, the nutrient–water relationship, photosynthesis, assimilation, and overall efficiency (Tao et al., 2016). Deliberate actions taken by plants in response to stress are species-specific and dependent on the plant's developmental status and environmental context (Sandhu et al., 2018). Enzyme activity and cell division in plants are altered by high temperatures (Rafique, 2019), and the duration and distribution of plant growth are also affected (Hussain et al., 2018).

6.9 PHYSIOLOGICAL EFFECTS OF WATER STRESS

6.9.1 Rate of Photosynthesis, Water Loss, and Stomatal Opening/Closing

Sorghum genotypes sensitive to water stress exhibited physiological alterations induced by respective stress, such as a reduced photosynthesis rate (Fracasso et al., 2016). The rates of photosynthesis, transpiration, efficiency of water usage, and stomatal conductivity are all significantly impacted by drought stress. The quantum yield of photosystem II at its maximum (Fv/Fm) is a useful indicator of how much drought affects photosynthesis. In sorghum cultivated under water deficit conditions, it serves as a sign of photosynthetic activity that is considerably reduced (Husen et al., 2014; Johnson et al., 2014). In sorghum, the photosynthetic activity is impacted by drought stress in several manners, such as a decrease in transpiration rate and stomatal conductivity (Zhang, Zhu et al., 2019), a decrease in quantum yield, and an increase in leaf temperature, a loss of Rubisco and chlorophyll, an enhancement in O_2 evolution, and a loss of phosphoenolpyruvate carboxylase (PEPCase) activity (Kapanigowda et al., 2014; Bao et al., 2017). The Fv/Fm and photosynthetic rate of sorghum varieties with the potential to tolerate water deficit stress is significantly higher in various studies conducted under drought stress conditions (Fracasso et al., 2016; Sukumaran et al., 2016). Photosynthetic recovery after rehydration has a significant role in determining drought tolerance in sorghum plants, avoiding a decline in grain yield, and the plant's potential to evade or endure water-deficit stress (Chaves et al., 2009). An important mechanism by which tolerant sorghum genotypes retain grain yield under water stress is a rise in photosynthetic rate, which delivers raw material and energy needed for plant growth and development (Getnet et al., 2015).

Studies on normal and drought-stressed sorghum have found substantial genetic variation in the net carbon accumulation rate, transpiration rate, carbon and transpiration rate ratio, and water use efficiency. According to several studies, the increase in the ratio of photosynthetic activity to transpiration activity and the efficiency of water usage has been shown to increase drought tolerance in sorghum pre-flowering stage, according to several studies (Vadez, Krishnamurthy et al., 2011). Transpiration rate was not varied between control and drought-tolerant sorghum cultivars exposed to water stress, but it has been found to vary significantly in cultivars sensitive to water stress. Water use efficiency was also significantly greater in cultivars tolerant to water stress than in cultivars sensitive to water stress (Fracasso et al., 2016). Significant correlations have been disclosed between sorghum grain yield, transpiration effectiveness, and water extraction (Vadez, Krishnamurthy et al., 2011). Heritability estimates for the ratio of photosynthetic rate to transpiration rate ranged from 0.8 to 0.9 at 80% and 40% field capacity, respectively, suggesting a strong trait of genetic component. Selecting valuable genotypes relying on this trait is crucial to develop sorghum varieties tolerant to water or drought stress, as the genetic component of the tolerance mechanisms plays a crucial role. Drought-tolerant genotypes reduce stomatal conductivity and transpiration activity during the vegetative stage, conserving water that can be deployed during grain filling in water-deficit environments. An intriguing study found that a QTL recognized for stomatal conductivity was linked to lower transpiration but not photosynthetic rate or biomass of shoot (Lopez et al., 2017).

6.9.2 Chlorophyll Content and Maintenance of Green Color

A plant's ability to absorb light for photosynthetic activity is inversely proportional to its overall chlorophyll content and chlorophyll a and b contents. One factor in a plant's resilience to drought is its ability to keep its chlorophyll content from dropping too low while under stress. According to several studies, sorghum grown in arid conditions has had its chlorophyll content significantly reduced significantly (Reddy et al., 2014; Chen, Liu et al., 2016; Fracasso et al., 2016; Fadoul et al.,

2018; Amoah and Antwi-Berko, 2020). For example, after growing in water stress conditions, senescent cultivars showed a 75% suppression in overall chlorophyll content, while stay-green cultivars showed a 23% decrease. One study found that the total chlorophyll content of stressed plants was 4.3% lower than control plants (Devnarain et al., 2016). Some carotenoids, like chlorophyll, have been shown to have their concentrations drop under extreme water deficit stress. It was found by Takele (2010) that in drought-tolerant sorghum, carotenoid, and chlorophyll contents were lower before and after flowering. Carotenoid levels are reduced in drought-sensitive genotypes likely because genes involved in the biosynthesis of carotenoid and terpenoid are repressed (Fracasso et al., 2016). Drought stress severely impacts the response of light and carbon dioxide fixation mechanisms because of the downregulation of genes contributing to the biosynthetic pathways for carotenoids and chlorophyll. The correlation between chlorophyll concentration, number of green leaves, and a fraction of green leaf area was highly significant at both the flowering and maturity stages. Grain yield was significantly correlated with leaf attributes at both developmental stages (Reddy et al., 2014).

Chlorophyll loss and a subsequent decrease in photosynthetic potential characterize the senescence of leaves. The well-characterized trait aids sorghum's ability to adapt to post-flowering water-deficit conditions to remain green, which postpones leaf senescence and enhances grain yield. Numerous researchers have investigated the physiological means of sorghum stay green trait. Early on, it was hypothesized and studied that an increased rate of cytokinin, nitrogen, and chlorophyll due to water-deficit stress is linked to the stay-green trait. Examples include a correlation between staying green and increased nitrogen concentration in leaves, specifically during the flowering stage. A different study demonstrated that remaining green genotypes of sorghum have higher cytokinin levels than their non-stay-green counterparts, suggesting that the stay-green cultivars old more slowly.

Moreover, the chlorophyll content of the remaining green genotypes is higher than that of the aging genotypes. Higher chlorophyll content in leaf, a lesser rate of losing green area in plant leaves exposed to water stress, reduced tillering, and smaller upper leaves are all characteristics of the stay-green trait of sorghum, which is a response to water deficit stress (Kassahun et al., 2010; Vadez, Deshpande et al., 2011; Borrell, Mullet et al., 2014; Borrell, van Oosterom et al., 2014; George-Jaeggli et al., 2017). Higher water extraction and transpiration efficiency are also associated with this trait. Higher relative grain yield was found in the post-flowering phases following the gene flow of stay green quantitative trait loci from the B35 to the senescent cultivar R 16. Higher chlorophyll rates of the leaf accompanied this at flowering. Canopy size is regulated by the Stg QTL, which, during inflorescence emergence, causes a suppression in tillering and a reduction in the size of upper leaves, an increase in the size of lower leaves, and, in some instances, a reduction in the number of leaves per culm. Because the canopy is smaller before flowering, there is more water for the plant to use during grain filling, increasing biomass production and grain yield (Borrell, Mullet et al., 2014; Borrell, van Oosterom et al., 2014). In stay-green lines, the senescence of lower leaves before flowering causes the loss of old leaves during the flowering process, contributing to the smaller canopy size. By conserving water before flowering, plants have more of it afterward, allowing them to keep producing food by photosynthesis for as long as possible and "staying green" while grains develop (George-Jaeggli et al., 2017). These and other studies suggest that staying green may function after flowering as a mechanism of water stress tolerance, making accessing the water needed for overall plant growth and grain productivity easier.

6.10 PHYSIOLOGICAL TRAITS RELATED TO WATER STRESS TOLERANCE

Prasad et al. (2019) provided an overview of sorghum's characteristics linked to its drought resistance. Stay-green is both a drought-adaptive and a characteristic trait in sorghum. The term "stay green" is commonly used to describe a plant's capacity to maintain chlorophyll content after flowering, which helps it to avoid lodging and gain well-filled grain (Jordan et al., 2012). Sorghum line B35, a descendant of Ethiopian durra and Nigerian cultivar, is the most viable line harboring the stay-green phenotype and is extensively employed as a trait source throughout areas. A few potential sources of stay-green traits in sorghum fodder have also been outlined as an additional perk. Those with drought-vulnerable genotypes experience a greater reduction in stay-green expression due to chlorophyll loss. The efficiency of stay-green is correlated with an interruption in the initiation of leaf senescence or the rate at which leaf senescence progresses. Under pre- and post-flowering water-deficit stress, the rate of leaf senescence was inversely related to yield (Prasad et al., 2021). Modulating the trait known as "stay green" are nitrogen requirements of the grain, nitrogen availability in the leaf via translocation and nitrogen uptake by the roots. Four QTLs, designated Stg1, Stg2, Stg3, and Stg4, were found to be accountable for about 54% of the phenotypic variability seen in stay-green genotypes of sorghum when the QTL related with the stay-green phenotype were targeted. QTLs Stg1 and Stg2 have been located on sorghum's third chromosome, while Stg3 and Stg4 have been identified on chromosomes 2 and 5, respectively.

Furthermore, it was found that the water availability and water loss phenotypes are linked to the stay-green trait (Borrell, Mullet et al., 2014). Stay-green QTL-carrying sorghum genotypes reduce tiller number, canopy size, and coverage, reducing water use during the vegetative stage and freeing up more water to be utilized in the post-anthesis water stress occurs during the grain filling duration (Jaegglia et al., 2017). To learn how plants respond to water stress, however, there is a need to investigate the stay green genotype root architecture in greater detail. According to a recent study (Liedtke et al., 2020), the relationship between canopy size and staying green is weak before but strong after flowering. That is why the stay-green phenotype of sorghum helps with grain yield in both normal and drought conditions.

Long, branched roots that can reach deep into the soil and draw up moisture are among the most important phenotypes for plant drought resistance (Fenta et al., 2014). For example, it was found that sorghum genotypes with more seminal roots and larger vessel diameters were better able to withstand drought. The genotype absorbed more soil water with longer roots and smaller diameters than with shorter roots and a larger diameter. According to these findings, sorghum's potential to tolerate water stress is more strongly associated with root length than root diameter. Also, a broad root angle decreases the amount of energy necessary for its penetration into the soil, which is a positive correlation between root angle and soil penetration capacity (Wasson et al., 2012). While a wide root angle with shallow roots can only access the soil surface moisture, a constricted root position with deep roots can access the soil's deeper layers of stored moisture. It has been shown in numerous studies that plants with deeper and healthier root systems produce higher yields (Liang et al., 2017; Wang et al., 2012; Rostamza et al., 2013). Genetic choice and irrigation scheduling can be used to alter the layout of the root system for improved performance in dry soil (Passioura, 2012). It is accepted that osmotic adjustment occurs when there is a net increase in solutes, which results in a decrease in the osmotic potential of tissues. Reduced osmotic potential shifts water concentration gradients from high (soil) to low (tissue), improving turgor. The body's ability to hold onto solutes improves when stress is reduced.

Leaf area, shoot expansion, canopy structure, and leaf surface characteristics can all be modified by sorghum in response to soil moisture (Prasad et al., 2021). Moreover, osmotic potential and accumulation of total solutes both change throughout the day. Sorghum leaves adapt osmotically at about 0.1 MPa per day, as the lowest leaf water potential drops by about 0.15 MPa per day. Under extreme water stress, the aperture of stomata closes at a leaf osmotic potential of 1.4–1.5 MPa. When the leaf potential is low, about 1.4 MPa, the stomata remain closed, leading to an increase in abscisic acid level. When the leaf potential drops below 3.7 MPa, the outer membrane of the chloroplast swells (Assefa et al., 2010). The osmotic adjustment of sorghum genotypes subjected to post-anthesis water stress was higher than that of the control, resulting in a 24% increase in yield. The grain size was the cause of the difference in yield, and variations in grain number are linked to a better harvest index (Amelework et al., 2015). By comparing sorghum with cereal crops like wheat, it had less research on its osmotic adjustment traits. There has been insufficient research on the QTLs and genes involved in osmoregulation and transgenics in sorghum, reflecting osmotic adaptation genes. Consequently, more research is required into the osmotic adaptation trait in summer fodder crops.

To a lesser extent, stomatal conductance can be inferred from the temperature of the canopy. Sorghum can keep its stomata open in a broad array of leaf turgor and water potential (Rebetzke et al., 2013). Plants with a high stomatal conductance have a lower canopy temperature because they transpire more. The variation in transpiration, stomatal conductivity, and water use efficiency among crop genotypes can be studied with the help of canopy temperature depression (Mutava, 2012). Grain sorghum drought resistance could be enhanced by investigating the links between canopy temperatures, water use efficiency, and grain yield. Drought tolerance is increased when water is conserved during the initial phases of crop development, even in the face of high evaporative demand. Under high vapor pressure deficit, sorghum genotypes were found to exhibit a trait known as limited transpiration (Gholipoor et al., 2012; Choudhary et al., 2013). Sorghum farmers can use this advantage by increasing crop resistance to drought.

Although cowpea is known as a drought-resistant crop, it is still affected by water scarcity. Significant traits for increasing grain yield of cowpea under drought stress include delay in leaf senescence, the greenness of plant stem, and deep rooting. The variability in drought onset and the effects of abiotic stresses on phenotypic expression have slowed the pace at which phenotype-based selection has been implemented. Efforts in recent years have centered on utilizing DNA markers established through QTL mapping. By manipulating cowpea stress responses using metabolite-based markers, it is hoped that drought tolerance can also be bestowed. Cowpea is remarkably drought-resistant because of its ability to conserve moisture. Stomatal closure and paraheliotropism, two drought-evasion mechanisms in cowpea, have been well described. However, the potential for moisture conservation by osmolyte accumulation has not been adequately explored (Gouto et al., 2017).

6.11 PHYSIOLOGICAL MECHANISMS ASSOCIATED WITH WATER STRESS TOLERANCE

Water deficit stress and plant tolerance result from intricate biological processes that modify the plant's physiology, biochemistry, genome, proteome, and metabolome (Ngara et al., 2021). Sorghum plant responses to drought stress are summarized in Figure 6.3. Plants have defense mechanisms to help them survive, recover from damage, and tolerate drought. Plants can avoid drying out in drought conditions by storing more water in their roots or drawing more moisture from the soil (Osmolovskaya et al., 2018). Plants can withstand drought by increasing their root depth, closing their stomata, rolling their leaves, waxing their stems to remain green in the face of arid conditions, and maintaining a high efficiency of transpiration (Badigannavar et al., 2018). Drought escape, in which plants avoid drought by maturing and reproducing before the onset of dry weather, is the superior mechanism (Manavalan and Nguyen, 2017). Mechanisms for enduring drought include early inflorescence and maturity, high nitrogen content in leaves, high potential for photosynthetic activities, and integration remobilization. The plant's potential to endure water deficits without compromising its physiological processes that maintain tissue and cellular viability and metabolic stability is drought tolerance. Among these are high stomatal conductivity, desiccation-tolerant enzymes, high proline levels, and osmoregulation (Badigannavar et al., 2018).

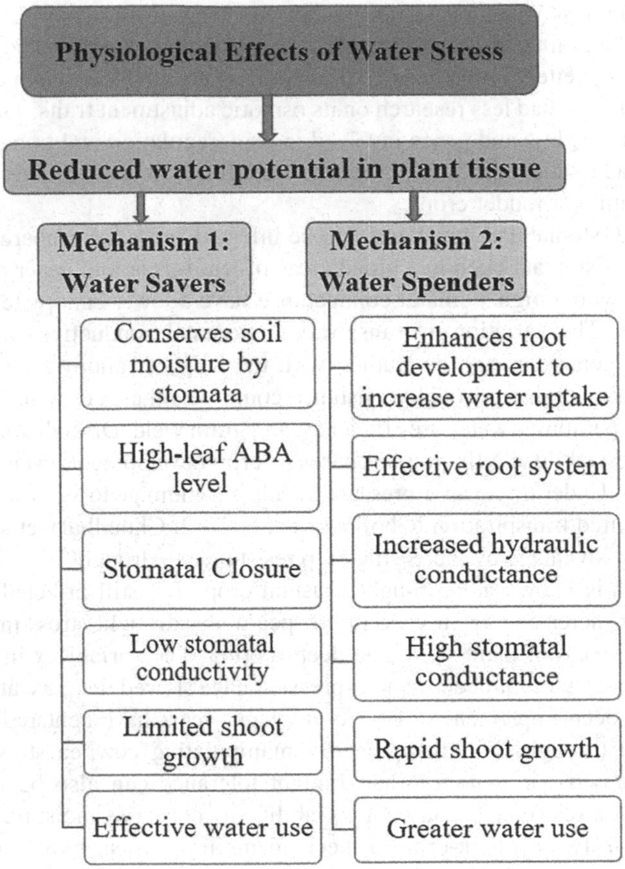

FIGURE 6.3 Physiological tolerance mechanisms involved in plants in response to water stress.

6.12 STAGES SENSITIVE TO WATER AND THERMAL STRESS

Compared to vegetative development stages, reproductive sorghum stages are more vulnerable to abiotic stresses (Prasad et al., 2019; Djanaguiraman, Prasad et al., 2020). The stages of panicle formation, flowering, and filling of grains are particularly vulnerable to water stress in sorghum. Similarly, most cereals are more vulnerable to thermal stress during the reproductive stage compared to the vegetative stage, with the former resulting in a larger loss of yield. The most sensitive periods for thermal stress in sorghum were 10 days to 5 days prior to anthesis (the micro- or megasporogenesis phase) and 5 days prior to and 5 days after anthesis, resulting in maximum reductions in floret fertility and a deprived seed set (Prasad et al., 2015). Moreover, more research is required to uncover the trade-off arrays linked with the more subtle phase vulnerability to water stress from seedling to maturity.

6.13 MINERAL STRESS

Mineral nutrition is a separate field from plant physiology. Arnon and Stout (1939) have recommended an essential criterion based on which different minerals are considered essential nutrients for crop growth and development. Distribute essential minerals into four categories on the basis of their biological mechanism and physiological activities. Scientists have observed varying degrees of mineral stress (deficiency or excess) in various plants. High nitrate concentrations (NO_3) suppress root development and reduce the proportion of root to shoot, and nitrate plays a crucial role in cytokinin biosynthesis and transport. Potassium-deficient plants are sensitive to lodging and air circulation, while phosphorous deficiency restricts the elongation of the primary roots, enhances the development of lateral roots, reduces the fraction of shoots and roots dry weight, and decreases the number of leaves. Reduced overall photosynthetic activity and root hydraulic conductivity, lower root-to-shoot dry weight ratio, altered carbohydrate metabolism, accompanied by induced starch accumulation, and so on are all effects of Sulphur deficiency (Singh and Chahal, 2020).

6.14 PHYSIOLOGICAL EFFECTS OF MINERAL STRESS

When sorghum plants lacked nitrogen, their leaf area (LA), chlorophyll content (Chl), and photon number (Pn) all dropped, leading to less dry matter (DM) accumulation. Nitrogen deficiency reduced the photosynthetic capacity of sorghum leaves, and this was most evident in a decline in stomatal conductance (g_s) and intercellular CO_2 (C_i) instead of a mesophilic activity. The leaf spectral reflectance of sorghum plants was extremely sensitive to their nitrogen status, particularly in the evident red-edge regions. Leaf reflectance was most affected by nitrogen deficiency at 555 and 715 nm. Leaf chlorophyll and nitrogen levels were linked to changes in reflectance at these wavelengths. The reflectance proportions R1075/R735 and R405/R715 had the highest r_2 values and were found to be highly associated with leaf chlorophyll and nitrogen concentrations (Zhao et al., 2005).

6.15 PHYSIOLOGICAL MECHANISM OF PLANT SYSTEM UNDER NITROGEN DEFICIENCY

Despite sorghum plants' greater tolerance to water deficit stress and high temperatures and ability to use nitrogen more effectively than most C3-type crops (Young and Long, 2000), Nitrogen deprivation inhibited the growth of plant and DM deposition and allocation. Sorghum's smaller leaf area was primarily responsible for the plant's reduced biomass production due to nitrogen deficiency, though a decrease in leaf photosynthetic capacity was also noted. Levels of chlorosis and photosynthetic rate in leaves were inversely proportional to leaf nitrogen levels, falling off linearly as nitrogen levels in leaves dropped (Zhao et al., 2005).

Several physiological and biochemical processes are required for successful leaf photosynthesis, including g_s, C_i, the photochemical potential of photosystem II, and the concentrations and activities of carbon fixation enzymes. The concentration of leaf chlorosis and the photosynthetic rate decreased linearly with decreasing leaf nitrogen, but the photosynthetic rate did so at a slower rate than chlorosis. For instance, a drop in leaf nitrogen from 50–20 mg kg^{-1} caused a 51% drop in leaf chlorosis and a 33% drop in leaf photosynthetic rate. Leaf photosynthesis in the nitrogen-deficient plants was expressed on a chlorosis basis, and while this was 12% (in response to 20% nitrogen treatment) or 20% (in response to 0% nitrogen treatment) lower as compared to that in the control plants, it was 12–38% higher in the former (data not shown). Losses in leaf area based on photosynthetic rate and leaf area because of nitrogen scarcity are the main factors restricting plant growth and production. However, this type of chlorophyll adaptation can partly reduce the negative impacts of nitrogen deficiency on leaf photosynthetic potential. The amount or activity of ribulose bisphosphate carboxylase/oxygenase (Rubisco) was found to be decreased by nitrogen deficiency in several other crops (Zhao et al., 2005). Nitrogen deficiency has recently been shown to decrease PEPCase and Rubisco levels in sorghum leaves.

Instead, it was found that under nitrogen deficiency, sorghum leaf photosynthetic rate was limited by a combination of a decrease in g_s and C_i, both of which occurred simultaneously with the reduction in the rate of leaf photosynthetic activity. In contrast to the differences in their photosynthetic light-response curves, the differences in leaf photosynthetic rate between the three nitrogen treatments were much smaller at the same C_i. As a result, it appears that under nitrogen-deficient conditions, the primary cause of the decrease in sorghum leaf photosynthetic rate is the resulting decrease in stomatal conductivity. Therefore, in sorghum, nitrogen deficiency impacts the photosynthetic leaf rate, transpiration rate, and water use efficiency (Zhao et al., 2005).

Nitrogen translocation occurs primarily from lower to upper leaves and occasionally from vegetative to reproductive organs in maize plants when nitrogen is scarce. Chlorosis in mature/lower leaves is a potential outcome of nitrogen re-translocation (Gaikpa et al., 2022). It takes much less nitrogen for maize, a C4 plant, to synthesize Rubisco protein because of its ability to increase carbon dioxide at the site where Rubisco activity occurs (Schlüter et al., 2012). This phenomenon may be responsible for the crops' ability to produce some yield even when the conditions for suitable bioavailability of nitrogen for the growth and development of plants are not met (Guo et al., 2019).

6.16 PHYSIOLOGICAL MECHANISMS OF PLANT SYSTEMS UNDER PHOSPHORUS DEFICIENCY

For crop production, phosphorous is the second-most limiting mineral, following nitrogen, and affects crop growth and development, significantly impacting international food security. The maize plant has physiological mechanisms of adaptation, in addition to morphological mechanisms, to flourish in low phosphorous conditions. The low phosphorous-tolerant maize plants experience a significant decrease in leaf area and shoot weight when grown in phosphorous-deficient soil. However, the crop's root development grew differently. Perhaps this means that in phosphorous-deficient conditions, the plant root zone receives a larger share of the photoassimilates (Gaikpa et al., 2022). However, in phosphorous-deficient plants, growth is not constrained by the availability of carbohydrates in developing organs. White clover showed similar morphological and physiological effects under phosphorous-deprived conditions, with less leaf area per plant and less whole-plant photosynthesis.

Under phosphorous deficiency, some maize hybrids have been shown to have low protein levels with a rise in proline levels. It has been shown that plants produce proline as an initiator molecule in response to drought, soil salinity, and nutrient deficiency (particularly phosphorous deficiency; Liang et al., 2013; Siddique et al., 2018; Chun et al., 2019). Under low-phosphorous conditions excluding the final growth phase, maize leaf cells were significantly longer than those in a maize plant with sufficient phosphorous (Amanullah et al., 2016). However, studies showed that the reduced cell production and leaf elongation rates in phosphorous-deficient maize plants were due to a regulatory mechanism involving cell division at the shorter zone of the epidermal layer and the production rate of the lower cell along the length of the leaves. It was revealed that the exact mechanism by which phosphorous deficiency influences leaf growth remains unknown (Gaikpa et al., 2022).

6.17 PHYSIOLOGICAL MECHANISMS OF PLANT SYSTEMS UNDER POTASSIUM DEFICIENCY

Plants have elicited multiple physiological responses to potassium deficiency. The effects of potassium deficiency on the development and growth of wheat seedlings were reported by Thornburg et al. (2020). It was proved that low potassium levels could alter miRNAs. After this major discovery, it was hypothesized that altering miRNAs and their targets could be a safe fortress to improve fertilizer efficacy, leading to higher biomass production and crop yield. Higher concentrations of calcium, magnesium, and sodium ions were mobilized into the plant cells of maize crops grown under potassium deficiency compared to maize plants grown under conditions having optimum potassium availability (Gaikpa et al., 2022). It was found that the higher concentration of calcium, magnesium, and sodium uptake by potassium-deficient maize plants did not fully adapt to the lowered molarity in response to the inadequate potassium content. Field experiments showed no significant difference in physiological maturity between foliar potassium-sprayed and control maize (Amanullah et al., 2016). The application of potassium causes maize to take in more nitrogen and phosphorus, leading to lush growth and an interruption of physiological maturity.

6.18 CONCLUSION

Thermal, mineral, and water stress harms crop growth, development, and productivity by altering plant physiology in numerous ways. Understanding the physiological underpinnings of fodders' tolerance or susceptibility can help breeders produce varieties and hybrids more resilient to abiotic stresses. The tolerance mechanism to thermal, mineral, and water stresses in summer fodders is less understood than those of rice and wheat. These abiotic stresses will have larger negative effects on growth and production if they occur during the reproductive periods of crop development when the plants are especially vulnerable. Reduced tissue water content, membrane stability, green leaf area, carbon uptake, and partitioning contribute to stunted development, biomass loss, and yield reductions due to water stress. To a similar extent, severe temperatures impact the membranes and photosynthetic activity the most. High temperatures reduce membrane stability, slow carbon assimilation, increase respiration, and reduce floret fertility, all of which negatively affect seed production (both in terms of quantity and quality). Stay-green, lowered canopy temperatures, reduced transpiration, increased reproductive success, and a strong root system are all hallmarks of drought-tolerant plants. In contrast, increased membrane stability, greater gamete viability, and reproductive success leading to a greater seed yield are hallmarks of plants that can withstand high-temperature stress. The response of summer fodders at mineral level (nitrogen, phosphorus, potassium) below optimum concentration can be understood by looking at changes in root and shoot structure and physiological properties. Therefore, a holistic contribution-based collaborative breeding team and the pipeline are necessary to increase stress tolerance and genetic advances in summer fodders.

REFERENCES

Alam, M. R.; Nakasathien, S.; Molla, M. S. H.; Islam, M. A.; Maniruzzaman, M.; Ali, M. A.; et al. Kernel water relations and kernel filling traits in maize (*Zea mays* L.) are influenced by water-deficit condition in a tropical environment. *Front. Plant Sci.* **2021**, *12*, 717178. doi: 10.3389/fpls.2021.717178.

Alemu, S. T. Photosynthesis limiting stresses under climate change scenarios and role of chlorophyll fluorescence: A review article. *Cogent Food Agric.* **2020**, *6*, 1785136.

Amanullah; Iqbal, A.; Irfanullah; Hidayat, Z. Potassium management for improving growth and grain yield of maize (*Zea mays* L.) under moisture stress condition. *Sci. Rep.* **2016**, *6*, 34627.

Amelework, B.; Shimelis, H.; Tongoona, P.; Laing, M. Physiological mechanisms of drought tolerance in sorghum, genetic basis and breeding methods: A review. *Afr. J. Agric. Res.* **2015**, *10*, 3029–3040.

Amoah, J. N.; Antwi-Berko, D. Comparative physiological, biochemical and transcript response to drought in sorghum genotypes. *Biotechnol J. Int.* **2020**. https://doi.org/10.9734/bji/2020/v24i330102

Arnon D. I.; Stout, P. R. The essentiality of certain elements in minute quantity for plants with special reference to copper. *Plant Physiol.* **1939**, *14*, 371–375.

Assefa, Y.; Staggenborg, S. A.; Prasad, P. V. V. Grain sorghum water requirement and responses to drought stress: A review. *Crop. Manag*. **2010**, *9*, 1–11.

Asseng, A.; Ewert, F.; Martre, P.; Rotter, P.; Lobell, D. B.; Cammarano, D.; Kimball, B. A.; Ottman, M. J.; Wall, G. W.; White, J. W.; et al. Rising temperatures reduce global wheat production. *Nat. Clim. Chang*. **2015**, *5*, 143–147.

Badigannavar, A.; Teme, N.; de Oliveira, A. C.; Li, G.; VaksMann, M.; Viana, V. E.; GanapaThi T.; Sarsu, F. Physiological, genetic and molecular basis of drought resilience in sorghum [*Sorghum bicolor* (L.) Moench]. *Indian J Plant Physiol*. **2018**, *23*(4), 670–688. https://doi.org/10.1007/s40502-018-0416-2

Bano, S.; Aslam, M.; Saleem, M.; Basra, S.; Aziz, K. Evaluation of maize accessions under low temperature stress at early growth stages. *J. Anim. Plant Sci*. **2015**, *25*, 392–400.

Bao, S. G.; Shi, J. X.; Luo, F.; Ding, B.; Hao, J. Y.; Xie, X. D.; Sun, S. J. Overexpression of sorghum WINL1 gene confers drought tolerance in *Arabidopsis thaliana* through the regulation of cuticular biosynthesis. *Plant Cell Tiss Org*. **2017**, 128(2), 347–356. https://doi.org/10.1007/s11240-016-1114-2

Bechtold, U.; Penfold, C. A.; Jenkins, D. J.; Legaie, R.; Moore, J. D.; Lawson, T.; et al. Time-series transcriptomics reveals that *AGAMOUS-LIKE22* affects primary metabolism and developmental processes in drought-stressed Arabidopsis. *Plant Cell*. **2016**, *28*, 345–366. doi: 10.1105/tpc.15.00910

Bilska-Kos, A.; Solecka, D.; Dziewulska, A.; Ochodzki, P.; Jo'nczyk, M.; Bilski, H.; Sowi'nski, P. Low temperature caused modifications in the arrangement of cell wall pectins due to changes of osmotic potential of cells of maize leaves (*Zea mays* L.). *Protoplasma*. **2017**, *254*, 713–724.

Borrell, A. K.; Mullet, J. E.; George-Jaeggli, B.; van Oosterom, E. J.; Hammer, G. L.; Klein, P. E.; Jordan, D. R. Drought adaptation of stay-green sorghum is associated with canopy development, leaf anatomy, root growth, and water uptake. *J. Exp. Bot*. **2014**, *65*(21), 6251–6263. https://doi.org/10.1093/jxb/eru232

Borrell, A. K.; van Oosterom, E. J.; Mullet, J. E.; George-Jaeggli, B.; Jordan, D. R.; Klein, P. E.; Hammer, G. L. Stay-green alleles individually enhance grain yield in sorghum under drought by modifying canopy development and water uptake patterns. New Phytol. **2014**, *203*(3), 817–830. https://doi.org/10.1111/nph.12869

Brestic, M.; Zivcak, M.; Kunderlikova, K.; Sytar, O.; Shao, H.; Kalaji, H. M.; Allakhverdiev, S. I. Low PSI content limits the photo-protection of PSI and PSII in early growth stages of chlorophyll b-deficient wheat mutant lines. *Photosynth. Res*. **2015**, *125*, 151–166.

Brestic, M.; Zivcak, M.; Olsovska, K.; Shao, H.-B.; Kalaji, H. M.; Allakhverdiev, S. I. Reduced glutamine synthetase activity plays a role in control of photosynthetic responses to high light in barley leaves. *Plant Physiol. Biochem*. **2014**, *81*, 74–83.

Chaves, M. M.; Flexas, J.; Pinheiro, C.; Photosynthesis under drought and salt stress: Regulation mechanisms from whole plant to cell. *Ann. Bot*. **2009**, *103*(4), 551–560. https://doi.org/10.1093/aob/mcn125

Chen, D.; Wang, S.; Cao, B.; Cao, D.; Leng, G.; Li, H.; Yin, L.; Shan, L.; Deng, X. Genotypic variation in growth and physiological response to drought stress and re-watering reveals the critical role of·recovery in drought adaptation in maize seedlings. *Front Plant Sci*. **2016**, *6*, 1241. https://doi. org/10.3389/fpls. 2015.01241

Chen, X.; Wu, Q.; Gao, Y.; Zhang, J.; Wang, Y.; Zhang, R.; Zhou, Y.; Xiao, M.; Xu, W.; Huang, R. The role of deep roots in sorghum yield production under drought conditions. *Agronomy*, **2020**, *10*, 611.

Chen, Y.; Zhang, Z.; Tao, F.; Palosuo, T.; Rötter, R. P. Impacts of heat stress on leaf area index and growth duration of winter wheat in the North China Plain. *Field Crop. Res*. **2017**, 230–237.

Chen, Y.-E.; Liu, W.-J.; Su, Y.-Q.; Cui, J.-M.; Zhang, Z.-W.; Yuan, M.; Zhang, H.-Y.; Yuan, S. Different response of photosystem II to short and long-term drought stress in *Arabidopsis thaliana*. *Physiol. Plant*. **2016**, *158*, 225–235.

Choudhary, S.; Sinclair, T. R.; Prasad, P. V. V. Hydraulic conductance of intact plants of two contrasting sorghum lines, SC 15 and SC1205. *Funct. Plant Biol*. **2013**, *40*, 730–738.

Chun, S. C.; Paramasivan, M.; Chandrasekaran, M. Proline accumulation influenced by osmotic stress in arbuscular mycorrhizal symbiotic plants. *Front. Microbiol*. **2019**, *9*, 2525.

Devnarain, N.; Crampton, B. G.; Chikwamba, R.; Becker, J. V.; O'Kennedy, M. M. Physiological responses of selected African sorghum landraces to progressive water stress and re-watering. *S. Afr. J. Bot*. **2016**, *103*, 61–69.

Djanaguiraman, M.; Boyle, D. L.; Welti, R.; Jagadish, S. V. K.; Prasad, P. V. V. Decreased photosynthetic rate under high temperature in wheat is due to lipid desaturation, oxidation, acylation, and damage of organelles. *BMC Plant Biol*. **2018**, *18*, 55.

Djanaguiraman, M.; Narayanan, S.; Erdayani, E.; Prasad, P. V. V. Effects of high temperature stress during anthesis and grain filling periods on photosynthesis, lipids and grain yield in wheat. *BMC Plant Biol.* **2020**, *20*, 268.

Djanaguiraman, M.; Perumal, R.; Jagadish, S. V. K.; Ciampitti, I. A.; Welti, R.; Prasad, P. V. V. Sensitivity of sorghum pollen and pistil to high-temperature stress. *Plant Cell Environ.* **2018**, *41*, 1065–1082.

Djanaguiraman, M.; Prasad, P. V. V.; Ciampitti, I. A.; Talwar, H. S. Impact of abiotic stress on sorghum physiology. In *Sorghum in the 21st Century: Food–Fodder–Feed–Fuel for a Rapidly Changing World.* Springer Nature, Singapore, **2020**; pp. 157–188.

Djanaguiraman, M.; Prasad, P. V. V.; Schapaugh, W. T. High day- or nighttime temperature alters leaf assimilation, reproductive success, and phosphotidic acid of pollen grain in soybean [*Glycine max* (L.) Merr.]. *Crop. Sci.* **2013**, *53*, 594–1604.

Eggen, M.; Ozdogan, M.; Zaitchick, B.; Ademe, D.; Foltz, J.; Simane, B. Vulnerability of sorghum production to extreme, sub-seasonal weather under climate change. *Environ. Res. Lett.* **2019**, *14*, 045005.

ElBasyoni, I.; Saadalla, M.; Baenziger, S.; Bockelman, H.; Morsy, S. Cell membrane stability and association mapping for drought and heat tolerance in a worldwide wheat collection. *Sustainability*. **2017**, *9*, 1606.

Erb, M.; Kliebenstein, D. J. Plant secondary metabolites as defenses, regulators, and primary metabolites: The blurred functional trichotomy. *Plant Physiol.* **2020**, *184*, 39–52. doi: 10.1104/pp.20.00433

Esim, N.; Atici, Ö. Relationships between some endogenous signal compounds and the antioxidantsystem in response to chilling stress in maize (*Zea mays* L.) seedlings. *Turk. J. Bot.* **2016**, *40*, 37–44.

Fadoul, H. E.; Siddig, M. A. E.; Abdalla, A. W. H.; Hussein, A. A. E. Physiological and proteomic analysis of two contrasting *Sorghum bicolor* genotypes in response to drought stress. *Aust J Crop Sci.* **2018**, *12*(09), 1543–1551. https://doi.org/10.21475/ajcs.18.12.09. PNE134

Farooq, M.; Aziz, T.; Wahid, A.; Lee, D.-J.; Siddique, K. H. Chilling tolerance in maize: Agronomic and physiological approaches. *Crop Pasture Sci.* **2009**, *60*, 501–516.

Fenta, B. A.; Beebe, S. E.; Kunert, K. J.; Burridge, J. D.; Barlow, K. M.; Lynch, P. J. Field phenotyping of soybean roots for drought stress tolerance. *Agronomy*. **2014**, *4*, 418–435.

Fracasso, A.; Trindade, L. M.; Amaducci, S. Drought stress tolerance strategies revealed by RNA-Seq in two sorghum genotypes with contrasting WUE. *BMC Plant Biol.* **2016**. https://doi.org/10.1186/s12870-016-0800-x.

Frank, D. A.; Reichstein, M.; Bahn, M.; Thonicke, K.; Frank., D.; Mahecha, M. D.; Smith, P.; van der Velde, M.; Vicca, S.; Babst, F.; et al. Effects of climate extremes on the terrestrial carbon cycle: Concepts, processes and potential future impacts. *Glob Chang. Biol.* **2015**, *21*, 2861–2880.

Gaikpa, D. S.; Opata, J.; Mpanga, I. K. Towards sustainable maize production: Understanding the morpho-physiological, genetics, and molecular mechanisms for tolerance to low soil nitrogen, phosphorus, and potassium. *Stresses*. **2022**, *2*(4), 395–404.

George-Jaeggli, B.; Mortlock, M. Y.; Borrell, A. K. Bigger is not always better: Reducing leaf area helps stay-green sorghum use soil water more slowly. *Environ Exp Bot.* **2017**, *138*, 119–129. https://doi.org/10.1016/j.envexpbot.2017.03.002

Getnet, Z.; Husen, A.; Fetene, M.; Yemata, G. Growth, water status, physiological, biochemical and yield response of stay green sorghum (*Sorghum bicolor* (L.) Moench) varieties—a feld trial under drought-prone area in Amhara Regional State, Ethiopia. *J. Agron.* **2015**, *14*(4), 188–202. https://doi.org/10.3923/ja.2015.188.202

Gholipoor, M.; Sinclair, T. R.; Prasad, P. V. V. Genotypic variation within sorghum for transpiration response to drying soil. *Plant Soil.* **2012**, *357*, 35–40.

Goufo, P.; Moutinho-Pereira, J. M.; Jorge, T. F.; Correia, C. M.; Oliveira, M. R.; Rosa, E. A., . . . Trindade, H. Cowpea (*Vigna unguiculata* L. Walp.) metabolomics: Osmoprotection as a physiological strategy for drought stress resistance and improved yield. *Frontiers in Plant Science*. **2017**, *8*, 586.

Guo, T.; Wang, D.; Fang, J.; Zhao, J.; Yuan, S.; Xiao, L.; Li, X. Mutations in the rice OsCHR4 gene, encoding a CHD3 family chromatin remodeler, induce narrow and rolled leaves with increased cuticular wax. *Int. J. Mol. Sci.* **2019**, *20*, 2567.

Gupta, N. K.; Agarwal, S.; Agarwal, V. P.; Nathawat, N. S.; Gupta, S.; Singh, G. Effect of short-term heat stress on growth, physiology and antioxidative defence system in wheat seedlings. *Acta Physiol. Plant.* **2013**, *35*, 1837–1842.

Hasanuzzaman, M.; Bhuyan, M.; Zulfiqar, F.; Raza, A.; Mohsin, S. M.; Mahmud, J. A.; et al. Reactive oxygen species and antioxidant defense in plants under abiotic stress: Revisiting the crucial role of a universal defense regulator. *Antioxidants*. **2020**, *9*, 681. doi: 10.3390/antiox9080681.

Hasanuzzaman, M.; Nahar, K.; Alam, M.; Roychowdhury, R. Physiological, biochemical, and molecular mechanisms of heat stress tolerance in plants. *Int. J. Mol. Sci.* **2013**, *14*, 9643–9684.

Haswell, E. S.; Verslues, P. E. The ongoing search for the molecular basis of plant osmosensing. *J. Gen. Physiol.* **2015**, *145*, 389–394. doi: 10.1085/jgp.201411295.

Hayashi, T. Varietal difference in the effects of low temperature on tassel development in hybrid maize. *Plant Prod. Sci.* **2016**, *19*, 230–237.

Hossain, M. A.; Bhattacharjee, S.; Armin, S. M.; Qian, P.; Xin, W.; Li, H. Y.; et al. Hydrogen peroxide priming modulates abiotic oxidative stress tolerance: Insights from ROS detoxification and scavenging. *Front. Plant Sci.* **2015**, *6*, 420. doi: 10.3389/fpls.2015.00420.

Husen, A.; Iqbal, M.; Aref, I. M. Growth, water status, and leaf characteristics of *Brassica carinata* under drought and rehydration conditions. *Braz J Bot.* **2014**, *37*(3), 217–227. https://doi.org/10.1007/s40415-014-0066-1.

Hussain, H. A.; Hussain, S.; Khaliq, A.; Ashraf, U.; Anjum, S. A.; Men, S.; Wang, L. Chilling and drought stresses in crop plants: Implications, cross talk, and potential management opportunities. *Front. Plant Sci.* **2018**, *9*, 393.

Hussain, H. A.; Shengnan, M.; Hussain, S.; Ashraf, U.; Zhang, Q.; Anjum, S. A.; Ali, I.; Wang, L. Individual and concurrent effects of drought and chilling stresses on morpho-physiological characteristics and oxidative metabolism of maize cultivars. *bioRxiv*. **2019**, 829309.

Hussain, S.; Khaliq, A.; Ali, B.; Hussain, H. A.; Qadir, T.; Hussain, S. Temperature extremes: Impact on rice growth and development. In *Plant Abiotic Stress Tolerance*; Springer: Cham, Switzerland, **2019**; pp. 153–171.

Hussain, S.; Khan, F.; Cao, W.; Wu, L.; Geng. M. Seed priming alters the production and detoxification of reactive oxygen intermediates in rice seedlings grown under sub-optimal temperature and nutrient supply. *Front. Plant Sci.* **2016**, *7*, 439.

IPCC, Intergovernmental Panel on Climate Change. Summary for policymakers. In *Climate Change 2013: The Physical Science Basis. Contribution of Working Group I to V Assessment Report of the Intergovernmental Panel on Climate Change*; Stocker, T. F.; Qin, G. K.; Plattner, M.; Tignor, S. K.; Allen, J.; Boschung, A.; Nauels, A.; Xia, Y.; Bex, V.; Midgley, P. M.; eds. Cambridge University Press: Cambridge, UK, **2013**.

IPCC. *Climate Change 2014: Synthesis Report. Contribution of Working Groups I, II and III to the Fifth Assessment Report of the Intergovernmental Panel on Climate Change*; Core Writing Team, Pachauri, R. K.; Meyer, L. A.; eds. IPCC: Geneva, Switzerland, 2014; 151.

Izaurralde, R. C.; Thomson, A. M.; Morgan, J.; Fay, P.; Polley, H.; Hatfield, J. L. Climate impacts on agriculture: Implications for forage and rangeland production. *Agron. J.* **2011**, *103*, 371–381.

Jaegglia, B. G.; Mortlockb, M. Y.; Borrell, A. Bigger is not always better: Reducing leaf area helps stay-green sorghum use soil water more slowly. *Environ. Exp. Bot.* **2017**, *138*, 119–129.

Johnson, S. M.; Lim, F.-L.; Finkler, A.; Fromm, H.; Slabas, A. R.; Knight, M. R. Transcriptomic analysis of *Sorghum bicolor* responding combined heat and drought stress. *BMC Genomics*. **2014**, *15*(1), 1–19. https://doi.org/10.1186/1471-2164-15-456

Jordan, D. R.; Hunt, C. H.; Cruickshank, A. W.; Borrell, A. K.; Henzell, R. G. The relationship between the stay-green trait and grain yield in elite sorghum hybrids grown in a range of environments. *Crop. Sci.* **2012**, *52*, 1153–1161

Joshi, J.; Hasnain, G.; Logue, T.; Lynch, M.; Wu, S.; Guan, J.-C.; et al. A core metabolome response of maize leaves subjected to long-duration abiotic stresses. *Metabolites*. **2021**, *11*, 797. doi: 10.3390/metabo11110797

Kapanigowda, M. H.; Payne, W. A.; Rooney, W. L.; Mullet, J. E.; Balota, M. Quantitative trait locus mapping of the transpiration ratio related to prefowering drought tolerance in sorghum (*Sorghum bicolor*). *Funct Plant Biol.* **2014**, *41*(10–11), 1049–1065. https://doi.org/10.1071/Fp13363

Kassahun, B.; Bidinger, F.; Hash, C.; Kuruvinashetti, M. Stay-green expression in early generation sorghum [*Sorghum bicolor* (L.) Moench] QTL introgression lines. *Euphytica*. **2010**, *172*(3), 351–362. https://doi.org/10.1007/s10681-009-0108-0

Kaushal, N.; Awasthi, R.; Gupta, K.; Gaur, P.; Siddique, K. H. M.; Nayyar, H. Heat-stress-induced reproductive failures in chickpea (*Cicer arietinum*) are associated with impaired sucrose metabolism in leaves and anthers. *Funct. Plant Biol.* **2013**, *40*, 1334–1349.

Khan, M. I. R.; Fatma, M.; Per, T. S.; Anjum, N. A.; Khan, N. A. Salicylic acid-induced abiotic stress tolerance and underlying mechanisms in plants. *Front. Plant Sci.* **2015**, *6*, 462. doi: 10.3389/fpls.2015.00462.

Kumar, A.; Kaushik, P. Heat stress and its impact on plant function: An update. *Preprints*. **2021**, *2021*, 2021080489.

Kumar, M.; Kumar Patel, M.; Kumar, N.; Bajpai, A. B.; Siddique, K. H. M. Metabolomics and molecular approaches reveal drought stress tolerance in plants. *Int. J. Mol. Sci.* **2021**, *22*, 9108. doi: 10.3390/ijms22179108

Li, Z.; Howell, S. H. Heat stress responses and thermotolerance in maize. *Int. J. Mol. Sci.* **2021**, *22*, 948. doi: 10.3390/ijms22020948

Li, Z. G. Synergistic effect of antioxidant system and osmolyte in hydrogen sulfide and salicylic acid crosstalk-induced heat tolerance in maize (*Zea mays* L.) seedlings. *Plant Signal. Behav.* **2015**, *10*, e1051278. doi: 10.1080/15592324.2015.1051278.

Li, Z. G.; Xie, L. R.; Li, X. J. Hydrogen sulfide acts as a downstream signal molecule in salicylic acid-induced heat tolerance in maize (*Zea mays* L.) seedlings. *J. Plant Physiol.* **2015**, *177*, 121–127. doi: 10.1016/j.jplph.2014.12.018.

Liang, X.; Erickson, J. E.; Vermerris, W.; Rowland, D. L.; Sollenberger, L. E.; Silveira, M. L. Root architecture of sorghum genotypes differing in root angles under different water regimes. *J. Crop. Improv.* **2017**, *31*, 39–55.

Liang, X.; Zhang, L.; Natarajan, S. K.; Becker, D. F. Proline mechanisms of stress survival. *Antioxid. Redox Signal.* **2013**, *19*, 998–1011.

Liedtke, J. D.; Hunt, C. H.; George-Jaeggli, B.; Laws, K.; Watson, J.; Potgieter, A. B.; Cruickshank, A.; Jordan, D. R. High-throughput phenotyping of dynamic canopy traits associated with stay-green in grain sorghum. *Plant Phenomics* **2020**, *2020*, 4635153.

Lieu, E. L.; Kelekar, N.; Bhalla, P.; Kim, J. Fructose and mannose in inborn errors of metabolism and cancer. *Metabolites*. **2021**, *11*, 479. doi: 10.3390/metabo11080479.

Liu, Q.; Hallerman, E.; Peng, Y.; Li, Y. Development of Bt rice and Bt maize in China and their efficacy in target pest control. *Int. J. Mol. Sci.* **2016**, *17*, 1561.

Lopez, J. R.; Erickson, J. E.; Munoz, P.; SAballos, A.; Felderhof, T. J.; Vermerris, W. QTLs associated with crown root angle, stomatal conductance, and maturity in sorghum. *Plant Genome*. **2017**. https://doi.org/10.3835/plantgenome2016.04.0038

Lu, D.; Cai, X.; Yan, F.; Sun, X.; Wang, X.; Lu, W. Effects of high temperature after pollination on physicochemical properties of waxy maize flour during grain development. *J. Sci. Food Agric.* **2014**, *94*, 1416–1421.

Ma, H.; Liu, C.; Li, Z.; Ran, Q.; Xie, G.; Wang, B.; et al. ZmbZIP4 contributes to stress resistance in maize by regulating ABA synthesis and root development. *Plant Physiol.* **2018**, *178*, 753–770. doi: 10.1104/pp.18.00436.

Manavalan, L. P.; Nguyen, H. T. Drought tolerance in crops: Physiology to genomics. *Plant Stress Physiol.* **2017**. https://doi.org/10.1079/9781780647296.0001

Medeiros, D. B.; Brotman, Y.; Fernie, A. R. The utility of metabolomics as a tool to inform maize biology. *Plant Commun.* **2021**, *2*, 100187. doi: 10.1016/j.xplc.2021.100187

Mishra, D.; Shekhar, S.; Agrawal, L.; Chakraborty, S.; Chakraborty, N. Cultivar-specific high temperature stress responses in bread wheat (*Triticum aestivum* L.) associated with physicochemical traits and defense pathways. *Food Chem.* **2017**, *221*, 1077–1087.

Mutava, R. N. Evaluation of Sorghum Genotypes for Variation in Canopy Temperature and Drought Tolerance. (Doctoral Dissertation), Kansas State University, Manhattan, KS, USA, **2012**.

Nadeem, M.; Li, J.; Wang, M.; Shah, L.; Lu, S.; Wang, X.; Ma, C. Unraveling field crops sensitivity to heat stress: Mechanisms, approaches, and future prospects. *Agronomy*. **2018**, *8*(7), 128.

Ngara, R.; Goche, T.; Swanevelder, D. Z. H.; Chivasa, S. Sorghum's whole-plant transcriptome and proteome responses to drought stress: A review. *Life*. **2021**. https://doi.org/10.3390/life11070704

Niu, S.; Du, X.; Wei, D.; Liu, S.; Tang, Q.; Bian, D.; et al. Heat stress after pollination reduces kernel number in maize by insufficient assimilates. *Front. Genet.* **2021**, *12*, 728166. doi: 10.3389/fgene.2021.728166

Osmolovskaya, N.; Shumilina, J.; Kim, A.; Didio, A.; Grishina, T.; Bilova, T.; Keltsieva, O. A.; Zhukov, V.; TIkhonovich, I.; Tarakhovskaya, E.; Frolov, A.; Wessjohann, L. A. Methodology of drought stress research: Experimental setup and physiological characterization. *Int J Mol Sci.* **2018**. https://doi.org/10.3390/ijms19124089

Osuna, D.; Prieto, P.; Aguilar, M. Control of seed germination and plant development by carbon and nitrogen availability. *Front. Plant Sci.* **2015**, *6*, 1023. doi: 10.3389/fpls.2015.01023.

Panison, F.; Sangoi, L.; Kolling, D. F.; Coelho, C. M. M. D.; Durli, M. M. Épocas de colheita e desempenho agron ômico de h íbridos de milho com ciclos de crescimento contrastantes. *Acta Sci. Agron.* **2016**, *38*, 219–226. [CrossRef]

Passioura, J. B. Phenotyping for drought tolerance in grain crops: When is it useful to breeders? *Funct. Plant Biol.* **2012**, *39*, 851–859.

Prasad, P. V. V.; Djanaguiraman, M. High night temperature decreases leaf photosynthesis and pollen function in grain sorghum. *Funct. Plant Biol.* **2011**, *38*, 993–1003.

Prasad, P. V. V.; Djanaguiraman, M.; Jagadish, S. V. K.; Ciampitti, I. A. drought and high temperature stress and traits associated with tolerance. In *Sorghum: A State of the Art and Future Perspectives*; Ciampitti, I. A.; Prasad, P. V. V.; eds. ASA, CSSA, SSSA: Madison, WI, **2019**; 58, 245–265.

Prasad, P. V. V.; Djanaguiraman, M.; Perumal, R.; Ciampitti, I. A. Impact of high temperature stress on floret fertility and individual grain weight of grain sorghum: Sensitive stages and thresholds for temperature and duration. *Front. Plant Sci.* **2015**, *6*, 820.

Prasad, V. R.; Govindaraj, M.; Djanaguiraman, M.; Djalovic, I.; Shailani, A.; Rawat, N. . . . Prasad, P. V. Drought and high temperature stress in sorghum: Physiological, genetic, and molecular insights and breeding approaches. *Int. J. Mol. Sci.* **2021**, *22*(18), 9826.

Rafique, S. Differential expression of leaf proteome of tolerant and susceptible maize (*Zea mays* L.) genotypes in response to multiple abiotic stresses. *Biochem. Cell Biol.* **2019**, *97*, 581–588.

Rebetzke, G. J.; Condon, A. G.; Rattey, A. R.; Farquhar, G. D.; Richards, R. A. Genomic regions for canopy temperature and their genetic association with stomatal conductance and grain yield in bread wheat (*Triticum aestivum* L.). *Funct. Plant Biol.* **2013**, *40*, 14–26.

Reddy, N. R. R.; Ragimasalawada, M.; Sabbavarapu, M. M.; Nadoor, S.; Patil, J. V. Detection and validation of stay-green QTL in postrainy sorghum involving widely adapted cultivar, M35-1 and a popular stay-green genotype B35. *BMC Genomics*. **2014**, *15*(1), 1–16. https://doi.org/10.1186/1471-2164-15-909

Riva-Roveda, L.; Escale, B.; Giauffret, C.; P érilleux, C. Maize plants can enter a standby mode to cope with chilling stress. *BMC Plant Biol.* **2016**, *16*, 212.

Rodziewicz, P.; Swarcewicz, B.; Chmielewska, K.; Wojakowska, A.; Stobiecki, M. Influence of abiotic stresses on plant proteome and metabolome changes. *Acta Physiol. Plant.* **2014**, *36*, 1–19. doi: 10.1007/s11738-013-1402-y

Rostamza, M.; Richards, R. A.; Watt, M. Response of millet and sorghum to a varying water supply around the primary and nodal roots. *Ann. Bot.* **2013**, *112*, 439–446.

Sanderson, M. A.; Stair, D. W.; Hussey, M. A. Physiological and morphological response of perennial forages to stress. *Advances in Agron.* **1997**, *59*, 171–224.

Sandhu, S.; Singh, J.; Kaur, P.; Gill, K. Heat stress in field crops: Impact and management approaches. In *Advances in Crop Environment Interaction*; Springer: Singapore, 2018; pp. 181–204.

Schlüter, U.; Mascher, M.; Colmsee, C.; Scholz, U.; Bräutigam, A.; Holger, F.; Sonnewald, U. Maize source leaf adaptation to nitrogen deficiency affects not only nitrogen and carbon metabolism but also control of phosphate homeostasis. *Plant Physiol.* **2012**, *160*, 1384–1406.

Sengupta, S.; Mukherjee, S.; Basak, P.; Majumder, A. L. Significance of galactinol and Raffinose family oligosaccharide synthesis in plants. *Front. Plant Sci.* **2015**, *6*, 656. doi: 10.3389/fpls.2015.00656

Sharma, A.; Shahzad, B.; Kumar, V.; Kohli, S. K.; Sidhu, G. P. S.; Bali, A. S.; et al. Phytohormones regulate accumulation of osmolytes under abiotic stress. *Biomolecules* **2019**, *9*, 285. doi: 10.3390/biom9070285.

Siddique, A.; Kandpal, G.; Kumar, P. Proline accumulation and its defensive role under diverse stress condition in plants: An overview. *J. Pure Appl. Microbiol.* **2018**, *12*, 1655–1659.

Siebers, M. H.; Slattery, R. A.; Yendrek, C. R.; Locke, A. M.; Drag, D.; Ainsworth, E. A.; Bernacchi, C. J.; Ort, D. R. Simulated heat waves during maize reproductive stages alter reproductive growth but have no lasting effect when applied during vegetative stages. *Agric. Ecosyst. Environ.* **2017**, *240*, 162–170.

Singh, A.; Chahal, H. S. Management of abiotic stress in forage crops. *Abiotic Stress in Plants*. **2020**, *13*.

Singh, V.; Nguyen, C. T.; van Oosterom, E. J.; Chapman, S. C.; Jordan, D. R.; Hammer, G. L. Sorghum genotypes differ in high temperature responses for seed set. *Field Crops Res.* **2015**, *171*, 32–40.

Sita, K.; Sehgal, A.; HanumanthaRao, B.; Nair, R. M.; Vara Prasad, P. V.; Kumar, S.; Gaur, P. M.; Farroq, M.; Siddique, K. H. M.; Varshney, R. K.; et al. Food legumes and rising temperatures: Effects, adaptive functional mechanisms specific to reproductive growth stage and strategies to improve heat tolerance. *Front. Plant Sci.* **2017**, *8*, 1–30.

Slama, I.; Abdelly, C.; Bouchereau, A.; Flowers, T.; Savouré, A. Diversity, distribution and roles of osmoprotective compounds accumulated in halophytes under abiotic stress. *Ann. Bot.* **2015**, *115*, 433–447. doi: 10.1093/aob/mcu239

Song, Y.; Chen, Q.; Ci, D.; Shao, X.; Zhang, D. Effects of high temperature on photosynthesis and related gene expression in poplar. *BMC Plant Biol.* **2014**, *14*, 111.

Stavridou, E.; Voulgari, G.; Michailidis, M.; Kostas, S.; Chronopoulou, E. G.; Labrou, N. E.; et al. Overexpression of a biotic stress-inducible *Pvgstu* gene activates early protective responses in tobacco under combined heat and drought. *Int. J. Mol. Sci.* **2021**, *22*, 2352. doi: 10.3390/ijms22052352

Su, X.; Wu, S.; Yang, L.; Xue, R.; Li, H.; Wang, Y.; Zhao, H. Exogenous progesterone alleviates heat and high light stress-induced inactivation of photosystem II in wheat by enhancing antioxidant defense and D1 protein stability. *Plant Growth Regul.* **2014**, *74*, 311–318.

Sukumaran, S.; Li, X.; Li, X.; Zhu, C.; Bai, G.; PeRumal, R.; Tuinstra, M. R.; Prasad, P. V.; Mitchell, S. E.; Tesso, T. T. QTL mapping for grain yield, flowering time, and stay-green traits in sorghum with genotyping-by-sequencing markers. *Crop Sci.* **2016**, *56*(4), 1429–1442. https://doi.org/10.2135/cropsci2015.02.0097

Sun, J.; Zheng, T.; Yu, J.; Wu, T.; Wang, X.; Chen, G.; Tian, Y.; Zhang, H.; Wang, Y.; Terzaghi, W. TSV, a putative plastidic oxidoreductase, protects rice chloroplasts from cold stress during development by interacting with plastidic thioredoxin Z. *New Phytol.* **2017**, *215*, 240–255.

Takele, A. Diferential responses of electrolyte leakage and pigment compositions in maize and sorghum after exposure to and recovery from pre-and post-fowering dehydration. *Agric Sci China.* **2010**, *9*(6), 813–824. https://doi.org/10.1016/S1671-2927(09)60159-0

Tao, Z.-Q.; Chen, Y.-Q.; Li, C.; Zou, J.-X.; Yan, P.; Yuan, S.-F.; Wu, X.; Sui, P. The causes and impacts for heat stress in spring maize during grain filling in the North China Plain—A review. *J. Integr. Agric.* **2016**, *15*, 2677–2687.

Thornburg, T. E.; Liu, J.; Li, Q.; Xue, H.; Wang, G.; Li, L.; Julia Elise, F.; Kyele, E. D.; Wanying, L.; Baohong, Z.; et al. Potassium deficiency significantly affected plant growth and development as well as microRNA-mediated mechanism in wheat (*Triticum aestivum* L.). *Front. Plant Sci.* **2020**, *11*, 1219.

Török, Z.; Crul, T.; Maresca, B.; Schütz, G. J.; Viana, F.; Dindia, L.; et al. Plasma membranes as heat stress sensors: From lipid-controlled molecular switches to therapeutic applications. *Biochim. Biophys. Acta.* **2014**, *1838*, 1594–1618. doi: 10.1016/j.bbamem.2013.12.015.

Vadez, V.; Deshpande, S. P.; Kholova, J.; Hammer, G. L.; Borrell, A. K.; Talwar, H. S.; Hash, C. T. Stay-green quantitative trait loci's effects on water extraction, transpiration efciency and seed yield depend on recipient parent background. *Funct Plant Biol.* **2011**, 38(7), 553–566. https://doi.org/10.1071/FP11073

Vadez, V.; Krishnamurthy, L.; Hash, C.; UpadHyaya, H.; Borrell, A. Yield, transpiration efciency, and water-use variations and their interrelationships in the sorghum reference collection. *Crop Pasture Sci.* **2011**, *62*(8), 645–655. https://doi.org/10.1071/CP11007

Wang, A.-Y.; Li, Y.; Zhang, C.-Q. QTL mapping for stay-green in maize (*Zea mays*). *Can. J. Plant Sci.* **2012**, *92*, 249–256.

Wang, D.; Heckathorn, S. A.; Mainali, K.; Tripathee, R. Timing effects of heat-stress on plant ecophysiological characteristics and growth. *Front. Plant Sci.* **2016**, *7*, 1–11.

Wang, L. C.; Tsai, M. C.; Chang, K. Y.; Fan, Y. S.; Yeh, C. H.; Wu, S. J. Involvement of the *Arabidopsis* HIT1/AtVPS53 tethering protein homologue in the acclimation of the plasma membrane to heat stress. *J. Exp. Bot.* **2011**, *62*, 3609–3620.

Wang, X.; Dinler, B. S.; Vignjevic, M.; Jacobsen, S.; Wollenweber, B. Physiological and proteome studies of responses to heat stress during grain filling in contrasting wheat cultivars. *Plant Sci.* **2015**, *230*, 33–50.

Waqas, M. A.; Khan, I.; Akhter, M. J.; Noor, M. A.; Ashraf, U. Exogenous application of plant growth regulators (PGRs) induces chilling tolerance in short-duration hybrid maize. *Environ. Sci. Pollut. Res.* **2017**, *24*, 11459–11471.

Wasson, A. P.; Richards, R. A.; Chatrath, R.; Misra, S. C.; Prasad, S. V.; Rebetzke, G. J.; Kirkegaard, J. A.; Christopher, J.; Watt, M. Traits and selection strategies to improve root systems and water uptake in water limited wheat crops. *J. Exp. Bot.* **2012**, *63*, 3485–3498.

Xalxo, R.; Yadu, B.; Chandra, J.; Chandrakar, V.; Keshavkant, S. Alteration in carbohydrate metabolism modulates thermotolerance of plant under heat stress. In *Heat Stress Tolerance in Plants: Physiological, Molecular and Genetic Perspectives*; Wani, S. H.; Kumar, V.; eds. Wiley: Hoboken, NJ, **2020**; 77–115. doi: 10.1016/j.jprot.2020.103968.

Xu, Q.; Xu, X.; Shi, Y.; Xu, J.; Huang, B. Transgenic tobacco plants overexpressing a grass *PpEXP1* gene exhibit enhanced tolerance to heat stress. *PLOS One*. **2014**, *9*, e100792.

Yan, G. U.; Cao, M. K.; Zhang, Y. Q.; Sun, Y.; Wen-He, H. U.; Chun-Sheng, W. U. Effects of low temperature on photosynthetic fluorescence and enzyme activity of carbon metabolism of maize at booting stage. *J. Plant Physiol.* **2015**, *51*, 941–948.

Yang, L.; Wang, Y.; Yang, K. *Klebsiella variicola* improves the antioxidant ability of maize seedlings under saline-alkali stress. *PeerJ*. **2021**, *9*, e11963. doi: 10.7717/peerj.11963.

Young, K. J.; Long, S. P. Crop ecosystem responses to climatic change: Maize and sorghum. In *Climate Change and Global Crop Productivity*; Reddy, K. R.; Hodges, H. F.; eds. CABI: Wallingford, UK, **2000**; 107–131. doi: 10.1079/9780851994390.0107.

Zhang, F.; Zhu, K.; Wang, Y.; Zhang, Z.; Lu, F.; Yu, H.; Zou, J. Changes in photosynthetic and chlorophyll fuorescence characteristics of sorghum under drought and waterlogging stress. *Photosynthetica*. **2019**, *57*(4), 1156–1164. https://doi.org/10.32615/ps.2019.136

Zhao, D.; Reddy, K. R.; Kakani, V. G.; Reddy, V. R. Nitrogen deficiency effects on plant growth, leaf photosynthesis, and hyperspectral reflectance properties of sorghum. *Eur. J. Agron.* **2005**, *22*(4), 391–403.

Zhou, Z. H.; Wang, Y.; Ye, X. Y.; Li, Z. G. Signaling molecule hydrogen sulfide improves seed germination and seedling growth of maize (*Zea mays* L.) under high temperature by inducing antioxidant system and osmolyte biosynthesis. *Front. Plant Sci.* **2018**, *9*, 1288. doi: 10.3389/fpls.2018.01288.

Zivcak, M.; Brestic, M.; Kalaji, H. M.; Govindjee. Photosynthetic responses of sun- and shade-grown barley leaves to high light: Is the lower PSII connectivity in shade leaves associated with protection against excess of light? *Photosynth. Res.* **2014**, *119*, 339–354.

7 Weeds of Summer Fodders and Their Control

Safdar Ali, Asif Iqbal, Muhammad Aamir Iqbal, Muhammad Abdullah Saleem, and Imran Ali

7.1 INTRODUCTION

7.1.1 Concept of Weeds

The most convenient definition of *weed* is "any plant grown where it is not wanted" or "a plant that is objectionable and interferes with man's activities."

7.1.2 Classification of Weeds Based on the Growing Season

7.1.2.1 Winter Weeds

Weeds that grow in the winter season, that is, from October to April–May.

7.1.2.2 Summer Weeds

Weeds that grow in the summer season, that is, April–May to October.

7.1.3 Classification of Weeds Based on Life Cycle or Total Length of Life or Duration

I. *Annual weeds*
Weeds complete their life cycle, that is, vegetative and reproductive life, in one season. These weeds grow rapidly. Some species produce two or more generations in a season (year). These are either winter or summer weeds. These weeds reproduce only from seed.

II. *Biennial weeds*
Weeds complete their life cycle in two seasons. Such weeds complete their vegetative growth in one season and reproductive growth in another, after which they die. Such weeds also reproduce from seeds.

III. *Perennial weeds*
They live for over two seasons; once fully developed, they usually produce seeds yearly. Such weeds reproduce from seed and storage organs (rhizomes).

7.1.4 Classification of Weeds Based on the Structure of the Leaf

Based on the structure of the leaf, weeds are classified into the following groups.

7.1.4.1 Broad-Leaved

These are weeds with net-veined leaves, tap root systems, and stiff stems.

DOI: 10.1201/b23394-7

7.1.4.2 Grasses or Narrow-Leaved Weeds

These weeds have two-ranked long, narrow, usually flat leaves with parallel vain and round hollow stems and fibrous root systems.

7.1.4.3 Sedges

These are similar to grasses but have three-ranked leaves and solid triangular stems. Most of these have modified rhizomes (root systems) for food storage and propagation.

7.2 ECONOMIC IMPORTANCE OF WEEDS

Many components influence Pakistan's poor seed production more than the average yield of other maize-growing countries worldwide. The main limitations restricting maize growth and yield are uneconomical smallholdings, bad economic prospects, and illiteracy. There are more limitation factors, such as inappropriate planting techniques, poor nutrient management, suboptimal biomass production, and weed infestations, because of unmanaged intervention in maize fields and late weed treatments causing poor seed yield of maize. Weed growth lowers yield and decreases the overall profitability of a farming sector. In Pakistan, yield losses of maize crops are about 45% due to weed infestation (Safdar *et al.*, 2015). Ahmad *et al.* (2016) suggested that removing weeds on time could result in a 100% increase in maize output. Weed infestation depletes domestic grains (wheat, rice, maize, sorghum, bajra), production by 3.4 million tons per year, resulting in a massive loss in the country's gross domestic product (Chovancova *et al.*, 2020).

7.2.1 Weeds Are More Efficient Than Crop Plants

Weeds impair the crop's efficiency in utilizing all existing resources for dry matter accumulation and thus affect grain yield and economic productivity. Weed infestation causes yield reductions ranging from 35% to severe crop failure (Little *et al.*, 2021). That is why selecting the appropriate herbicide is useful in decreasing weed infestations and resistance (Kundu *et al.*, 2020). Plenty of herbicides are available in the market to prevent weed growth. Herbicides are considered the most effective source to control perennial and annual weeds. However, it is critical to choose the appropriate herbicide and apply it at the right time; otherwise, the crops may be harmed (Merritt *et al.*, 2020).

Crop plants and weeds compete for natural and artificial resources. Due to their better traits compared to crops, such as seed viability, seed dormancy, C4-mechanism, competition ability, prolific seed production, dormancy mechanism, competitive ability, and greater temperature compensation point, weeds have significantly reduced crop productivity and quality throughout the history of agriculture. Weeds cause more economic losses than any other crop pest (Gharde et al., 2018).

Undoubtedly, agricultural and weed plants compete and, to varied degrees, obstruct growth processes. Competition between weeds and crops for various growing resources, including light, water, nutrients, thermal energy, and CO_2, has been noted. This competition also causes four harvesting operations impediments (Galon *et al.*, 2018). The longer the weeds directly conflict with the crop, the more harm they do. The crucial period of weed elimination is a vital component of an integrated weed management system, according to Knezevic *et al.* (2019). Generally, it has been accepted that weeds compete for different resources like light, space, soil nutrients, carbon dioxide, and water (Horvath *et al.*, 2023). Both the timing and the actual act of getting rid of weeds are crucial since the competitiveness of weeds relies on how long they interfere with the crop (Iqbal *et al.*, 2020). According to Reddy (2018), maize is sensitive to weeds during the early growth period primarily because it grows slowly. Operations carried out before and following crucial times are less productive and do not produce significant benefits (Maitra *et al.*, 2020).

Weeds consume more amount of CO_2 in comparison to cultivated crops. It is due to the relatively more leaf area of weeds; for example, wild mustard and wheat leaf areas at the blooming stage are 7300 and 140 cm^2, respectively. Weeds compete with crop plants for space; for example, a plant carpet weed covers a 3.5-m^2 area. They use relatively more light and respond better than the cultivated crops to high-light intensities. Weeds like fat-hen and pigweed consume double the quality of N and P and triple the dose of K as compared to those crops with which they compete.

7.2.2 Crop Losses Due to Weeds

One of the main biological factors limiting agricultural output is weeds. It is said that with each kilogram of weeds produced, 1 kilogram of wheat yield is reduced. Weeds can spoil the quality of a crop and so lower its value, for example, wild onion, which may also taint milk when eaten by cows. Weeds are also host plants for various pests and diseases of crop plants; for example, *Chenopodium* weed is the host of cowpea mosaic virus, which causes mosaic disease in cowpea. Some weeds excrete poisonous substances that adversely affect crop plants. Baru grass infestation in tomato crops reduces seed germination.

Some weeds live parasitically on other plants for certain dissolved minerals, organic substances, water, and others. *Cuscuta* spp. can affect tomatoes, cucurbits, onions, peppermint, peppers, tobacco, alfalfa, clover, and several ornamental plants. Weeds like baru grass, wild safflower, puncture vine, and field bindweed make land preparation and crop harvesting difficult. Eradication of weeds is a laborious job and increases the cost of crop production. Most of the water absorbed by weeds is lost through transpiration. Weeds growing in the water channel decrease the efficiency of the water channel because weeds act as obstacles to water flow. Weed has more potential for seed production than field crops, for example, fat hen, wild oat, and safflower produce 8635, 2688, and 582 seeds/plant, respectively. A single wheat plant can produce a maximum of 200 seeds/plant. On average, weeds reduce crop yield by up to 42%, depending upon the nature and weed density.

7.2.3 Weed Competition with Main Crops

Weeds compete with main crops for regarding vital plant nutrients, light, moisture, and space. Weeds take up double the nitrogen (N) and phosphorus (P) amount while triple the potassium (K) amount when competing with crop plants. It is estimated that weeds use 225 liters of water to produce 0.5 kg of dry matter. They reduce not only seed and fodder yields but also the quality of the fodder. Normal yield reductions for green fodder and seed are 25–35% due to weed infestation. To increase productivity, it is crucial to control weeds.

Fodder crops like Egyptian clover are also affected by weeds at the earliest, and thus, an early weed-free period in Berseem is important (Wasnik *et al.*, 2017).

In Egyptian clover, the infestation of weed flora reduces green fodder by up to 23–28% and seed yield by up to 38–44% (Wasnik *et al.*, 2017). In addition, weeds like *Rumex dentatus*, *Cichorium intybus*, and *Sonchus asper* make it difficult to harvest Egyptian clover.

7.2.3.1 Allelopathic Effects of Weeds

Allelopathy refers to a plant or microorganism's direct or indirect stimulation or inhibition of another by releasing chemicals into the environment. Compounds are released into the environment due to root exudation, leaching, and volatilization of decaying plant tissue (Sturm *et al.*, 2018). The allelopathic phenomenon is plant-specific, and various parts of a plant differ in their effect on the seedling growth of other plants (Shinde and Salve, 2019). Likewise, weeds also secrete different chemicals (allelo-chemicals) from their leaves, stems, roots, flowers, and fruit which significantly affect the growth and development of crop (Scrivanti *et al.*, 2011). Even at low concentrations, allelopathic effects can impair seedling growth due to the intricate interactions of various chemical classes, including phenolic compounds, alkaloids, flavonoids, steroids, terpenoids, and amino acids (Saha *et al.*, 2018).

Numerous studies have investigated the interactions between weeds and crops or allelopathy (Li *et al.*, 2021). Numerous investigations have demonstrated the presence of compounds with allelopathic potential in different plant parts. According to Xie et al. (2010), *Alternanthera philoxeroides* has the potential to be allelopathic in its effective invasion of new places. Various approaches have also been proposed to investigate the allelopathic effect (Sturm *et al.*, 2018). When the rice was treated with aqueous extracts of the leaf, petiole, and root of *Eichhornia crassipes*, Scavo *et al.* (2018) discovered that the leaf and root extracts had a substantial impact on rice germination.

7.2.4 Major Fodder Crops and Their Weeds in Pakistan

Mainly, there are two seasons for fodder cultivation in Pakistan during which crops are sown. The first season starts in April and ends in September, called the summer or kharif season, while the second season is the winter or rabi season, starting in October and ending in April. The duration from June–July and December–January is known as the lean period or fodder scarcity period because, in these periods, one-season crops are being harvested and the upcoming season crops are being cultivated.

The major fodder crops and their weeds are given in Table 7.1.

7.2.5 A Detailed Review of Some Important Weeds in Pakistan

7.2.5.1 *Cichorium intybus*

7.2.5.1.1 Common Names

Kasni, Blue sailors, coffee chicory, French endive, succory, witloof, and Belgium endive.

7.2.5.1.1.1 Origin Native to Asia (Afghanistan, Iraq, Lebanon, Iran, Syria, Turkey, Armenia, Georgia, Kazakhstan, Tajikistan, Uzbekistan, Kyrgyzstan, Turkmenistan, China, India, Azerbaijan, and Pakistan), Northern Africa (Algeria and Tunisia), and Europe (Denmark, Sweden, Switzerland, United Kingdom, Hungary, Belgium, France, Germany, Netherlands, Spain, Poland, Albania, Bulgaria, Italy, Romania, Czechoslovakia, and Portugal).

7.2.5.1.2 Widespread Naturalization

This plant has become widely naturalized in Australia, especially in the south and east. It has become widely naturalized in Queensland, Australian Capital Territory (ACT), Victoria, New South Wales, and Tasmania. Additionally, it is common in South Australia's southern, eastern, and southeast regions as well as in the southern sections of Western Australia (including Kangaroo Island).

7.2.5.2 *Parthenium hysterophorus*

Parthenium hysterophorus L., a foreign invasive weed, is quickly taking over cropped and uncultivated lands in Pakistan. It was recently brought to the Indian subcontinent by an unidentified source; it was first noted in 1956 in Poona, Maharashtra, India, and it is claimed to have come from the United States in wheat grains. It was initially noted in Pakistan in the 1980s. It was only

TABLE 7.1
Major Weeds Found in Fodder Crops
Winter or Rabi Season Weeds

Sr. No.	Common Name	English Name	Botanical Name	Family
1	Jangli palak	Sour dock	*Rumex dentatus*	Polygonaceae
2	Bathu	Common lambs quarter	*Chenopodium album*	Chenopodiaceae
3	Pohli	Wild safflower	*Carthamus oxyacantha*	Asteraceae
4	Loombar gha	Beard grass	*Polypogen monspeliensis*	Poaceae
5	Kaurgandal	Leafy spurge	*Euphorbia simplex*	Euphorbiaceae
6	Milkweed	Perennial sow thistle	*Sonchus arvensis*	Asteraceae
7	Ryegrass	Poison ryegrass/ Ivary	*Lolium temulentum*	Poaceae
8	–	Common Chickweed	*Stellaria media*	Caryophyllaceae
9	Guien/Buien	Sweetgrass/Annual bluegrass	*Poa annua*	Poaceae
10	Billi booti	Blue pimpernel	*Anagallis arvensis*	Primulaceae
11	Khandi or wild senji	Yellow sweet clover	*Melilotus indica*	Fabaceae
12	Maina	Toothed bur clover	*Medicago denticulata*	Fabaceae
13	Chotta takla	Forked catchfly	*Silene conoidea*	Caryophyllaceae
14	Rari or rewari (narrow-leaved)	Hairy vetch	*Vicia hirsuta*	Fabaceae
15	Karund	Nettle leaf	*Chenopodium murale/ Chenopodiastrum murale*	Chenopodiaceae
16	Jangli matar	Grass pea	*Lathyrus sativus*	Fabaceae
17	Dokanni/Pili mattri	Meadow pea	*Lathyrus aphaca*	Fabaceae
18	Rari or rewari (broad-leaved)	Vetch	*Vicia sativa*	Fabaceae
19	Kandiali palak	Dock/Sorrel	*Rumex spinosus*	Polygonaceae
20	Pitpapra	Fumatory	*Fumaria parviflora*	Fumariaceae
21	Bhang	Indian hemp	*Cannabis sativa*	Cannabaceae
22	Khandi or wild senji	White sweet clover	*Melilotus alba*	Fabaceae
23	Jangli halon	Garden cress	*Coronopus didymus*	Brassicaceae
24.	Jaundhar	Wild oats	*Avena ludoviciana*	Poaceae
25	Gulli danda	Little seed canarygrass	*Phalaris minor*	Poaceae
26	Jangli sarson	London rocket/Wild mustard	*Sisymbrium irio*	Brassicaceae
27	Bara takla	Cow cockle	*Saponaria vaccaria*	Caryophyllaceae
28	Piazi	Wild onion	*Asphodelus tenuifolius*	Liliaceae
29	Jangli dhania	Corn spurry	*Spergula arvensis*	Caryophyllaceae
30	Maini	Wild fenugreek	*Trigonella polycerata*	Fabaceae
31	Kashni	Blue daisy	*Cichorium intybus*	Asteraceae

(Continued)

TABLE 7.1 ***(Continued)***
Major Weeds Found in Fodder Crops
Summer or Kharif Weeds

Sr. No.	Common Name	English Name	Botanical Name	Family
1	Madhana	Crowfoot grass/ Torpedo grass	*Dactyloctenium aegyptium*	Poaceae
2	Bari dodak	Pill pod spurge/ Snake weed	*Euphorbia hirta*	Euphorbiaceae
3	Kutta ghas	Sandbur	*Cenchrus biflorus*	Poaceae
4	Loomer grass	Foxtail	*Setaria verticillata*	Poaceae
5	Swanki	Water grass	*Echinochloa colona*	Poaceae
6	Takri ghas	Crabgrass	*Digitaria sanguinalis*	Poaceae
7	Jangli jhona/ Lal jhona	Wild rice	*Oryza* spp.	Poaceae
8	Swank	Barnyard grass	*Echinochloa crusgalli*	Poaceae
9	Ghuien	Grasslike fimbry	*Fimbristylis tenera*	Cyperaceae
10	Kandiali chulai	Spiny pigweed	*Amaranthus spinosus*	Amaranthaceae
11	Puthkanda	Prickly chaff flower	*Achyranthus aspera*	Amaranthaceae
12	Bhambola	Ground cherry	*Physalis minima*	Solanaceae
13	Rice motha	Common sedge	*Cyperus difformis*	Cyperaceae
14	Daryaibooti	Erect horseweed	*Conyza stricta*	Astereceae
15	Khatti booti	Indian sorrel	*Oxalis corniculata*	Oxalidaceae
16	Datura	Jimson weed	*Datura stramonium*	Solanaceae
17	Chibber	Wild melon/Native gooseberry	*Cucumis callosus*	Cucurbitaceae
18	Itsit/Chupatti	Horse purselane	*Trianthema portulacastrum*	Aizoaceae
19	Mothi	Hedgehog sedge	*Cyperus compressus*	Cyperaceae
20	Chulai	Smooth pigweed	*Amaranthus viridis*	Amaranthaceae
21	Lendhra/Kutta ghas	Sandbur	*Cenchrus catharticus*	Poaceae
22	Kaonmakki	Day flower	*Commelina benghalensis*	Commelinaceae
23	Jangli jute	Wild jute	*Corchorus tridens*	Tiliaceae
24	Tandla	False amaranth	*Digera arvensis*	Amaranthaceae
25	Khatti booti	Pink wood sorrel	*Oxalis martiana*	Oxalidaceae
26	Lunak	Purslane	*Portulaca quadrifida*	Portulaceae
27	Malancha	Alligator weed	*Alternanthera philoxeroides*	Amaranthaceae
28	Chhoti dodak	Shrubby spurge	*Euphorbia microphylla*	Euphorbiaceae
29	Hazardani	Leafflower	*Phyllanthus niruri*	Euphorbiaceae
30	Dodak/Chandni/ Umbrella milkweed	Sun spurge	*Euphorbia helioscopia*	Euphorbiaceae
31	Makoh	Black nightshade	*Solanum nigrum*	Solanaceae
32	Chhoti dodak	Chickenweed	*Euphorbia thymifolia*	Euphorbiaceae
33	Lunak	Purslane	*Portulaca meridiana*	Portulaceae
34	Oonthchra	Heliotrope	*Heliotropium eichwaldii*	Boraginaceae
35	Jangli mirch	Common spurge	*Croton sparsiflorus*	Euphorbiaceae
36	Bhang	Indian hemp	*Cannabis sativa*	Cannabaceae
37	Lunak	Purslane	*Portulaca oleracea*	Portulaceae
38	Bhakhra	Puncture vine	*Tribulus terrestris*	Zygophylaceae

TABLE 7.1 ***(Continued)***

Perennial Weeds (Lawn + Aquatic)

Sr. No.	Common Name	English Name	Botanical Name	Family
1	Chhittar thor	Prickly pear	*Opuntia dillenii*	Cactaceae
2	Kans or Kahi	Tiger grass/ Kans grass	*Saccharum spontaneum*	Poaceae
3	Siru/Dhabi/ Bharavai	Congo grass/ Lalang grass	*Imperata cylinderica*	Poaceae
4	Aak	Maddar root	*Calotropis procera*	Asolepiadaceae
5	Dab ghas	Halfa grass/ Big cordgrass/Salt reed-grass	*Desmostachya bipinata*	Poaceae
6	Amar bel (on zizyphus)	Dodder	*Cuscuta reflexa*	Convulvulaceae
7	Gajar ghas	Wild carrot weed	*Parthenium hysterophorus*	Asteraceae
8	Deela	Nut grass	*Cyperus rotundus*	Cyperaceae
9	Palwan gha	Marvel grass	*Dicanthium annulatum*	Poaceae
10	Pahari akk	Blush morning glory	*Ipomoea carnea*	Convulvulaceae
11	Amar bel (on lucerne)	Dodder	*Cuscuta chinensis*	Convulvulaceae
12	Lehli or hirankhuri	Field bind weed	*Convolvulus arvensis*	Convulvulaceae
13	Roadsided Itsit/ Biskhapra	Spiderling	*Boerhavia diffusa*	Nyctaginaceae
14	Baru	Johnsongrass	*Sorghum halepense*	Poaceae
15	Leh	Canada thistle	*Cirsium arvense*	Asteraceae
16	Khabal grass	Bermuda grass	*Cynodon dactylon*	Poaceae

introduced in Islamabad and now expands to other Punjab and Khyber Pakhtunkhwa regions. In addition to competing with beautiful pasture and agricultural species, parthenium weed exposes farmers and stock animals to an allergic skin condition.

A portion of the parthenium weed is allelopathic, which means it competes with nearby plant species for soil moisture and nutrients while preventing its neighbors' seedlings from sprouting and growing.

Additionally, parthenium weed is the root cause of significant economic, health, and environmental issues. Parthenium weed in India lowered crop productivity up to 40% and pasture production up to 90%. Parthenium weed had taken over 5.6% of Australia's total land area by 2000, demonstrating the problem's importance on a national scale. This weed's extreme invasiveness suggests it may have spread substantially wider since 2000 than reported in the literature. Parthenium weed is a significant issue in central Queensland's perennial grasslands, where it is expected to cause a US$129 million-a-year decline in beef production by the year 2050. The parthenium weed hinders pasture production by displacing useful forage plants and is estimated to be US$109 million in annual costs (Adamson, 1996). Given the US$4 million loss of sunflower harvests caused by the tobacco streak virus, it is anticipated that cropping systems will be affected to the tune of

US$10 million a year. The Einasleigh Uplands bioregion's most significant threat to biodiversity is thought to be parthenium weed. Even though parthenium weed is generally unappealing, sheep and cattle will consume it during feed shortages. Large doses will cause mutton to become tainted. Owing to Parthenium, Australia currently spends US$6.90 person^{-1} year^{-1} on human health expenses. The spread of *Parthenium* in Pakistan severely threatens the nation's agricultural productivity.

Parthenium hysterophorus is an alien invasive weed, and it is projected to spread further due to the agricultural department and the farming community's failure to control it. It has spread erratically throughout Pakistan, infesting huge regions like the Peshawar Valley. It replaced priceless native flora and imposed a serious threat to biodiversity because of its capacity to adapt to a variety of environmental factors and the lack of their natural enemies. According to studies, the area and density of *Parthenium* infection have greatly expanded during the last 5 years. Tillage and hand weeding are the main methods used to control parthenium weed in the valley. According to the study, *Parthenium* can spread in Pakistan dramatically by reducing the national economy. Other problems include loss of crop productivity, forage production, biodiversity, animal production, and health issues. Till the entire seed bank is used up, consistent effort should be made to manage *P. hysterophorus*. Quarantine measures should be implemented to prevent the introduction and spread of *Parthenium* to uninfected areas. Vehicles and equipment going through places where *Parthenium* is present should be watered down. Farmers nationwide should receive training on how to stop and manage *P. hysterophorus* introduction and spread. More research is needed to determine the effects of *Parthenium* on human and animal health and biodiversity. For the management of *Parthenium* to be sustainable, effective control measures must be developed.

7.2.5.3 *Medicago polymorpha*

A herbaceous legume, commonly known as maithi and bur medic, with a native distribution in western and central Asia and nations bordering the Mediterranean, *Medicago polymorpha* has been widely imported worldwide. Although not limited to these locations, it is particularly common in areas with a Mediterranean climate. Accidental movement of the spiny seed pods and possibly intentional introduction as a fodder plant were the two main causes of introduction; it is impossible to assess the relative importance of the two dispersal methods. The species can be an advantageous pasture plant, especially given its capacity to fix nitrogen, but in other regions, it is regarded as an invasive weed. It can occasionally harm livestock, and the seed pods can seriously contaminate wool.

7.2.5.4 *Lepidium didymum*

7.2.5.4.1 Description

Lepidium didymum is an annual or biennial herb with up to 40 cm (16 in.) long, glabrous, green stems that branch out from a central point. The fruits comprise two spherical valves with a short style and an apex notch. The four white petals are very short or absent, there are only two (rarely four), and there are 54 stamens in the unremarkable flowers. They have 1–5 mm long orange or reddish-brown and wrinkled seeds.

7.2.5.4.2 Distribution

Lepidium didymum's origin is uncertain, but it is widely said that South America, specifically Argentina, Bolivia, Brazil, Chile, Paraguay, Peru, Uruguay, and Venezuela, is where it originated. It is a weed variety that has spread to other areas. It has naturalized worldwide, from Africa to Europe, Asia to Australasia to North America and South America. It had been discovered in the wild in Britain by 1778, mostly in the southern parts of Ireland and England, on cultivated and abandoned ground, in gardens and lawns, and by pathways.

7.2.5.5 *Spergula arvensis*

7.2.5.5.1 Description

S. arvensis is an erect annual herb, ascending or spreading. It has stems up to 60 cm tall, slender, jointed, branched, and somewhat sticky. The leaves are green, threadlike, round, and up to 4 cm long. About 1.5 mm in diameter, thick, lens-shaped, dull black seeds with a pronounced, thin, light-colored wing on the periphery and a surface frequently roughened by tiny, rounded, projecting bodies.

7.2.5.5.2 Distribution

S. arvensis is a weed most common in temperate areas, although it can also be found in tropical areas. It can be found on every continent and is a weed in Tasmania, Australia, Finland, Alaska, and the United States.

7.2.5.5.3 Crops Affected

In 33 countries, *S. arvensis* has been identified in 25 different crops. It thrives in wheat, oats, root crops, and flax. It has been noted in cereals in practically every region of the world.

7.2.5.6 *Chenopodium album*

7.2.5.6.1 Description

At first, it usually grows upright; unless supported by other plants, it usually goes recumbent because of the weight of the leaf and seeds. The leaves are alternating and provide a variety of looks. The plant's earliest leaves are serrated, roughly diamond-shaped, 3–7 cm long, and 3–6 cm wide. They are located close to the plant's base. The leaves are 1–5 cm long, 0.4–2 cm wide, unwettable, and mealy in appearance, and they are on the upper portion of the flowering stems. On a 10–40 cm long, densely branched inflorescence, the tiny flowers, which are radially symmetrical and grow in little cymes, are present. The flowers have five tepals, five of which are mealy on the outside and quickly merge at the base.

7.2.5.6.2 Distribution

The origin of the cosmopolitan weed *C. album* is masked by its widespread distribution. It is equally widespread in Australia, Asia, Europe, Pakistan, India, North America, South Africa, and South America and can be found all over North America. It is primarily found at higher altitudes in tropical areas. It is grown as a grain crop in the Himalayan region, where it was domesticated, and as a vegetable in India and Pakistan. Archaeological evidence indicates that it was grown as a pseudocereal in prehistoric Europe.

7.2.5.6.3 Crops Affected

Although *C. album* is a potentially dangerous weed in practically all crops seeded in the tropics and subtropics during the winter, it appears to thrive most aggressively in temperate and subtemperate climates. It is a typical weed in about 40 crops in 47 nations, with the highest prevalence of sugar beet, potatoes, corn, and cereals. It is among the six most dangerous weeds in India, New Zealand, and Pakistan. It is one of the main weeds in Canada and Europe. It affects practically all crops sown in the summer and winter in temperate areas.

It is particularly prevalent in maize, sunflower, soybean, horticultural gardens, wheat, chickpea, barley, winter vegetables, and subtropical climates. It is also a significant weed of upland rice and tea in Japan, grapes and citrus groves in Spain, soybeans, cotton, strawberries, pastures and peanuts, tobacco, and rice in the United States, Canada, and Mexico. It is a problematic weed in main crops in Europe and America.

7.2.5.6.4 Crop Losses

C. album directly competes with crops for nutrients and light, which lowers agricultural productivity. Significant losses from *C. album* have been documented in field and greenhouse tests on various crops, including maize, tomato, lucerne, soybeans, sugar beets, and oat. In field tests conducted in Canada, *C. album* was found to reduce maize yields between 6% and 58% at densities of 172–300 plants/m^2. A 22.3% decrease in maize yield occurred in Spain's irrigated field testing when maize competed with *C. album* at equal densities. Field testing discovered that unchecked populations of *C. album* were to blame for 59% of US losses in maize production. Thirty-six percent of the fruit was lost in tomato field experiments conducted in the US using 64 *C. album* plants/m of row. Oat grain yield was found to be lowered by about 60% by *C. album* interference in field tests conducted in Canada when the crop was allowed to compete with the weed for the entire growing season. Losses of 23–36% of the American barley crop were attributed to *C. album* competition. The yield of lucerne grown under greenhouse conditions in Canada was lowered by *C. album* competition by about 23%.

In field studies conducted in Colorado, United States, sugar beet root yield was reduced by 48% when competing with about one *C. album* plant per meter of row. When sugar beet was grown in Japan with 13 *C. album* plants per meter of a row that emerged 10 days after the crop, root yield was reduced by 72%. In contrast, when 22 *C. album* plants/m^2 were allowed to thrive alongside sugar beet, yield losses as high as 93% were achieved in Wageningen, the Netherlands. When 170 *C. album* plants were grown per square meter, the yield of sugar beet root was 86% lower. The crop losses were economically significant in each case, even though the yield losses brought on by *C. album* vary according to the crop, weed density, and location.

In addition to aiding in weed growth, crop seed contamination by weed seeds also results in significant losses in crop seed quality and value. Very tiny *C. album* seeds frequently contaminate crop seeds gathered from weed-infested fields. For instance, *C. album* seeds are regularly discovered in numerous grain seeds as contaminants.

Farmers benefit from the allelopathic effects, but the opposite could result in significant financial losses. Crop plants such as maize, soybeans, carrots, onions, tomatoes, squash, sunflowers, cucumbers, lettuce, and oats have all been documented to be affected by *C. album*'s allelopathic effects. *C. album* residues caused field-grown maize and soybean growth losses of 15–30% and glasshouse soybean production losses of 16–20% in the United States.

C. album is to blame for significant agricultural losses because it is an alternate host to several economically significant pests and illnesses. In New Zealand, it was discovered that *C. album* was host to the fungus *Stagonospora atriplicis*. *C. album* was reportedly *Polymyxa betae*'s host in Japan. Several crop viruses also have *C. album* as an alternative host. Up to 47% of sugar beet losses result from this disease, predominantly spread by the aphid *Myzus persicae*. According to reports, *C. album* is a productive host for the peanut stunt cucumovirus in the US. In southwest Quebec, this nematode poses a significant barrier to the production of carrots. In South Africa, it has been demonstrated that *C. album* is infested by the potato root-knot worm *Ditylenchus destructor*, allowing it to live between crop seasons. This nematode is a significant groundnut pest in South Africa.

7.2.5.7 *Avena fatua*

A grass belonging to the oat genus is called *Avena fatua*. It goes by the name of common wild oat. Although this oat is native to Eurasia, it has spread to most temperate zones. It is considered a harmful weed in some places but has become naturalized in others.

A green grass with upright, hollow stems 30–120 cm tall and bearing spikelets, *Avena fatua* resembles an oat. Due to tiny hairs, dark green leaves can be up to a centimeter wide and rough.

The seedlings have hair as well. Compared to the commonly farmed oat seed, the seed kernel is hairier, longer, darker, and thinner (*Avena sativa*). When it invades, a few wild oat plants can significantly reduce wheat or farmed oat field output. Degrading a field crop or competing with crop plants for nutrients, this species and other wild oats can cause problems in prairie agriculture.

7.2.5.7.1 Crop Losses

The significance of *A. fatua*, one of the deadliest agricultural weeds in the world, is still growing. There are 11 million ha of cultivated area infested with *A. fatua* in the United States. *A. fatua* was referred to as an intractable weed because of how its life cycle coincides with crop expansion. A computer program has been created to evaluate the economics of controlling *A. fatua* in oilseeds and cereals. *A. fatua* exhibits a strong capacity for competition and is frequently more competitive than *Alopecurus myosuroides*, *Galium aparine*, or wheat. Plowing and heavy nitrogen and phosphorus fertilization increase shoot biomass and competitiveness. Most of the underground competition between cereals and *A. fatua* occurs. The time of emergence and density have an impact on crop production response. Cereal competition begins mostly at the two-node stage, which lowers crop tillers.

For many crops, numerous damage threshold levels have been estimated. The following information has been reported for cereals: Eight *A. fatua* plants per square meter resulted in a 14% decrease in production and a 5.5% decrease in protein content in wheat, while 100 panicles per square meter decreased the output of winter wheat by 34%. Wheat yield loss was less remained at 1–3% owing to *A. fatua* (5 plants m^{-2}), increasing to 60% with an increasing *A. fatua* population. Herbicide resistance has frequently developed due to the ongoing and extensive use of herbicides to control *A. fatua*, and it is currently ranked as the second-most herbicide-resistant weed globally. Approximately 2 million acres of crops in Canada, where the issue is worst, are infested with herbicide-resistant *A. fatua*.

7.2.5.8 *Cyperus rotundus*

It is a sedge native to Europe, Southern Asia, and Africa. It is also known as coco-grass, Java grass, nut grass, red nut sedge, purple nut sedge, and Khmer kravanh chruk.

7.2.5.8.1 Description

A perennial plant called *Cyperus rotundus* can grow up to 140 cm tall (55 in.). The names "nut grass" and "nut sedge," which it shares with a closely related species called *Cyperus esculentus*, are derived from its tubers, which resemble nuts in appearance but are not related to nuts botanically.

The base of the plant bears three-ranked leaves, approximately 5–20 cm (2–8 in.) long, like other Cyperaceae. The flower stalks have triangular cross sections. Three stamens, a pistil containing three stigmas, and three to eight uneven spikes make up the inflorescence of the bisexual flower. The fruit is a three-angled achene.

Rhizomes that are up to 25 mm (1.0 in.) long, white, and meaty are first produced by young plants. Before forming a bulb-like structure from which new branches and roots emerge, certain rhizomes grow upward in the soil. The new roots then sprout new rhizomes. Other rhizomes create horizontally or downward-growing chains of dark reddish-brown tubers or tuber-like structures.

7.2.5.9 *Cynodon dactylon*

Cynodon dactylon, or Bermuda grass or scutch grass, is widespread. It is indigenous to much of Asia, Australia, Africa, and Europe, and it has been introduced in the Americas. It is a widespread invasive species in Bermuda, even though it is not native to the island. It has been referred to as "crabgrass" in Bermuda.

7.2.5.9.1 Description

The small, gray-green blades typically measure 2–15 cm (0.79–5.91 in.) in length and have rough edges. The upright stems have a height range of 0.39 in. to 10.0 cm. The stems have a slight flattening and are frequently colored purple. The seed heads are produced at the top of the stem in two to six spikes, each measuring 2 to 5 cm (0.79 to 1.97 in.). It has a deep root system that, under drought and permeable soil conditions, can extend to a depth of more than 2 m (6.6 ft), although most of the root mass is located less than 60 cm (24 in.) below the surface. The grass roots anywhere a node meets the ground as it crawls along the surface with its stolons and forms a thick mat. Seeds, stolons, and rhizomes are used by *C. dactylon* to reproduce.

7.2.5.10 *Dactyloctenium aegyptium*

African-native *Dactyloctenium aegyptium*, often known as Egyptian crowfoot grass, is a member of the *Poaceae* family. The plant primarily flourishes in dense, moist soils.

7.2.5.11 *Pennisetum*

The grass family includes the widely distributed genus *Pennisetum*, native to tropical and warm temperate parts of the world. The line dividing the genus *Pennisetum* from *Cenchrus* is not clearly defined. *Pennisetum* gave rise to *Cenchrus*, which is classified together in a single monophyletic clade. Some species currently found in *Pennisetum* were originally found in *Cenchrus*, while others have been relocated. The degree of bristle fusion in the inflorescence is a key morphological characteristic used to differentiate them, but it is frequently unreliable and commonly referred to as fountaingrasses (fountain grasses).

7.2.5.11.1 Description

Currently, there are between 80 and 140 species in the genus *Pennisetum*. The species are indigenous to Africa, Asia, Australia, and Latin America, with some having established extensive natural populations in Europe and North America and on several marine islands.

They are grasses, either annual or perennial. Some are small, but others can grow stems as tall as 8 m. The inflorescence is a narrow, dense panicle with fascicles of spikelets and bristles scattered throughout. There are three different types of bristles; some species have all three, and others do not. Some bristles have hairs covering them, sometimes long, showy hairs that resemble plumes and are the source of the genus name, *Latin penna.*

7.2.5.12 Alligator Weed

Alligator weed is a mat-forming perennial weed. It is a significant aquatic weed issue in the coastal plain of Carolina. Except for the nodes, alligator weed stems have distinct joins and are hollow. Light green stems with faint parallel green lines. The nodes are where roots form. Rhizomes of two different forms are produced by terrestrial infestations: fleshy, white, root-like rhizomes and horizontal, purplish rhizomes that look like stems and have very short internodes. Terrestrial stems are typically not hollow but rather pithy or even solid. Along the stem, the leaves are arranged in an opposing pattern and range in form from oval to lance. Near the ends of the stems, tiny, white clover-like blooms are produced. From late April through October, flowers bloom. Due to the seldom formation of seeds, reproduction is vegetative.

7.2.5.13 Lambsquarter

Annual weed *lambsquarter* has many branches and grows upright between 1 and 4 ft tall. Young leaves and the undersides of older leaves have a light gray or white mealy coating with a wavy or coarsely serrated margin. The inconspicuous gray-green flowers are arranged in dense clusters and branch terminals. It spreads quickly, consumes water, and competes fiercely in most environments.

7.2.5.14 Pigweed

Annual weeds include amaranth and pigweed. With foliage that ranges from purple and crimson to gold, there are about 60 species. They have a long germination period, rapid growth, and high seed production rates. If used repeatedly, they may develop resistance to herbicides that prevent acetolactate synthase (ALS). They can grow up to 612 ft tall and have crimson stems and alternating egg-shaped leaves with wavy margins. Leaves and stems may or may not have hairs on them. They have a shallow, frequently reddish taproot. They produce dense, compact panicles between 2 and 8 in. long and about three-quarters of an inch broad, with tiny, inconspicuous green flowers. These weeds are frequently mistaken for lambsquarter. Pigweed can be extremely harmful if consumed in large amounts due to the probable nitrate buildup. The two most prevalent species that flourish on disturbed soils are redroot pigweed and spiny amaranth.

7.2.5.15 Bahiagrass

A typical perennial grass, bahiagrass spreads by rhizomes and may be identified by its distinctive Y-shaped black seedhead. This invasive grass spreads quickly and develops into a dense sod with short, narrow leaves. A forage grass alternative is bahiagrass.

7.2.5.16 Barnyard Grass

Annual barnyard grass can grow as tall as 5 ft and has sturdy stems. One of the few grass weeds with no ligules is barnyard grass. Hairs and auricles are absent from leaves. The base of the leaf sheaths frequently has a red or maroon hue. The leaves are coiled in the shoot and are silky. The size of a leaf can vary from 0.2 to 1.2 in. in width and 4 to 20 in. in length. Each leaf has a prominent midvein that becomes keeled as it approaches the base. The bottoms of the leaves may have a few small hairs.

7.2.5.17 Broomsedge

Virginia bluestem, or broom straw, are other names for broomsedge, which is a clump-forming perennial grass. It frequently remains unseen until it develops into a clump of stalks that resemble a broom. In the late summer, the plant has slender stems that can grow to 3 ft. The stalks produce numerous white wind-borne seeds. It thrives on less fertile soils and is particularly resistant to N, P, and low pH. The production of allelopathic compounds by broomsedge suppresses the emergence and expansion of rival species. In essence, it produces natural weed killers on its own.

7.2.5.18 Foxtail

Three distinct annual plant species are referred to as foxtail. Since they are all grasses, it is not easy to distinguish them before they reach maturity. A grass that forms clumps and has a seedhead that resembles a fox's tail is called foxtail. Giant foxtail has many hairs on the upper leaf surface and typical foxtail-like seedheads that droop as they develop. In contrast to other foxtails, it is often bigger and has a nodding seedhead. When fully grown, the yellow foxtail's distinctive seedhead resembles a foxtail and has long and silky hairs just at the base of its leaves. Green foxtail has hairless leaves and a distinctive seedhead resembling a foxtail.

7.2.5.19 Goosegrass

An annual grass called goosegrass needs both light and moisture to sprout. Goosegrass has a prostrate growth pattern, is frequently white in the center, and resembles a wagon wheel. It grows well in compacted soils but does not root at the nodes. The sheaths on the leaves are smooth, flattened, and change color from white to silver toward the base. It has a fibrous root structure. The seedhead comprises clusters of 2 to 13 spikes, each measuring between 112 and 6 in. long and 0.1 and 0.3 in. in diameter. Along each spike are two rows of spikelets with three to six 0.04–0.08-in.-long light brown to black seeds. Plants frequently have a compacted appearance to the earth, as if they have been trampled on repeatedly. Its characteristic white center distinguishes goosegrass from most other grass weeds.

7.2.5.20 Johnsongrass

Johnsongrass is a virulent perennial grass that reproduces both vegetatively and asexually. This grass has thick rhizomes and is very tall and coarse. It may grow to 8 ft tall and form tense bunches as it expands. The leaves have a noticeable white midvein, are smooth, and range in length from 6 to 20 in. Near the base, the smooth, pink to rust-colored stems turn red. Large, loosely branching, hairy, and purple panicles are seen. Each spikelet has an evident awn, usually appearing in pairs or threes. The reddish-brown seeds are almost 18 in. long. Fibrous roots are present. Rhizomes near the soil's surface are thick and covered with scales and purple dots. It can harm livestock when Johnsongrass is wilted by frost or develops new growth after a drought or pruning. Johnsongrass can have enough prussic acid in it during hot, dry conditions to kill livestock when taken up. It may result in nitrate overdose and death.

7.2.5.21 Nutsedge

Although a sedge, nutsedge (yellow and purple) is occasionally misidentified as nut grass. The stem helps to differentiate sedges from grasses. Sedge stems are always triangular or three-sided. Native to North America, yellow nutsedge is a sedge that can withstand freezing temperatures. Native to India, purple nutsedge is more frequently an issue for warm season grasses. The color of seedheads and tuber placement can be used to distinguish between the two species. Purple nutsedge has a reddish-purple seedhead, while yellow nutsedge has a yellowish-tan seedhead. Purple nutsedge has hairy tubers that are charcoal (black) in hue, while yellow nutsedge has chestnut-colored (tan) tubers. Identification is crucial since purple nutsedge is typically harder to manage.

7.2.5.22 Prickly Pear

A low-growing cactus, prickly pear has bright yellow blooms and pads with several individual prickles made up of spines. It is indigenous to the majority of the United States. There are no actual leaves on the prickly pear. Its rootstock is substantial. This species can spread thanks to the ability of the pads to take root. Each pad is a thick, succulent stem with several little spines. Individual fruits mature to a reddish-maroon color and are fleshy, growing to 1 to 112 in. The flowers are 2 to 3 in. in diameter and bright yellow and may have a crimson core.

7.2.5.23 Wild Mustard

Wild mustard is an annual weed that blooms with typical mustard-colored flowers. It features a fibrous root system in addition to a taproot. Cotyledons range in shape from kidney- to heart-shaped. The initial genuine leaves have wavy margins and are alternating, hairy, and elliptic. The alternating leaves have an ovate to an egg-shaped form. The bottom leaves grow in a rosette, are petiolated, have irregular lobed edges, and are toothed. Upper leaves gradually get smaller up the stem, are not lobed, and either clasp the stem or have short petioles. The plant's stems are upright, branching at the top, and typically covered in hair, at least at the base. At the tips of the branches, the blooms grow in groups. Each flower is about 1/4 in. large and has four yellow petals. The fruit has two valves that separate, leaving a center division. It is dry and elongated.

7.2.5.24 Annual Bluegrass

Annual bluegrass, also known as *Poa annua*, forms clumps and can withstand close mowing; if not, it can grow as tall as 11 in. It grows in pastures, but it is predominantly a weed of lawns. The foliage is pale green. The leaf blades are folded in the bud up to 12 in. long and hairless on both surfaces. The tip of the leaves has a distinctive boat form. The ligule is membranous and has a small tip. Leaf sheaths lack hair and are considerably compressed and flattened. A 34- to 212-in.-long open panicle makes up the seedhead. It has fibrous roots.

7.2.5.25 *Anagallis arvensis*

Anagallis arvensis is a species of low-growing annual plants with brightly colored flowers, most frequently scarlet but also occasionally bright blue and pink. It is also known as the scarlet pimpernel, red pimpernel, red chickweed, poor man's barometer, poor man's weatherglass, shepherd's weatherglass, or shepherd's clock.

7.2.5.25.1 Origin and Geographic Range

The native range of the species is Western Asia, North Africa, and Europe. With a distribution that includes the Americas, Central and East Asia, the Indian subcontinent, Malesia, the Pacific Islands, Australasia, and Southern Africa, *A. arvensis* has now practically naturalized everywhere. Humans have unintentionally or intentionally spread the species far as attractive flowers.

7.2.5.25.2 Description

The red pimpernel grows slowly and creeps when it is a summer annual, but when it is a winter annual, it creates a half-rosette with an upright stem. Its weak, sprawling stems have a square cross-section and can reach lengths of 5–30 cm (2–12 in.). They produce opposing pairs of oval, sessile, brilliant green leaves. From spring to autumn, the radially symmetric, orange, red, or blue flowers have a diameter of 10–15 mm (0.4–0.6 in.) and are produced singly in the leaf axils. The margins of the petals are somewhat crenate and covered in tiny glandular hairs. Although the flowers can self-pollinate, the lollipop hairs on the stamens make them attractive to various pollinators, especially flies. The dehiscent capsule fruits ripen from August through October in the northern hemisphere. The weight of the fruiting body bends the stalk, and the wind or rain carries the seeds to their destination. Plants with blue flowers (*A. arvensis Forma azurea*) are widespread in some places.

7.2.5.25.3 Crop Losses

Numerous research conducted in various nations supports the idea that *A. arvensis* is not a competitive weed in most crops because of its slow growth and limited root structure. It might, however, emerge early in the spring before other weeds (and crops) establish themselves, grow into dense masses, and stifle the early development of slow-growing crops. Little field evidence supports the widespread belief that *A. arvensis* is poisonous to livestock. However, as it is only seldom left in pastures by grazing animals, it is probably unappealing. Indoor feeding tests on some animals have revealed potential harm. After feeding on *A. arvensis* in the field, buffalo and cattle have recently perished in India. After handling the plant, cases of human dermatitis have been documented. Lucerne and clovers, two small-seeded field crops, are contaminated by *A. arvensis* seeds.

7.2.5.26 *Asphodelus tenuifolius*

7.2.5.26.1 Distribution

Although it appears to be a taproot system of a dicotyledon, the root of the upright annual monocotyledonous herb *A. tenuifolius* is a firm, compacted bundle of fibrous roots that occasionally twists to resemble a rope; the root is a yellowish in young plants and dark brown at maturity. Numerous, straightforward, sparsely dichotomous branching in the top section, sturdy, 3 mm in diameter, and up to 60 cm long scapes; Campanulate, white with a pink- or purple-striped bloom, in loose racemes; jointed bracteate, pedicellate, and short pedicel; leaves range in length from 10 to 40 cm, are numerous, all basal, hollow, and slender; the underlayment flowers typically do not bloom until late afternoon, and unless the weather is dull and chilly, they will wither and close before the next day; six perianth segments with six 1.5 cm long petals, six stamens, a simple, superior, three-carpeted, three-loculated ovary, and blossoming that progresses upward over weeks. The fruit is a three-valved globular capsule with transverse wrinkles, deep irregular dents on the front and back, and three-angled, dark-colored seeds that dehisce at partitions into the hollow.

7.2.5.26.2 Crop Losses

The biggest yield loss is brought on by *A. tenuifolius*'s interference in Pakistan and India. It has been found that a variety of crops are damaged. *A. tenuifolius* competes more ferociously with the crop than *Chenopodium album*, and infestations caused 42% yield losses in chickpea areas. *Sclerotinia sclerotiorum* has been isolated from *A. tenuifolius* in mustard fields and serves as a secondary host for *Macrophomina phaseoli*, a fungus that causes root rot, in Pakistan.

7.2.5.27 *Bromus japonicus*

7.2.5.27.1 Distribution

In addition to Northern Africa (Egypt), tropical (Indian subcontinent), and temperate (Caucasus, China, eastern, middle, and western Asia, Mongolia, Siberia) Asia, *B. japonicus* is a native of eastern, central, southeastern, and portions of southern (France) Europe. In Australasia, North America, and southern South America (Argentina), *B. japonicus* is an alien species. *B. japonicus* is a frequent foreign invasive species in mixed-grass grasslands in North America. According to the Alberta Weed Monitoring Network, it can also be found throughout Mexico, the southern US, and most of Canada.

7.2.5.27.2 Description

This plant is 40–90 cm tall and erect with annual culms. The leaf blades measure 12–30 cm, are 4–8 mm thick, and are pubescent on both surfaces. The panicle is effuse, 20–30 cm long by 5–10 cm broad and nodding; branches range from two to eight and are thin and long, between 5 and 10 cm, with one to four spikelets on each. Lemmas are elliptic and 7–9 mm long in a side view, herbaceous, nine-veined, typically glabrous with margins membranous that have a prominent angle at maturity, scabrid, and anthers that are about 1 mm long. Glumes are subequal and keel scabrid, have margins that are membranous, have a lower glume of 5–7 mm, are three- to five-veined, and have an upper glume 5–7 mm and a caryopsis of 7–8 mm length. Fl. and fr. is 2n = 14 from May through July. *B. japonicus* has significantly more developed roots than *B. tectorum*.

7.2.5.27.3 Crop Losses

B. japonicus is an invasive species that is tenacious and aggressive. When fresh, it is somewhat pleasant to livestock, but as it ages, it quickly turns indigestible. It can change seasonal patterns of rangeland feed supply and quality and livestock performance, harming perennial grass biodiversity. *B. japonicus* is one of the most prevalent species of grass weeds discovered in the winter wheat production system in the North China Plain. Wheat competes with *B. japonicus*, which could cut wheat output by 30%. In Canada, it is a weed that affects crops and rangelands and frequently signals that a range is in bad condition. *B. japonicus* poses a problem in overgrazed pastures in Saskatchewan and is a worry in fodder and reclaimed lands. The maize dwarf mosaic virus has been identified as having *B. japonicus* as a host plant.

7.2.5.28 *Carthamus oxyacantha*

7.2.5.28.1 Distribution

India, Iran, Iraq, Kyrgyzstan, Pakistan, Tajikistan, Turkmenistan, Afghanistan, Azerbaijan, and Iran.

7.2.5.28.2 Description

The fruit is an achene, broadly elliptic to slightly four-sided in cross section, obovate or elliptic, 3–5.5 mm long, 2–3.5 mm wide, 1.5–2 mm thick, +/– truncate at apex, with a marginal notch at the base. It is bone white to ivory, less commonly beige; glabrous; smooth; and glossy, with +/– thickly dispersed blotches and speckles in shades of brown. Scar subbasal is a rough, vertical ridge-filled cavity with an outline resembling a diamond. Early deciduous pappus is not present. The style base is deciduous, and the apex is a circular, rough, flat-to-uneven area surrounded by a black ring with an irregular edge. Endosperm is not present in the spatulate, broad cotyledon embryo.

7.2.5.28.3 Crop Losses

Wheat, barley, corn, chickpeas, and lentils are among the agricultural systems where *Carthamus oxyacantha* is a weed. It reduces the production of wheat, chickpeas, and field peas.

7.2.5.29 *Convolvulus arvensis*

A persistent weed of Eurasian origin, *Convolvulus arvensis* is a concern worldwide in agricultural and nonagricultural areas. *C. arvensis* L. has at least 84 common names and is regarded by horticulturists and farmers as one of the worst weeds in the world. Bindweed is a native of the Old World's Mediterranean region, where it is still used medicinally along with the morphologically related species scammony (*C. scammonia* L.). Although the initial research on *C. arvensis* refers to medicine usage, it was also spread accidentally and for adornment. The earliest occurrences of *C. arvensis* were mapped using herbarium vouchers and literature in every state, arvensis. Bindweed was initially discovered in North America (Virginia) in the 1730s, and by 1807, Pennsylvania was the location of commercial mail-order sales of the plant. The plant is standing as a cure-all diminished over time. *C. arvensis* did not emerge until the 1880s. In North America, *C. arvensis* became regarded as a native plant, but it was not until the early 1900s that efforts were made to eradicate this twining herb. After almost a century of conflict, the species has not been effectively subdued.

7.2.5.29.1 Distribution

Originally from Eurasia, *C. arvensis* is now widely dispersed in temperate and tropical areas of the globe. The range of *C. arvensis* ranges from 60°N to 45°S. It should be noted that a country's lack of inclusion in the accompanying "Distribution Table" does not guarantee that it does not exist there. Instead, it shows that no mention of its existence has been made in the literature searched for in this compendium.

7.2.5.29.2 Description

The herbaceous perennial *C. arvensis* has extremely deep roots. At nearly any depth down to 1 m, shoots emerge from adventitious buds on the deep root system. The stems twine or trail upward above the earth. Slender, 1.5-m-long, clockwise twining stems that are glabrous or finely pubescent. The leaves are alternating, petiolate, lanceolate to narrow-oblong shape, 1.2–5.0 cm long, acute at the apex, whole but frequently hastate-sagittate at the base, and glabrous or pubescent with a few crisped hairs dispersed throughout. Bracteoles are linear and 2–4 mm long, and the flowers are axillary, solitary, or in cymes of 2–3 on peduncles equal to the leaf below. Free of sepals, obtuse, and 2.5–4.5 mm long. Corolla funnel-shaped, pentamerous, white or pink, 10–25 mm long, 10–25 mm diameter, with five radial pubescent bands without distinct lobes. Corolla tube with five inserted stamens. Solitary stigmata in two oblong shapes. Two-celled ovary fruit is a capsule that varies in shape from globular to ovoid and has a persistent style base. Typically, there are four compressed-globose, 3–5 mm diameter, granular, dark-brown or black seeds per plant.

7.2.5.29.3 Crop Losses

C. arvensis can result in financial losses by lowering crop yields in horticulture and agriculture. This is accomplished by making crops compete for nutrients and light, even suffocating them as they grow quickly. With 20–80% yield reductions documented, annual crops, including cereals and grain legumes, appear particularly vulnerable to *C. arvensis*–related yield loss. Because the crop gets tangled up in the twining stems, *C. arvensis* can substantially hamper the harvesting of annual crops. It is frequently mentioned as a pesky weed in vineyards.

7.2.5.30 *Euphorbia prostrata*

7.2.5.30.1 Origin

Although it is endemic to the West Indies, *Euphorbia prostrata* has spread widely throughout the tropics and subtropics. It can be found all over tropical Africa and on the islands in the Indian Ocean.

7.2.5.30.2 Description

This is a monoecious, prostrate, annual herb with latex-coated stems and branches up to 20 cm long and colored purplish. The blades of the leaves are ovate, up to 5–8 mm, with an unequal base, one side cuneate and the other rounded; a rounded apex; shallowly toothed margins; and glabrous above and sparsely hairy beneath. The stipules are triangular, about 1 mm long with two teeth, and the petiole can be up to 1 mm long. Cyathia are almost sessile, 0.5–1 mm, with a barrel-shaped involucre, lobes triangular, minute, margin hairy; glands are 4 glands are present which are transversely elliptical, red, with very small pink or white appendages, and each involucre contains one female flower surrounded by a few male flowers. This type of inflorescence is known as a "cyathium," and it grows on short Flowers are unisexual; male flowers are sessile, with bracteoles that resemble hair, and perianths are absent. Female flowers have pedicels about 1.5 mm long and reflexed in the fruit, with perianths that have rims. The ovary is superior, glabrous, three-celled, and has three small, two-fid styles. A three-seeded, sharply three-lobed capsule with a truncate base and purplish and hairy sutures. Oblong-conical, apically 4-angled, transversely wrinkled, gray-brown, and without caruncle seeds, measuring approximately 1 mm by 0.5 mm.

7.2.5.30.3 Crop Losses

Numerous investigations of the phytotoxicity of extracts from Euphorbia species have been made. Wheat germination and seedling growth of wheat, peas, and several other species were inhibited by aqueous extracts of the stems, leaves, and roots of leafy spurge. The distribution and concentration of allelochemicals in various areas of weeds determine the variation in the allelopathic response of different plants. Lowered wheat and maize seedling growth, delayed germination, low chlorophyll, and dependence. It encouraged the development of maize seedlings at low concentrations. According to the application dose, this extract could be utilized to either impede or stimulate crop growth.

Cicer arietinum, Cajanus cajan, Vigna radiata, germination, seedling growth, and weight were all considerably decreased by an aqueous extract of *E. thiamifolia*'s leaf, stem, root, and inflorescence. Crops affected by *E. thiamifolia* plant parts varied.

Chickpea, lentil, and wheat seeds' germination were inhibited by extracts from the root, stem, leaf, and fruit of *E. helioscopia*, with the leaf having the strongest inhibitory effect. However, this weed's leaf, stem, and root extracts increased seed germination at low concentrations compared to the control in the examined crops.

Conclusion: Allelochemicals that at certain concentrations prevented the germination of some species may, at lower quantities, stimulate the germination of the same or a different species.

7.2.6 Weed Control Methods

One of the most important aspects of fodder agriculture, especially for seed production, is weed control. In addition to lowering the amount and quality of agricultural yield, weeds compete with crops for natural and applied resources. Numerous "weed management" techniques have been created to slow the growth of weeds. In Pakistan, hand weeding has historically been the main method of weed control. However, farmers are encouraged to implement labor- and cost-saving measures. The growth habits and life cycles of weeds vary. As a result, there is no guarantee that any weed control technique will be effective. Many different weed management techniques may be used in the field, and each one is more crucial than the others. Any weed management strategy should be chosen based on its efficiency and cost. Three major categories can be used to classify the numerous weed management techniques: cultural and preventive control, physical or mechanical control, and chemical weed control. Following is a discussion of these practices:

7.2.7 Sanitation-Preventive Control

The best and most economical way to eradicate weeds is to stop their spread before germination. This can be done in the following ways:

a. Use of clean crop seed
b. Clean farm machinery
c. Clean water channels
d. Weed-free manure
e. Clean feed and hay to animals
f. Removal of weeds before maturity: "An ounce of prevention is worth a pound of cure"

Once weeds are established, control becomes more difficult as they often spread and create more of a problem, which takes time, effort, and money to resolve. Preventive methods in controlling weeds have recently been fruitful in eliminating weeds in fodder crops. Deep plowing and soil inversion through soil tillage and soil inversion exposes weed seeds to direct sunlight, effecting their germination and stand establishment ability. Some cultural practices are also helpful in controlling weeds, including crop rotation, sowing, fertilizer application, dose and time, and irrigation time and method. Using clean seed or weed-free seed also reduces weed incidence in fields. Weed-free seed facilitates the uniform germination of fodder crops and a reduced weed population. Seed treatment of small-seeded forages such as Egyptian clover with 10% salt solution has been observed to be the most common practice to eliminate the *Cichorium intybus. Cichorium intybus* seeds float over the salted water because of the weight difference from Egyptian clover and thus can be separated easily.

7.2.8 Cultural Control

a. *Dab method*
When the seedbed comes into "Wattar" (proper moisture condition) after a round (soaking irrigation), it is plowed and planked twice. Then the field is left as such for 8 to 10 days so that weed seeds lying dormant in the soil can germinate. After the emergence of weed seedlings, the field is plowed and planked for final seedbed preparation. During this operation, the young weed seedlings are uprooted and buried.

b. *Fertilizer application*
Fertilizer should preferably be applied after weed removal. It is better to apply fertilizer along the row of crops to minimize the chances of its uptake by weeds away from crop rows.

c. *Irrigation*
Continuous standing of 6–7 cm of water for 1.5 months after rice seedling transplanting reduces weed germination and growth considerably.

d. *Crop rotation*
Rotation of crops with different production requirements keeps down weed growth by not allowing a buildup of weeds ecologically adapted to one crop, for example,

i. Growing rice in maize or cotton fields where It-sit is a problem.
ii. Growing of jute in the rice fields.
iii. In rice–wheat rotation, the inclusion of mungbean, sunflower, and soybean can check the growth of rice weeds to a considerable extent.
iv. Egyptian clover or alfalfa should be grown instead of wheat in wheat fields infested with foxtails or wild oat.

Crop rotation is a crucial part of integrated weed control. Under monoculture, weeds with the same life cycle as the crop typically grow more. In berseem-based farming systems, using good crop rotation techniques will assist farmers in combatting the dominance of some weeds and lessen weed competition. Some economically significant annual weeds can have their life cycles broken by the insertion of a crop with a different sowing and maturity timing. Depleting the soil's weed seed bank is a key factor in crop rotation. Soil weed seed banks decrease to low levels where they can be more readily controlled when other crops are grown in place of berseem for 2 or more years. Crop rotation has been discovered to be a very efficient cultural method for destroying the bonds between troublesome weeds. Blue daisy populations can be decreased by rotating berseem fields with other crops like wheat, chickpea, and mustard.

7.2.9 Sowing Methods

In crops like wheat, more seed rate should be used. Under such conditions, crop plants compete more effectively with weeds and provide less space for weeds. Line sowing should be adopted because it is easy to control weeds in a line-sown crop, and the time taken for weeding is also reduced. Berseem seeds, on the other hand, are disseminated in 5–6 cm of standing water. This facilitates young seedlings' quick germination and simple establishment. A seed drill is useful because it allows for line sowing, which aids in optimal seed placement and uniform plant population. Additionally, it facilitates simple inter-cultivation and is advised for seed production.

7.2.10 Physical and Mechanical Methods

Crop weeding by hand dates back to the beginning of agriculture. It entails uprooting and removing weeds using various tools and instruments, including by hand. Even while manual weeding is successful, it takes much time and labor. Its viability is very low because labor is expensive and in short supply. Additionally challenging is mechanical weeding in crops seeded widely. However, the line planting of an Egyptian clover seed crop is an effective application of mechanical control. A field experiment discovered that the three most common weeds in Egyptian clover were *Cornopus didymus*, *Medicago denticulata*, and *Cichorium intybus*. The most efficient method of weed control was hand hoeing, which resulted in higher yields of green fodder (612 q. ha^{-1}) and seed (3.74 q. ha^{-1}) than in unweeded plots (266 and 1.57 q. ha^{-1}, respectively). Two-hand weeding should be done 3 and 5 weeks after seeding because of the initial slow development period. After 3 and 5 weeks following sowing, it was discovered that hand weeding reduced weed density.

Chemical-approach herbicides are useful weapons in humans' ongoing battle with weeds. Herbicides can safely and successfully achieve their goal when used as directed. Since chemical weed treatment is more effective, less expensive, and time-consuming, it is favored. Additionally, it avoids crop mechanical harm from manual weeding. Additionally, manual weeding was more effective in controlling weeds than the herbicides Pendimethalin/oxyfluorfen at 0.1 kg ha^{-1} as a pre-emergence treatment andImazethapyr at 0.1 kg ai ha^{-1} as a post-emergence treatment.

Manual weeding has traditionally been the main method of weed control. Herbicides must be chosen correctly based on the type of weed flora infesting the crop in order to control weeds effectively. They must also be given at the ideal dose and timing using the right application technique. Furthermore, the control is more effective since the weeds that usually escape due to morphological similarity to the crop are killed even within the rows during mechanical control.

Using reliable and efficient chemical herbicides has been a popular weed management technique worldwide in recent decades. A pre-emergence treatment of oxyfluorfen at 0.100 kg ai ha^{-1} and Imazethapyr at 0.15 kg ai ha^{-1} could selectively and inexpensively reduce the threat of broadleaved weeds in Egyptian clover (immediately after harvest of the first cut). Like pre-emergence applications, post-emergence (3 WAS) applications of butachlor at 1.5 kg ha^{-1} and Imazethapyr at 0.10 kg ai ha^{-1} were successful in controlling weeds. After the first cutting, an increased dose of pendimethalin (0.5 kg ai ha^{-1}) and Imazethapyr (0.1 kg ai ha^{-1}) slows the growth of Egyptian clover. Pendimethalin's phytotoxic effects, when applied as pre-emergence or 7 days after sowing, resulted in a 50% decrease in the population of Egyptian clover. Nevertheless, post-emergence spraying 14 days after sowing was secure and generated the highest yields of green fodder. In the Pune region of Maharashtra, pre-emergence applications of oxyfluorfen at 0.1 kg/ha and imazethapyr at 0.10 kg ha^{-1} after the first cut of Egyptian clover were found to be the most fruitful and lucrative. Imazethapyr applied pre-emergence at 0.1 kg ai ha^{-1}, decreased the number of weeds in Jhansi, including *Cichorium intybus*. Imazethapyr post-emergence spraying, however, was discovered to be successful in controlling weeds and increasing production after the first and second cuts.

7.2.10.1 Integrated Weed Management

Integrated weed management aims to keep the weed population below the point at which it causes crop economic harm by employing all weed control methods (cultural, physical, biological, chemical) sequentially.

REFERENCES

Adamson, D.C. 1996. Determining the economic impact of *Parthenium* on the Australian beef industry: A comparison of static and dynamic approaches. Report Submitted for AG870 Research Project, Department of agriculture, University of Adelaide, Australia.

Ahmad, Z., S.M. Khan, S. Ali, I.U. Rahman, H. Ara, I. Noreen and A. Khan. 2016. Indicator species analyses of weed communities of maize crop in district Mardan, Pakistan. Pakistan J. Weed Sci. Res. 22:227–238.

Chovancova, S., F. Illek and J. Winkler. 2020. The effect of three tillage treatments on weed infestation in maize monoculture. Pak. J. Bot. 52:697–701.

Galon, L., M.A.M. Bagnara, R.L. Gabiatti, F.W. Reichert Júnior, F.J.M. Basso, F. Nonemacher and C.T. Forte. 2018. Interference periods of weeds infesting maize crop. J. Agric. Sci. 10:197–205.

Gharde, Y., P.K. Singh, R.P. Dubey and P.K. Gupta. 2018. Assessment of yield and economic losses in agriculture due to weeds in India. Crop Prod. 7:12–18.

Horvath, D.P., S.A. Clay, C J. Swanton, J.V. Anderson and W.S. Chao. 2023. Weed-induced crop yield loss: A new paradigm and new challenges. Trends Plant Sci. 28(5):567–582.

Iqbal, S., S. Tahir, A. Dass, M.A. Bhat and Z. Rashid. 2020. Bio-efficacy of pre-emergent herbicides for weed control in maize: A review on weed dynamics evaluation. J. Exp. Agric. Int. 42:13–23.

Knezevic, S.Z., P. Pavlovic, O.A. Osipitan, E.R. Barnes, C. Beiermann, M.C. Oliveira and A. Jhala. 2019. Critical time for weed removal in glyphosate-resistant soybean as influenced by pre-emergence herbicides. Weed Technol. 33:393–399.

Kundu, R., M. Mondal, S. Garai, H. Banerjee, D. Ghosh, A. Majumder and R. Poddar. 2020. Efficacy of herbicides on weed control, rhizospheric microorganisms, soil properties and leaf qualities in tea plantation. Indian J. Weed Sci. 52:160–168.

Li, J., L. Chen, Q. Chen, Y. Miao, Z. Peng, B. Huang and H. Du. 2021. Allelopathic effect of *Artemisia argyi* on the germination and growth of various weeds. Scientific Reports. 11(1):1–15.

Little, N.G., A. DiTommaso, A.S. Westbrook, Q.M. Kettering and C.L. Mohler. 2021. Effects of fertility amendments on weed growth and weed–crop competition: A review. Weed Sci. 69:132–146.

Maitra, S., T. Shankar and P. Banerjee. 2020. Potential and advantages of maize-legume intercropping system. Maize-Prod Use. 1–14.

Merritt, L.H., J.C. Ferguson, A.E. Brown-Johnson, D.B. Reynolds, T.M. Tseng and J.W. Lowe. 2020. Reduced herbicide antagonism of grass weed control through spray application technique. Agron. 10:1131–1145.

Reddy, C. 2018. A study on crop weed competition in field crops. J. Pharmacogn. Phytochem. 7:3235–3240.

Safdar, M.E., A. Tanveer, A. Khaliq and M.A. Riaz. 2015. Yield losses in maize (*Zea mays*) infested with parthenium weed (*Parthenium hysterophorus* L.). Crop Prot. 70:77–82.

Saha, D., S.C. Marble and B.J. Pearson. 2018. Allelopathic effects of common landscape and nursery mulch materials on weed control. Front. Plant Sci. 9: https://doi.org/10.3389/fpls.2018.00733

Scavo, A., A. Restuccia, G. Pandino, A. Onofri and G. Mauromicale. 2018. Allelopathic effects of *Cynara cardunculus* L. leaf aqueous extracts on seed germination of some Mediterranean weed species. Ital. J. Agron. 13:119–125.

Scrivanti, L.R., A.M. Anton and Z.A. Julio. 2011. Allelochemical potential of south *American Bothriochloa* species (Poaceae: Andropogoneae). Allelopathy 28:189–200.

Shinde, M.A. and J.T. Salve. 2019. Allelopathic effects of weeds on *Triticum aestivum*. Int. J. Eng. Sci. 9:19873–19876.

Sturm, D.J., G. Peteinatos and R. Gerhards. 2018. Contribution of allelopathic effects to the overall weed suppression by different cover crops. Weed Res. 58:331–337.

Wasnik, V.K., A. Maity, D. Vijay, S.R. Kantwa, C.K. Gupta and V. Kumar. 2017. Efficiency of different herbicides on weed flora of barseem (*Trifolium alexandrinum* L.). Range Manag. Agrofor. 38:221–226.

Xie, L.J., R.S. Zeng, H.H. Bi, Y.Y. Song, R.L. Wang, Y.J. Su, M. Chen, S. Chen and Y.H. Liu. 2010. Allelochemical mediated invasion of exotic plants in China. Allelopathy J. 25(1):31–50.

8 Insect Pests of Fodder Grains of Summer Season

Rashad Rasool Khan, Muhammad Umair Sial, Muhammad Arshad, Aqsa Riaz, Umm E. Ummara, Ayesha Parveen, and Tehrim Liaqat

8.1 INTRODUCTION

Fodder is an agricultural product only used to feed domesticated animals like cattle, rabbits, sheep, and horses. "Fodder" refers to food given to animals, like plants cut and brought to them. This differs from food that animals find on their own (called forage). Fodder comprises hay, straw, silage, compressed and pelleted feeds, oils, mixed rations, sprouting grains, and legumes (such as bean sprouts, fresh malt, or spent malt). In contrast to food and commercial crops, the fodders group is farmed primarily for fresh green vegetative biomass (Eskandari *et al.*, 2009). Because of their substantial dry matter output and low price, fodders such as maize, pearl millet, guar, sorghum, and cowpea are significant in feeding ruminating animals. Because of their substantial protein content, legume forages, including cowpea, guar, and lima beans, can be utilized in livestock nutrition to help reduce the high feed costs of protein supplements (Asangla and Gohain, 2016).

8.1.2 Evergreen Fodder

It consists of mott grass, lucerne grass, and Sudan grass, and it just needs to be planted once to provide food for many years (Jha and Tiwari, 2018).

8.1.3 Winter Fodder Crops

This kind of fodder is planted at the beginning of winter so that it may offer nutrition throughout the winter and mature in the summer. Examples of winter food crops include berseem, oats, ryegrass, mustard, barley, and ryegrass (ul Haq and Ijaz, 2021).

8.1.4 Summer Fodder Crops

Forage refers to using plants in their fresh, succulent state to feed domesticated animals, while fodders are crops that are collected and processed as hay or silage for use during starvation or other times of shortage on different continents. In the summer, forages can yield much food from a relatively small plot of land. Forage millet, sorghum, *Pennisetum*, cowpea, lablab, and brassicas are all viable choices (Iqbal *et al.*, 2015).

8.1.4.1 Sorghum

The genus *Sorghum* contains roughly 25 species of flowering plants in the grass family. Some of these species are raised as cereals for human use, while others are used to provide pasture for livestock. One species is raised for grain, while others are utilized as fodder plants. These are

DOI: 10.1201/b23394-8

farmed during warm temperatures worldwide or naturally occur in pasture lands (Ogunlakin *et al.*, 2021).

8.1.4.2 Millet

Millets are a diverse genus of small-seeded grasses commonly cultivated as cereal crops or grains for human and animal sustenance worldwide. Most species commonly referred to as millets are members of the Paniceae tribe, but some are also members of other taxa. In the semiarid tropics of Asia and Africa, millets are significant crops. 97% of the world's millet is produced in developing nations (Mcdonough *et al.*, 2000). This crop is preferred because of its yield and short growing season in hot, dry circumstances. There are several regions where millets are native (Saxena *et al.*, 2018).

8.1.4.3 Maize

As a result of its widespread adoption as a primary source of nutrition, maize now accounts for a greater proportion of global grain production than either wheat or rice. Maize is utilized for human consumption, producing biofuels, livestock feed, and other maize-based goods such as cornstarch and syrup. Dent corn, flint corn, pod corn, popcorn, flour corn, and sweet corn are the six primary varieties of maize (Linda, 2013).

8.1.4.4 Guar

Guar gum comes from an annual legume called the cluster bean (*Cyamopsis tetragonoloba*), sometimes known as guar. Guar gum is used in a variety of applications. There are other names for this bean, including Gavar, Gawar, and Guvar bean. Since guar has never been discovered growing in its natural environment, its ancestry remains a mystery. Guar has several applications for human and animal nutrition, but the most significant is the gelling agent that can be extracted from its seeds, called guar gum (Mudgil *et al.*, 2014).

8.1.4.5 Cowpea

The cowpea is a member of the genus *Vigna*, an annual herb bean. Because it can thrive on sandy soil and receives only moderate rainfall, it has become a significant crop in the semiarid regions that span Africa and Asia. Because the nodules of roots can fix atmospheric nitrogen, it needs very little input, making it a desirable crop for farmers short on resources. The entire plant is harvested for use as fodder for livestock, and it is likely this practice that gave rise to the plant's common name (Priyanka *et al.*, 2022)

8.2 STORED GRAIN PESTS

8.2.1 Lesser Grain Borer (*Rhyzopertha dominica*)

8.2.1.1 Introduction

The monotypic genus *Rhyzopertha* belongs to the Bostrichidae family of false powder post beetles. The smaller grain borer is the common name for the single species, *Rhyzopertha dominica*. It is a beetle that is often found in store-bought goods and is a pest of dried cereal grains all around the world. *R. dominica* is typically elongated and cylindrical and ranges in color from reddish brown to dark brown (Edde, 2012).

8.2.1.2 Description

The lesser grain borer is a global pest affecting many types of stored grains. Its reddish-brown color can easily recognize it. *R. dominica* typically measures 2.1–3 mm in length. Almost 11 antennal segments with a three-segmented antennal club are attached to a reddish-brown body. Their body is usually slim and cylindrical. The shield conceals the head with a rounded neck and hood-like appearance. Male and female individuals of *R. dominica* have little to no externally discernible difference. *R. dominica* resembles other species in the family Bostrichidae morphologically, especially those in the subfamily Dinoderinae (Park *et al.*, 2015).

8.2.1.3 Distribution

Their exact origins are still unknown, although experts believe they are native to the Indian subcontinent because of other *Bostrichid* species in that area. *R. dominica* is now found all over the globe; however, it is commonly found in temperate regions between latitudes 40° North and South of the equator. It is primarily found in grain storage areas and woodland areas. Therefore, via grain transport, human contact has contributed to the spread of *R. dominica* (Edde, 2012).

8.2.1.4 Life Cycle

Within 3 weeks, the female lays approximately 300 to 500 eggs. They are often attached to the grain and are deposited individually or in clusters of 2 to 30. Within 7 to 18 days, depending on the temperature, the eggs hatch into white larvae having brown heads and rather short legs. These pierce the grains they feed, growing into fleshy structures with a recognizable "C" shape. At least five molts may occur before the pupal stage in the grain. Pupation lasts for roughly a week. The complete life cycle lasts between 24 and 133 days, depending on the temperature. The life cycle lasts 45 days at 26°C and 70% relative humidity (RH). Adults have a 10-month life span (Win and Rolania, 2020).

8.2.1.5 Damage

R. dominica feeds on a variety of stored grains. Their preferred food is grain products such as rice, wheat, sorghum, oats, pearl millet, and guar. It seems that *R. dominica* mainly relies on dry grains. Since lesser grain borers are the main pests of grain, they will target intact grain, making it vulnerable to subsequent pest attacks. Both adults and the larvae consume the grain, producing floury dust and maybe little more than empty husks. The larvae dig into the kernels and grow within the grain while the adult is active, and it may infest a considerable number of kernels (Buss and Foltz, 2009).

8.2.2 Rice Weevil (*Sitophilus oryzae*)

8.2.2.1 Introduction

Internal feeder *Sitophilus oryzae* L. (Coleoptera: Curculionidae), a widespread pest of economic significance, bores into grains that have been stored. The stored product pest known as the rice weevil feeds on the seeds of several crops, including wheat, rice, maize, and millet. The larvae prefer to feed on the germ layer rather than the endosperm, which results in a significant loss of protein and vitamins. Adult weevils mostly consume the endosperm, which lowers the carbohydrate level. A larger loss in germination will be caused by insects that target the germ than by others (Dal Bello *et al.*, 2000).

8.2.2.2 Description

The adults have a long snout, typically between 3 and 4.6 mm long. Their body looks dark or black, but closer inspection reveals four orange or red dots in a cross on the wing coverings. It may be confused with the maize weevil, which has a similar appearance. Both species are capable of hybridization (Hong *et al.*, 2018). A few characteristics can distinguish most adults, but the only ones consistently accurate are those found in genitalia.

8.2.2.3 Distribution

These worldwide-distributed insects have their origins in the Far East. These weevils are found in all warm, tropical regions of the earth, giving them an almost global range (CABI, 2015). They may occur in any place if the grain is left undisturbed for a while, and the physical circumstances favor development (Ramadan *et al.*, 2020; Plague *et al.*, 2010).

8.2.2.4 Life Cycle

Between 300 and 400 eggs are laid by female rice weevils, and the life cycle takes 32 days to complete. The pupa matures after 6 days, and the larva after 18 days. Depending on the temperature, the life cycle lasts 30 to 40 days in the summer and 123–148 days in the winter. Adults have a life span of 7–8 months. Adult rice weevils may fly and have a life span of up to 2 years (Koehler, 1999; Singh, 2017).

8.2.2.5 Damage

Rice weevils are found in all stored grains, such as wheat, maize, oats, sorghum, millet, cowpea, and others (Felicia *et al.*, 2012). They could infest the field grain as well. Grain kernels are often attacked by larvae, which hollow up the kernels. Mature and emerging adults make holes in the sides of grains and multiply in stored grains. The female uses strong mandibles for chewing a hole into the grain, then drops one egg inside the hole, and seals it with ovipositor secretions. As it grows within the grain and feeds, the larva hollows out the grain. After 2–4 days, it pupates within the grain kernel and emerges as an adult (Srivastava and Subramanian, 2016; Ahmed and Raza, 2010).

8.2.3 Rust-Red Flour Beetle (*Tribolium castaneum*)

8.2.3.1 Introduction

The red flour beetle (*Tribolium castaneum*) is a beetle species in the darkling beetle family Tenebrionidae. It is a global pest of stored items, especially food grains, and a model organism for ethological and food safety studies (Grunwald *et al.*, 2013).

8.2.3.2 Description

Adult beetles are 3–4 mm long, black or brown, and uniformly rust. The pronotum and head are occasionally darker than the rest of the body. The adult has a flattish curved-sided reddish-brown body. The head and upper thorax have minute holes, and the wing casings (elytra) are ridged throughout their length. The tips of the antennae are expanded, with the last three segments broader than the previous segments. The eyes are dark reddish black (Devi and Devi, 2015).

8.2.3.3 Distribution

The red flour beetle is of Indo-Australian origin. Although traditionally considered a sedentary insect, the genetic and ecological study has revealed that it may travel long distances through flight (Ridley *et al.*, 2011). It has a limited ability to survive outdoors than its closely related species. As a result, it has a greater southern range, yet both species are found globally in hot climates. Adults can live for more than 3 years.

8.2.3.4 Life Cycle

Females deposit 300–400 eggs throughout their 5–8-month adult life. The development duration for each stage is generally 3 days for eggs, 16 days for larvae, and 5 days for pupae under ideal circumstances of 35°C and 60–80% relative humidity. Both phases inflict severe harm to grains that have previously been damaged physically or by the activities of other pests. When disturbed, adults react fast and flee for protection. They can be found on the surface and deep beneath grain storage facilities. By flying, adults scatter over short distances (Pai *et al.*, 2005).

8.2.3.5 Damage

The red flour beetle causes loss and damage to stored grain such as maize, guar, barley, sorghum, millet, rice, and bean. According to the United Nations, *T. castaneum* and *T. confusum* are the two most prevalent secondary pests of all plant commodities in storage around the globe (Sallam, 1999).

8.2.4 Sawtoothed Grain Beetle (*Oryzaephilus surinamensis*)

8.2.4.1 Introduction

Oryzaephilus surinamensis, commonly called the sawtoothed grain beetle, is a type of beetle in the superfamily Cucujoidea. It is a common pest of grain and grain products worldwide, including chocolate, medicated drugs, and nicotine. Carl Linnaeus came up with the binomial name for the species, which means "rice-lover from Suriname." He did this after getting beetles from Surinam. The malt beetle is another name for it (Khairi *et al.*, 2010).

8.2.4.2 Description

O. surinamensis is a 2.4–3 mm long, dark brown beetle with "teeth" running along the prothorax's side. It looks almost exactly like the merchant grain beetle, called *O. ercator*. However, *O. surinamensis* has smaller eyes and a wider, more triangular head, which makes it different from other species (Mortazavi *et al.*, 2010; Abbas *et al.*, 2011).

8.2.4.3 Distribution

Sawtoothed grain beetle is a common pest of stored products; it also very familiar in the food industry and can be found in food production, storage, retail, and home pantries. *O. surinamensis* is found everywhere in the world. *O. surinamensis* is less common in colder places like Canada and the northern United States (Aldawood *et al.*, 2013).

8.2.4.4 Life Cycle

A female can lay between 43 and 285 eggs on a food mass. The best temperature range for larvae to grow in eggs is between 27°C and 29°C. In this range, they hatch within 3–5 days (Hashem *et al.*, 2019). The larvae are yellowish white and have brown heads. They can grow to be 3 mm long. They move freely around the food mass and eat broken pieces of grain or grain kernels that other insects have damaged. Bigger larvae may bore into kernels. Most of the damage done to grain comes from larvae. Before becoming adults, larvae molt two to four times (Marouf *et al.*, 2013). The larvae change into adults by making cocoons out of broken parts of the grain. After about a week, adults start to show up (Marouf *et al.*, 2013). Adults can live between 6 and 10 months on average, but they can live up to 3 years. At 85–95°F (29–35°C), the life cycle takes 27–51 days (Al-Dosary, 2009).

8.2.4.5 Damage

One of the most common insects of grains, pet foods, and seeds. When insects feed on a product, the dry mass of the product shrinks, and the amount of water in the product goes up because of the insect's metabolism. This can lead to mold growth. Insect damage to grain lowers its value and can make it unusable. The buyer may reject the grain if there are enough insect parts or live insects (EI-Orabi *et al.*, 2007; Radek *et al.*, 2017).

8.2.5 Rusty Grain Beetle (*Cryptolestes ferrugineus*)

8.2.5.1 Introduction

The rusty grain beetle, *Cryptolestes ferrugineus*, and its relatives, *Cryptolestes* spp., eat a wide range of cereal grains and other foods. Mostly it hurts grain, especially wheat. It also eats beans, oats, rye, wheat, rice, bran, corn, citrus pulp, and sunflower seeds. In grains, the embryo is often attacked. The larva gets in under the pericarp, eats the embryo, and then finishes growing. This beetle is a major secondary pest of cereal grains. It often appears after other insects have already caused damage (Golam *et al.*, 2011).

8.2.5.2 Description

Adults are just around 2 mm long, have a flattened reddish-brown color, and have long bead-like antennae. Adults have a ridge (sublateral carina) between their eyes and on their prothorax. The legs of a larva are highly developed and long, and flat. A darkish head capsule is present in the larvae; otherwise, it is a pure white-to-yellow color (Baltaci *et al.*, 2008).

8.2.5.3 Distribution

C. ferrugineus is found worldwide, from the tropics to the subarctic. It can withstand the winter in temperate areas, unlike some other *Cryptolestes* species. Flour mill beetles are a common Canadian pest; however, this species is commonly misidentified.

8.2.5.4 Life Cycle

Growth and survival are influenced by environmental factors, including temperature and humidity, as well as the availability and quality of food. In the cracks or randomly scattered amid the food, the females deposit 100–400 eggs. Eggs hatch within 4–5 days. Their life cycle is completed within 3 weeks. Adults have a life span of 6–9 months and are more cold-tolerant than other stored pests (Collins *et al.*, 2002; Abdelghany and Fields, 2017).

8.2.5.5 Damage

Damage is usually limited to grain stored in a place with much moisture. Both larvae and adults feed on the germ and endosperm of seeds. Grains are completely damaged when they are present in large numbers (Jagadeesan *et al.*, 2012).

8.2.6 Khapra Beetle (*Trogoderma granarium*)

8.2.6.1 Introduction

Originating in South Asia, the khapra beetle (*Trogoderma granarium*), often known as the cabinet beetle, is one of the most devastating pests of grain products and seeds worldwide. It is one of the 100 worst invasive species in the world. Its resistance to many pesticides, its preference for dry circumstances and low-moisture food, and the length of time it can go without eating all contribute to the difficulty of controlling infestations (Day and White, 2016).

8.2.6.2 Description

The full-grown beetles' size ranges between 1.6 to 3 mm with a brownish-reddish color. Males are black or dark brown, whereas females are paler overall and significantly bigger (Ghimire *et al.*, 2017). The larvae may grow up to 5 mm long and have a thick coat of reddish-brown hair. The larval stage typically lasts between 4 and 6 weeks but may last as long as 7 years in certain cases.

8.2.6.3 Distribution

The khapra beetle is a synanthropic, meaning it lives mostly close to people. The beetle thrives in warm, dry places like pantries, malthouses, grain and fodder processing industries, and warehouses housing discarded grain sacks or boxes. Native distribution of this species includes India, Burma, and Western Africa. Little is known about the beetle's behavior in natural settings (Kavallieratos, Athanassiou and Boukouvala, 2017).

8.2.6.4 Life Cycle

Adult khapra beetles usually live for between 5 and 10 days. The khapra beetle lays eggs ranging from 0.7 mm to about 0.25 mm broad and 0.02 mg in weight. One end is more rounded, and the other end is more pointed. The pointy end has several protrusions that look like spines. At first, the eggs are milky white, but they turn a pale yellow after a few hours.

The food the khapra beetle eats greatly affects how it works. Borzoi and his colleagues found that rye is the best place for individuals to breed and grow (Borzoui *et al.*, 2015). By comparison, the walnut and rice diets made the females less fertile and the adults smaller. They also made the larval stage last longer (Kavallieratos, Athanassiou, Diamantis *et al.*, 2017).

8.2.6.5 Damage

Khapra beetles can eat almost any dried animal or plant matter. It can eat things with as little as 2% moisture and grow on dead mice, dried blood, and dried insects, among other things. It prefers wheat, barley, oats, rye, maize, rice, flour, malt, and noodles as its grains and cereal products (Eliopoulos, 2013; Emery *et al.*, 2010). In addition to these foods, it consumes maize, dried orange pulp, ground rice, and various dry fruits (Myers and Hagstrum, 2012).

8.2.7 Angoumois Grain Moth (*Sitotroga cerealella*)

8.2.7.1 Introduction

The Angoumois grain moth (*Sitotroga cerealella*) species belong to the Gelechiidae family, usually considered "rice grain moth." Mostly it is associated as a pest of stored grains and cereals because they make dens within the grains of kernel crops, making them useless for consumption by a human. *S. cerealella* lays eggs between the spaces within grains, which hatch later during transport and storage. Hence, this moth can be transferred to a country or household where it is not present. Therefore, continuous protection is needed against Angoumois grain moth until consumption of grains. The source of origin is still unknown (Shah *et al.*, 2014).

8.2.7.2 Description

Whitish oval eggs are laid initially, but eggs turn in a red shade after some time. A mature egg is about 2 mm long and hatches after a maximum of one month if a favorable temperature is available. The wingspan of an adult Angoumois grain moth is 10–15 mm, while the length of the body is approximately 5–10 mm (Demissie *et al.*, 2014). The body color is yellowish brown or gray brownish. They have dark dots on the edges of the forewings. Unlike other moths, their hind wings are curved distinctively. Yellow-golden-colored forewings and gray-colored hind wings give them brown color. Hairs present on the edge of wings also give them a unique characteristic. Males have pointed and thin abdomens, while females exhibit long, bulky abdomens with no color (Bukero *et al.*, 2015).

8.2.7.3 Distribution

Mostly prefer to live in warmer climatic conditions. However, smaller populations are now reported in colder areas like Russia. It has also been reported in the UK from imported food products but has not settled there. The optimum temperature for survival is to about 86°F, and 75% moisture content is required for hatching. These are mostly distributed in developmental areas of agriculture (Muthukumar *et al.*, 2016).

8.2.7.4 Life Cycle

The Angoumois grain moth's life cycle starts with an egg, which develops into burrowing larvae that eat cereal grains or seeds. The female moths typically lay 40 eggs. Following pupation, the larva emerges as an adult moth after 10 days or as few as 5 days of being protected in a silk cocoon inside a grain. Under ideal circumstances, the life cycle takes 35–40 days to become an adult moth (Nadeem *et al.*, 2010). The Angoumois grain moth relies on environmental conditions, particularly humidity, and temperature, for egg hatching, fertility, and survival rate (Hamed and Nadeem, 2012).

8.2.7.5 Damage

Because it is ranked as the third-most prevalent pest globally, *S. cerealella* is a destructive moth pest that can cause significant yield loss in many different parts of the world (Keneni *et al.*, 2011). This is one of the worst and most damaging pests of stored grains, especially rice, wheat, sorghum, and maize. According to certain research, *S. cerealella* produced cereal weight loss of roughly 13% and a 76% increase in the percentage of damaged grains during 8 months of preservation (Giga *et al.*, 1991). This translates to a waste of 183.3 million tonnes annually due to *S. cerealella* worldwide.

8.2.8 Rice Moth (*Corcyra cephalonica*)

8.2.8.1 Introduction

In tropical regions, *Corcyra cephalonica* pest poses a major threat to numerous essential stored food products, including wheat, maize, sorghum, cottonseed, coffee, cocoa beans, rice, and groundnuts. *C. cephalonica* belongs to the family Pyralidae. A little moth can develop into a big problem. Its caterpillars consume cereals and other dry plant materials like seeds (Bhandari *et al.*, 2014).

8.2.8.2 Description

The rice moth's color is buff-brown or pale throughout the body without any marking or patch; dark veins may appear on wings, and the hind wings are mostly glassy. The wingspan is approximately 15–25 mm, with a body length of 15 mm. There is a clump of scales on the head. The larvae have a whitish or yellow-colored body with a dark brown–colored head. The body is covered by long hair (Jagadish *et al.*, 2009).

8.2.8.3 Distribution

This pest prefers a warm climate yet must survive in warm stores in temperate areas. It is found in Europe, Africa, and North America. It is a major pest of stored grains in Pakistan and India, particularly in the larval stage. It has spread widely, even in Northern Europe, via imported food (Mitcham *et al.*, 2006).

8.2.8.4 Life Cycle

The life span of a larva is approximately 26–31 days, and the life span of a pupa is about 9–16 days. The ages of adults that hatched were 4–6 days for males and 8–13 days for females. During the ovipositor stage, which lasts 6–8 days, each female lays 90 to 200 eggs. The deposited eggs hatch at 26–28°C after 4–7 days. Following the appearance of adults, courtship and spawning take place right away. It takes a gestation period from 33 to 52 days to complete. Six generations can pass annually (Nathan *et al.*, 2006).

8.2.8.5 Damage

This moth can cause a devaluation of the market value of grains. By clumping grains together and creating a web, caterpillars cause damage and feed inside it. Most damage occurs at the larval stage because, at this stage, they are more active, motile, and voracious feeders. The entire supply turns into a webby heap with a nasty odor due to high contamination, making it useless for human utilization. Grains, dry fruits, and oilseeds are also affected and damaged, resulting in significant harm. It causes about 5–10% loss in stored cereals and grains in India. Approximately 5 million tons of food grains are lost yearly in Indian stores due to pest damage (Dwivedi and Garg, 2003).

8.2.9 Almond Moth (*Ephestia cautella*)

8.2.9.1 Introduction

An almond moth or tropical storehouse moth is a little pest that attacks stored goods. In addition to dried fruits, flour, oats, bran, and other grains are infested by almond moths. It is a member of the Pyralidae family of snout moths. The closely related Indian meal moth and the Mediterranean flour moth, frequent pantry pests within the same subfamily, may be mingled with this species. Like the raisin moth, the almond moth possesses practically a global distribution as a result of the accidental transportation of food items in its larval state (Arbogast, 2002).

8.2.9.2 Description

Most adult almond moths are light brown, with smaller, generally gray hind wings. Its wingspan varies from 14–22 mm when expanded. Short fringe lines are present at the back side of the wings corner. Larvae are mostly dark-headed with gray color. The 12–15 mm long caterpillar can be recognized by the pattern of spots running around its back (Faruki, 2004).

8.2.9.3 Distribution

Almond moths are found all over the world. Although it does optimally in tropical climes, because of its propensity to invade dry objects that are moved worldwide, it has been transported to many places of the world. For instance, it has recently been shipped with cargoes of coconuts across Polynesia. Since the almond moth is mainly known as a pest, its habitation is frequently in some dry foods kept in a storehouse or other industrial setting. They have been discovered in various foods, notably dried fruits, beans, nuts, and other cereals (Phillips and Throne, 2010).

8.2.9.4 Life Cycle

Eggs hatch in around 3 1/2 days under ideal circumstances, and larvae proceed through 5 instars within 17–37 days. The pupa stage remains roughly 7 days in ideal humidity and temperature (El-Naggar and Mikhaiel, 2011). Adult females typically live for 10 days, whereas adult males live for 6–7 days.

8.2.9.5 Damage

When they hatch, several dried food items are utilized as the principal food source for larval almond moths. Even though moths damage various foods, goods made from wheat and maize provide the fastest growth for larvae. Furthermore, since larvae cannot pierce the hulls or shells of whole seeds or grains, broken or crushed seeds and cereal goods are more suitable for larvae than whole seeds or grains (Ryne *et al.*, 2006).

8.3 MANAGEMENT OF STORED GRAIN PESTS

Cultural, biological, and chemical methods manage stored grain pests.

8.3.1 Cultural

Good store hygiene can help to reduce the severity of pest infestation by cleaning the store between harvests, burning, and removing infested residues. Dipping the grain sacks into hot water and removing wood from stores. Moreover, fumigating the store to discard the residual infestations (Dent and Binks, 2020). The temperature inside storage spaces must be closely monitored since it affects the insect population. Insect infestations may be located and measured using trapping techniques. It may be destroyed at any stage of growth by freezing infected food at or below −18°C (0°F) for 3 days or by heating it to 60°C (140°F) for 15 minutes (Buatone, 2010; Lorenz and Hardke, 2013; Patel and Vekaria, 2013).

8.3.2 Biological

Natural enemies like predators, parasites, and fungi can control the pests in the store. Due to their small populations relative to insect pest fecundity, arthropod species that prey on it are inadequate as biological control agents. Additionally, it is extremely detrimental for natural predators and parasitoids to become victims to other species. Its deep burrowing ability makes successful proliferation possible, enabling them to evade predators and danger effectively (Ahmed, 1996; edde, 2012). *Anisopteromalus calandrae*, a parasitoid, and *Lariophagus distinguendus* are important in the management of stored pests (Lucas and Riudavets, 2002). The entomopathogenic fungi *Isaria fumosorosea*, *Beauveria bassiana*, *Metarhizium anisopliae*, *Simplicillium lamellicola*, and *Lecanicillium muscarium* are used to eradicate the insect pests (Storm *et al.*, 2016; Ak, 2019).

8.3.3 Chemical

Most insecticides fail to control the pest population because the pest has become resistant (Collins *et al.*, 2017). Organophosphorus pesticides are among the protective agents. Fumigation is a recommended management method when infestations become serious (Shi *et al.*, 2012). Some treat grains before storage using dust and sprays of synergized pyrethrins labeled to provide further protection, following the warnings and safety instructions. If the issue becomes serious and pervasive, contact a certified pest control specialist with the skills, tools, and pesticides to handle the situation safely (Buatone, 2010).

Pesticides are poisons; thus, adhering to all safety recommendations listed on labels is critical. Stored grain pests have been controlled with pyrethroids, organophosphates, and fumigants. The most lasting pesticide employed against these pests in Pakistan was tetrachlorvinphos, followed by diazinon, malathion, and azinphos-methyl. In most countries, these pests have evolved resistance to several pesticides. Currently, fumigation with phosphine is a primary method of controlling this species worldwide (Isman, 2006; Jagadeesan *et al.*, 2015).

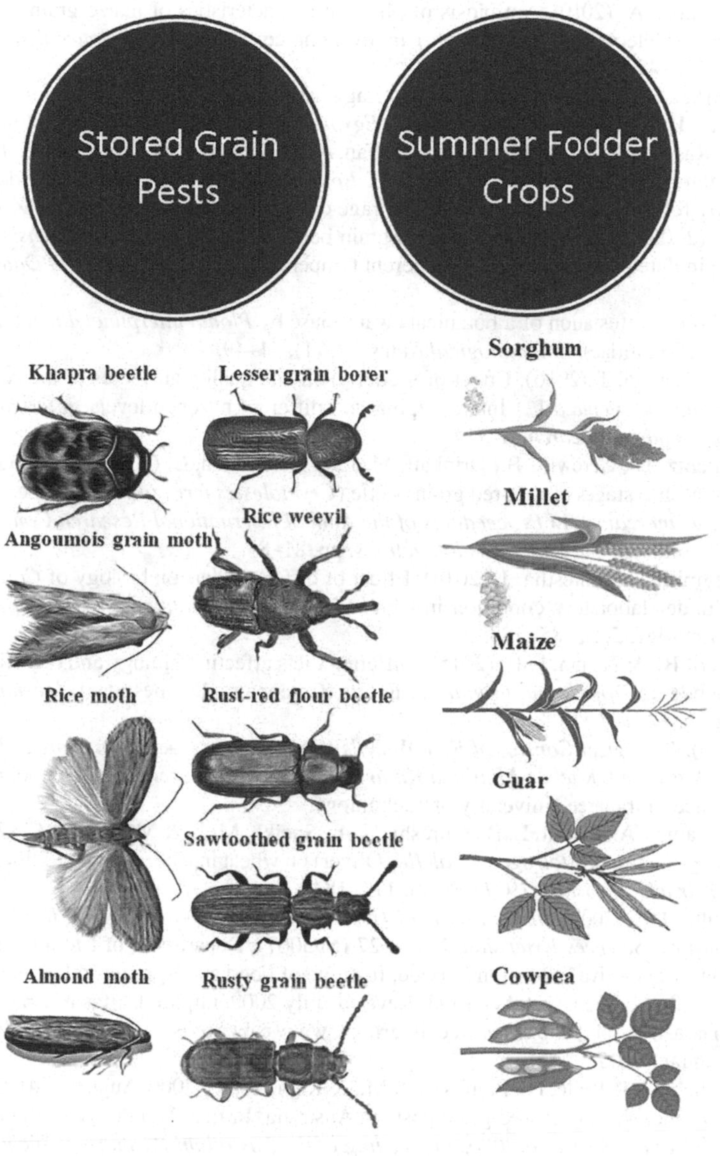

FIGURE 8.1 Main summer fodder crops and stored grain pests.

REFERENCES

Abbas, H., Nouraddin, S., Reza, Z. H., Iraj, B., Mohammad, B., Hasan, Z., Hossein, A. M., & Hadi, F., 2011. Effect of gamma radiation on different stages of Indian meal moth *Plodia interpunctella* Hübner (Lepidoptera: Pyralidae). *African Journal of Biotechnology*, *10*, 4259–4264.

Abdelghany, A. Y., & Fields, P. G. (2017). Mortality and movement of *Cryptolestes ferrugineus* and *Rhyzopertha dominica* in response to cooling in 300-kg grain bulks. *Journal of Stored Products Research*, *71*, 119–124.

Ahmed, K. S. (1996). Studies on the ectoparasitoid, *Anisopteromalus calandrae* How. (Hymenoptera: Pteromalidae) as a biocontrol agent against the lesser grain borer, *Rhyzopertha dominica* (Fab.) in Saudi Arabia. *Journal of Stored Products Research*, *32*(2), 137–140.

Ahmed, K. S., & Raza, A. (2010). Antibiosis of physical characteristics of maize grains to *Sitotroga cerealella* (Oliv.) (Gelechiidae: Lepidoptera) in free choice test. *Pakistan Journal of Life and Social Sciences*, *8*(2), 142–147.

Ak, K. (2019). Efficacy of entomopathogenic fungi against the stored-grain pests, *Sitophilus granarius* L. and *S. oryzae* L. (Coleoptera: Curculionidae). *Egyptian Journal of Biological Pest Control*, *29*(1), 1–7.

Aldawood, A. S., Rasool, K. G., Alrukban, A. H., Soffan, A., Husain, M., Sutanto, K. D., & Tufail, M. (2013). Effects of temperature on the development of *Ephestia cautella* (Walker) (Pyralidae: Lepidoptera): A case study for its possible control under storage conditions. *Pakistan Journal of Zoology*, *45*(6).

Al-Dosary, N. H. (2009). Role of the saw-toothed grain beetle (*Oryzaephilus surinamensis* L.) (Coleoptera: Silvanidae) in date palm fruit decay at different temperatures. *Basrah Journal of Date Palm Research*, *8*(2), 1–14.

Arbogast, R. T. (2002). Infestation of a botanicals warehouse by *Plodia interpunctella* and *Ephestia elutella* (Lepidoptera: Pyralidae). *Entomological News*, *113*(1), 41–49.

Asangla, K. H., & Gohain, T. (2016). Effect of fodder yield and quality attributes of maize (*Zea mays* L.) + cowpea (*Vigna unguiculata* L.) Intercropping and different nitrogen levels. *International Journal of Agricultural Science Research*, 6, 349–356.

Baltaci, D., Klementz, D., Gerowitt, B., Drinkall, M. J., & Reichmuth, C. (2008, November). Sulfuryl fluoride against all life stages of rust-red grain beetle (*Cryptolestes ferrugineus*) and merchant grain beetle (*Oryzaephilus mercator*). In *Proceedings of the Annual International Research Conference on Methyl Bromide Alternatives and Emissions Reductions*, pp. 81–83.

Bhandari, G., Regmi, R., & Shrestha, J. (2014). Effect of different diets on biology of *Corcyra cephalonica* (Stainton) under laboratory condition in Chitwan, Nepal. *International Journal of Applied Sciences and Biotechnology*, *2*(4), 585–588.

Borzoui, E., Naseri, B., & Namin, F. R. (2015). Different diets affecting biology and digestive physiology of the Khapra beetle, *Trogoderma granarium* Everts (Coleoptera: Dermestidae). *Journal of Stored Products Research*, *62*, 1–7.

Buatone, S. (2010). *Biological Control of Rice Weevils* (Sitophilus oryzae L.) *In Stored Milled Rice by the Extracts of Mintweed, Kitchen Mint and Kaffir Lime* (Doctoral dissertation), School of Biology. Institute of Science. Suranaree University of Technology.

Bukero, A., Nizamani, I. A., Rajput, L. B., Qureshi, N. A., Sheikh, M. J., & Murtaza, J. G. (2015). Biology of Angoumois grain moth, *Sitotroga cerealella* (Oliver) on wheat in laboratory conditions. *Entomological Society of Karachi, Pakistan (1971)*, *30*(2), 179–184.

Buss, E. A., & Foltz, J. L. (2009). *Insect Borers of Trees and Shrubs1. University of Florida Institute of Food and Agricultural Sciences Extension*. ENY-327 (MG007), Department of Entomology and Nematology, Florida Cooperative Extension Service, Institute of Food and Agricultural Sciences, University of Florida. Date first printed: October 1993. Revised: July 2009. http://edis.ifas.ufl.edu.

CABI. (2015). Datasheet of *Sitophilus* weevil group. www.cabi.org/isc/datasheet/10850; 10887; 10926; accessed January 2017.

Collins, P. J., Daglish, G. J., Pavic, H., Lambkin, T. M., & Kapittke, R. (2000, August 1–4). Combating strong resistance to phosphine in stored grain pests in Australia. In E. J. Wright, H. J. Banks, & E. Highley, eds. *Stored Grain in Australia 2000, Proceeding of the Australian Postharvest Technical Conference, Adelaide*, pp. 109–112. Canberra, Australia: CSIRIO Stored Grain Research Laboratory, 2002.

Collins, P. J., Falk, M. G., Nayak, M. K., Emery, R. N., & Holloway, J. C. (2017). Monitoring resistance to phosphine in the lesser grain borer, *Rhyzopertha dominica*, in Australia: A national analysis of trends, storage types and geography in relation to resistance detections. *Journal of Stored Products Research*, *70*, 25–36.

Dal Bello, G., Padin, S., Lastra, C. L., & Fabrizio, M. (2000). Laboratory evaluation of chemical-biological control of the rice weevil (*Sitophilus oryzae* L.) In stored grains. *Journal of Stored Products Research*, *37*(1), 77–84.

Day, C., & White, B. (2016). Khapra beetle, *Trogoderma granarium* interceptions and eradications in Australia and around the world (No. 1784–2016–141903). *SARE Working Paper 1609*, School of Agricultural and Resource Economics, University of Western Australia, Crawley, Australia.

Demissie, G., Rajamani, S., & Ameta, O. P. (2014). Effect of temperature and relative humidity on development and survival of Angoumois grain moth, *Sitotroga cerealella* (Olivier) (Lepidoptera: Gelechiidae) on stored maize. *International Journal of Sciences: Basic and Applied Research (IJSBAR)*, *15*(2), 9–21.

Dent, D., & Binks, R. H. (2020). *Insect Pest Management*. 3rd ed. CABI, Nosworthy Way Wallingford Oxfordshire OX10 8DE UK.

Devi, M. B., & Devi, N. V. (2015). Biology of rust-red flour beetle, *Tribolium castaneum* (Herbst) (Coleoptera: Tenebrionidae). In *Biological Forum*, Vol. 7, No. 1, pp. 12–15. New Delhi: Satya Prakashan.

Dwivedi, S. C., & Garg, S. (2003). Toxicology laboratory of flower extract of *Lantana camara* on the life cycle of Corcyra cephalonica. *Indian Journal of Entomology (India)*, *65*(3), 330–334.

Edde, P. A. (2012). A review of the biology and control of *Rhyzopertha dominica* (F.) The lesser grain borer. *Journal of Stored Products Research*, *48*, 1–18.

EI-Orabi, M. N., Sawires, S. G., Antonious, A. G., & Salama, S. I. (2007). Gamma irradiation effects on some biological aspects of *Ephestia kuehniella* (Zell.), inherited sterility and mating competitiveness. *Arab Journal of Nuclear Sciences and Applications*, *40*(2), 241–250.

Eliopoulos, P. A. (2013). New approaches for tackling the khapra beetle. *CAB Reviews*, *8*, 012.

El-Naggar, S. M., & Mikhaiel, A. A. (2011). Disinfestation of stored wheat grain and flour using gamma rays and microwave heating. *Journal of Stored Products Research*, *47*(3), 191–196.

Emery, R. E., Chami, M., Garel, N., Kostas, E., & Hardie, D. C. (2010). The use of hand-held computers (pdas) to audit and validate eradication of a post-border detection of khapra beetle, *Trogoderma Granarium*, in Western Australia. In *Proceedings of the 10th International Working Conference on Stored Product Protection*, pp. 1031–1037.

Eskandari, H., Ghanbari, A., & Javanmard, A. (2009). Intercropping of cereals and legumes for forage production. *Notulae Scientia Biologicae*, *1*, 7–13.

Faruki, S. I. (2004). Effect of insecticide on the irradiated tropical warehouse moth, *Cadra cautella* (Walker) (Lepidoptera: Phycitidae). *Journal of Biological Sciences (Faisalabad)*, *4*(5), 681–686.

Felicia, J., Gervais, N. K., Kouassi, S., & Kouahou, F.-B. (2012). Overview of storage problems and impact of insects on stored rice and maize in rural area: Case of Bouafle region, *Ivory Coast. European Journal of Scientific Research*, *83*(3), 349–363.

Ghimire, M. N., Myers, S. W., Arthur, F. H., & Phillips, T. W. (2017). Susceptibility of *Trogoderma granarium* Everts and *Trogoderma inclusum* LeConte (Coleoptera: Dermestidae) to residual contact insecticides. *Journal of Stored Products Research*, *72*, 75–82.

Giga, D. P., Mutemerewa, S., Moyo, G., & Neeley, D. (1991). Assessment and control of losses caused by insect pests in small farmers' stores in Zimbabwe. *Crop Protection*, *10*(4), 287–292.

Golam, R. S., Ali, A. P., & Mohammad, H. S. (2011). Effects of nitrogen and phosphine mixtures on stored-product insects mortality. *African Journal of Biotechnology*, *10*(32), 6133–6144.

Grunwald, S., Adam, I. V., Gurmai, A. M., Bauer, L., Boll, M., & Wenzel, U. (2013). The red flour beetle *Tribolium castaneum* as a model to monitor food safety and functionality. *Advances in Biochemical Engineering/Biotechnology*, *135*, 111–122.

Hamed, M., & Nadeem, S. (2012). Effect of cereals on the development of *Sitotroga cerealella* (Olivier) (Lepidoptera: Gelechiidae) and subsequent quality of the egg parasitoid, *Trichogramma chilonis* (Ishii) (Hymenoptera: Trichogrammatidae). *Pakistan Journal of Zoology*, *44*(4), 923–929.

Hashem, M. Y., Ahmed, A. A. I., Ahmed, S. S., Mahmoud, Y. A., & Khalil, Sh S. H. (2019). Impact of modified atmospheres on respiration of last instar larvae of the rice moth, *Corcyra cephalonica* (Lepidoptera: Pyralidae). *Archives of Phytopathology and Plant Protection*, *51*(3), 1–16.

Hong, K. J., Lee, W., Park, Y. J., & Yang, J. O. (2018). First confirmation of the distribution of rice weevil, *Sitophilus oryzae*, in South Korea. *Journal of Asia-Pacific Biodiversity*, *11*(1), 69–75.

Iqbal, M. A., Bilal, A., Shah, M. H. U., & Kashif, A. (2015). A study on forage sorghum (*Sorghum bicolor* L.) Production in perspectives of white revolution in Punjab, Pakistan: Issues and future options. *Agriculture and Environment Science*, 15(4), 640–647.

Isman, M. B. (2006). Botanical insecticides, deterrents, and repellents in modern agriculture and an increasingly regulated world. *Annual Review of Entomology*, *51*(1), 45–66.

Jagadeesan, R., Collins, P. J., Daglish, G. J., Ebert, P. R., & Schlipalius, D. I. (2012). Phosphine resistance in the rust red flour beetle, *Tribolium castaneum* (Coleoptera: Tenebrionidae): Inheritance, gene interactions and fitness costs. *PLOS One*, *7*(2), e31582.

Jagadeesan, R., Nayak, M. K., Pavic, H., Chandra, K., & Collins, P. J. (2015). Susceptibility to sulfuryl fluoride and lack of cross-resistance to phosphine in developmental stages of the red flour beetle, *Tribolium castaneum* (Coleoptera: Tenebrionidae). *Pest Management Science*, *71*(10), 1379–1386.

Jagadish, P. S., Nirmala, P., Rashmi, M. A., Hedge, J. N., & Neelu, N. (2009). Biology of rice moth, *Corcyra cephalonica* Stainton on foxtail millet. *Karnataka Journal of Agricultural Sciences*, *22*(3), 674–675.

Jha, S. K., & Tiwari, N. I. T. I. S. H. (2018). Evaluation of intensive fodder cropping systems for round the year green fodder production in Chhattisgarh. *Forage Research*, *44*(2), 115–118.

Kavallieratos, N. G., Athanassiou, C. G., & Boukouvala, M. C. (2017). Invader competition with local competitors: Displacement or coexistence among the invasive khapra beetle, *Trogoderma granarium* everts (Coleoptera: Dermestidae), and two other major stored-grain beetles? *Frontiers in Plant Science*, *8*, 1837.

Kavallieratos, N. G., Athanassiou, C. G., Diamantis, G. C., Gioukari, H. G., & Boukouvala, M. C. (2017). Evaluation of six insecticides against adults and larvae of *Trogoderma granarium* Everts (Coleoptera: Dermestidae) on wheat, barley, maize and rough rice. *Journal of Stored Products Research*, *71*, 81–92.

Keneni, G., Bekele, E., Getu, E., Imtiaz, M., Damte, T., Mulatu, B., & Dagne, K. (2011). Breeding food legumes for resistance to storage insect pests: Potential and limitations. *Sustainability*, *3*(9), 1399–1415.

Khairi, M. M. A., Elhassan, M. I., & Bashab, F. A. (2010, March). The status of date palm cultivation and date production in Sudan. In *IV International Date Palm Conference 882* (pp. 37–42).

Koehler, P. G. (1999). *Rice Weevil, Sitophilus Oryzae (Coleoptera curculionidae)*. ENY261, The Entomology and Nematology Department, UF/IFAS Extension. Original publication date March 1994. Revised January 2012 and February 2022. https://edis.ifas.ufl.edu

Linda Campbell Franklin. (2013). Corn. In A. F. Smith, ed. *The Oxford Encyclopedia of Food and Drink in America*, 2nd ed., pp. 551–558, p. 553. Oxford: Oxford University Press.

Lorenz, G., & Hardke, J. (2013). Insect management in rice. *Arkansas Rice Production Handbook*, 139–162.

Lucas, E., & Riudavets, J. (2002). Biological and mechanical control of *Sitophilus oryzae* (Coleoptera: Curculionidae) in rice. *Journal of Stored Products Research*, *38*(3), 293–304.

Marouf, A., Amir-Maafi, M., & Shayesteh, N. (2013). Two-sex life table analysis of population characteristics of almond moth, *Cadra cautella* (Lepidoptera: Pyralidae) on dry and semi-dry date palm varieties. *Journal of Crop Protection*, 2(2), 171–181.

Mcdonough, C. M., Rooney, L. W., & Serna-Saldivar, S. O. (2000). *The Millets: Food Science and Technology*. New York: Handbook of Cereal Science and Technology.

Mitcham, E., Martin, T., & Zhou, S. (2006). The mode of action of insecticidal controlled atmospheres. *Bulletin of Entomological Research*, *96*(3), 213–222.

Mortazavi, S. M. H., Arzani, K., & Arujalian, A. A. (2010, March). Modified atmosphere packaging of date fruit (*Phoenix dactylifera* L.) Cultivar Barhee in Khalal Stage. In *IV International Date Palm Conference 882* (pp. 1063–1069).

Mudgil, D., Barak, S., & Khatkar, B. S. (2014). Guar gum: Processing, properties and food applications—a review. *Journal of Food Science and Technology*, *51*(3), 409–418.

Muthukumar, M., Ragumoorthi, K. N., Balasubramani, V., & Vijayakumar, A. (2016). Impact of different Maize cultivars on pre harvest infestation by *Sitotroga cerealella* (Oliver) (Lepidoptera-Gelechiidae). *Life Sciences Leaflets*, *71*, 20–28.

Myers, S. W., & Hagstrum, D. H. (2012). Quarantine. In D. H. Hagstrum, T. W. Philips, & G. Cuperus, eds. *Stored Product Protection*, pp. 297–304. Manhattan, KS: Kansas State University.

Nadeem, S., Ashfaq, M., Hamed, M., & Ahmed, S. (2010). Optimization of short and long term storage duration for *Trichogramma chilonis* (Ishii) (Hymenoptera: Trichogrammatidae) at low temperatures. *Pakistan Journal of Zoology*, *42*(1).

Nathan, S. S., Kalaivani, K., Mankin, R. W., & Murugan, K. (2006). Effects of millet, wheat, rice, and sorghum diets on development of *Corcyra cephalonica* (Stainton) (Lepidoptera: Galleriidae) and its suitability as a host for *Trichogramma chilonis* Ishii (Hymenoptera: Trichogrammatidae). *Environmental Entomology*, *35*(3), 784–788.

Ogunlakin, G. O., Onibokun, E. H., & Fashogbon, B. (2021). Effect of germination period on some functional and engineering properties of sorghum flour. *Asian Journal of Advances in Agricultural Research*, *17*(3), 12–22. https://doi.org/10.9734/ajaar/2021/v17i330196

Pai, A., Bennett, L., & Yan, G. (2005). Female multiple mating for fertility assurance in red flour beetles (*Tribolium castaneum*). *Canadian Journal of Zoology*, *83*(7), 913–919.

Park, S., Lee, S., & Hong, K. J. (2015). Review of the family Bostrichidae (Coleoptera) of Korea. *Journal of Asia-Pacific Biodiversity*, *8*(4), 298–304.

Patel, A. V., & Vekaria, M. V. (2013). Management of lesser grain borer, *Rhyzopertha dominica* (F.) On wheat. *Indian Journal of Entomology*, *75*(4), 347–348.

Phillips, T. W., & Throne, J. E. (2010). Biorational approaches to managing stored-product insects. *Annual Review of Entomology*, *55*(1), 375–397.

Plague, G. R. P., Gaelle, V., Bridget, E. W., & Kevin, M. D. (2010). Rice weevils and maize weevils (Coleoptera: Curculionidae) respond differently to disturbance of stored grain. *Annals of the Entomological Society of America*, *103*(4), 683–687.

Priyanka, A. K. M., Mathur, A. C., Bagri, R. K., & Sharma, R. S. (2022). Current status and prospect of web blight of cowpea: A review. *Legume Research—An International Journal*, *1*, 7.

Radek, A., Vlastimil, K., Jan, P., & Vaclav, S. (2017). Field efficacy of brief exposure of adults of six storage pests to nitrogen-controlled atmospheres. *Plant Protection Science*, *53*(3), 169–176.

Ramadan, M. M., Abdel-Hady, A. A., Guedes, R. N. C., & Hashem, A. S. (2020). Low temperature shock and chill-coma consequences for the red flour beetle (Tribolium castaneum) and the rice weevil (Sitophilus oryzae). *Journal of Thermal Biology*, *94*, 102774.

Ridley, A. W., Hereward, J. P., Daglish, G. J., Raghu, S., Collins, P. J., & Walter, G. H. (2011). The spatiotemporal dynamics of *Tribolium castaneum* (Herbst): Adult flight and gene flow. *Molecular Ecology*, *20*(8), 1635–1646.

Ryne, C., Ekeberg, M., Jonzén, N., Oehlschlager, C., Löfstedt, C., & Anderbrant, O. (2006). Reduction in an almond moth *Ephestia cautella* (Lepidoptera: Pyralidae) population by means of mating disruption. *Pest Management Science: Formerly Pesticide Science*, *62*(10), 912–918.

Sallam, M. N. (1999). Insect damage: Damage on post-harvest. *Food and Agriculture Organization of the United Nations (FAO) and International Centre of Insect Physiology and Ecology (ICIPE)*, 137.

Saxena, R., Vanga, S. K., Wang, J., Orsat, V., & Raghavan, V. (2018). Millets for food security in the context of climate change: A review. *Sustainability*, *10*(7), 2228.

Shah, J., Chaturvedi, R., Chowdhury, Z., Venables, B., & Petros, R. A. (2014). Signaling by small metabolites in systemic acquired resistance. *The Plant Journal*, *79*(4), 645–658.

Shi, M., Collins, P. J., Ridsdill-Smith, J., & Renton, M. (2012). Individual-based modelling of the efficacy of fumigation tactics to control lesser grain borer (*Rhyzopertha dominica*) in stored grain. *Journal of Stored Products Research*, *51*, 23–32.

Singh, B. K. P. (2017). Study on the life cycle of *Sitophilus oryzae* on rice cultivar Pusa 2–21 in laboratory condition. *International Journal of Education & Applied Sciences Research*, *4*(2), 37–42.

Srivastava, C., & Subramanian, S. (2016). Storage insect pests and their damage symptoms: An overview. *Indian Journal of Entomology*, *78*(special), 53–58.

Storm, C., Scoates, F., Nunn, A., Potin, O., & Dillon, A. (2016). Improving efficacy of *Beauveria bassiana* against stored grain beetles with a synergistic co-formulant. *Insects*, *7*(3), 42.

Ul Haq, I., & Ijaz, S. (eds.). (2021). *Sustainable Winter Fodder: Production, Challenges, and Prospects*. Boca Raton, FL. ISBN: 9780367518363, 448 Pages 13 B/W Illustrations, CRC Press.

Win, N., & Rolania, K. (2020). Influence of seasonal variation on the biology of lesser grain borer, *Rhyzopertha dominica* (Fabricius) on wheat. *Journal of Entomology and Zoology*, *8*(5), 285–290.

9 Insect Pests of Summer Fodder Crops

Field Identification and Control

Muhammad Dildar Gogi, Muhammad Jalal Arif, Rabia Ramzan, Ahmad Nawaz, Muhammad Hamid Bashir, Abid Ali, Muhammad Sufian, Oscar Emanuel Liburd, and Ashfaq A. Sial

9.1 INTRODUCTION

Fodder and forage crops are the prime nutritional source for most of the livestock in developed and developing countries. Successful and sustainable cultivation of quality fodder can ensure a substantial increase in the production of livestock (Fuglie *et al.*, 2021), which furnishes about 34% protein food and has been reported as the paramount custodian of food security at the globe (FAO, 2018; Mehmood *et al.*, 2022). The livestock production system is facing severe crises in producing quality fodder around the globe. According to an estimate, over 1.3 billion people depend on livestock, and approximately 0.82 billion global population and 13% of people from developing countries are seriously undernourished (Mehmood *et al.*, 2022). To fill this livestock-associated food-security gap around the globe, sustainable and quality production of livestock and its products/by-products need to ensure merely depends on the availability of quality fodder and forages on a sustainable basis (Mehmood *et al.*, 2022).

Pakistan's livestock industry is showing a substantial growth rate of about 4.2% per annum, emphasizing a parallel increase in feed and fodder production for satisfactory production and growth of livestock in the country. In Pakistan and around the globe, the shortage of fodder, especially during November–January (winter) and May–June (summer), has become a worse and prime limiting factor for the growth of the livestock industry. In Pakistan, about 2.32 million hectares of area, with a small quantity of fluctuation, remains under cultivation of various fodder crops with approximate annual fodder production of 51.92 million tons and average fodder production of 22.5 tonnes ha^{-1} which is about 50 times less than the demand of the livestock industry in the Pakistan (Sajjad, 2019).

Summer (kharif) fodders include maize, sorghum, millet, guar, Sudan grass, Napier grass, and cowpeas. These fodder crops are the prime nutritious and quality feed sources for our livestock industry (Sajjad, 2019). Fodder crops are damaged by many insect pests, including both the sucking and chewing insect pest categories. These insect pests are major biotic constraints that limit the production of fodder crops in Pakistan and worldwide (Atwal, 1976). The attack of these insect pests not only deteriorates the fodder quality but also reduces the yield significantly. Low-quality fodder infested and damaged by insects is not preferred for silage and feed to livestock. The summer fodder crops are damaged and deteriorated qualitatively and quantitatively by both the chewing (stem borers, pod borer, armyworms, hairy caterpillar, cutworms, grasshoppers, shoot fly, foliage beetles) and sucking (pyrilla, aphids, jassid, sucking bugs, hoppers) insect pests.

DOI: 10.1201/b23394-9

Farmers mostly practice blind and careless sprays of toxic and highly persistent insecticides on these fodder crops against these insect pests to make the fodder free of pests' attack. Injudicious use of persistent and toxic insecticides on fodder crops contaminates fodder with their toxic residues. When these insecticide-contaminated fodders are fed to livestock, they result in mammalian toxicity, which magnifies through the food chain into the body of a human and causes many health complications.

Despite new technological innovations and modern agricultural reforms, the agricultural system is under various biotic stresses, including fodder cultivation and livestock production. Climate change and deteriorated environment have made the agroecosystem more and more vulnerable to pest outbreaks and infestation in crops on a large scale (Wang, 2022). In this changing and challenging scenario, accurate identification of the pests (insects and diseases), biological knowledge of pests, knowledge of technology, and strategic implementation of compatible control tactics in an integrated way are the key components of any pest management program (Arif *et al.*, 2017; Gogi *et al.*, 2017; Wang, 2022). This chapter focuses on the field identification of insect pests of summer fodders and their eco-friendly management with the least use of safer insecticides.

9.2 INSECT PESTS OF FODDER MAIZE, SORGHUM, AND MILLET

9.2.1 Stem Borer

Insect pests are major biotic constraints that limit the production of maize, sorghum, and millet production in Pakistan and worldwide, and stem borers are economically the most important (Atwal, 1976). In different African countries, almost three species of stem/stalk borers, including the African maize stalk borer, the spotted stem borer, and the African pink stem borer, attack maize, sorghum millet, rice, and sugarcane (Chernoh, 2014). The spotted stem borer, *Chilo partellus* Swinhoe (Pyralidae: Lepidoptera), is a major constraint in the productivity of maize and sorghum crops worldwide (Sharma and Gautam, 2010; Thakur *et al.*, 2018). The spotted stem borer is documented as a major biotic constraint in eastern and southern Africa (Ahmed *et al.*, 2003). It attacks all stages (growth) of maize crops, from seedlings to harvesting, by causing severe losses in fodder crops (Tamiru *et al.*, 2011). Several authors have reported almost 47 species of genus *Chilo* from diverse regions where maize is being produced (De-Prins and De-Prins, 2016). The production of maize crops is severely infested by maize stem borer, up to 15–60%, while this pest can cause losses from 24–75% alone in maize crops (Kumar, 2002). In Pakistan, 10–50% losses are observed by maize stem borer in the regions where it has been produced (Farid *et al.*, 2007). The maize yield losses due to *C. partellus* are 80% in Africa alone, while in Kenya, 18% of losses in maize and 88% in sorghum crops have been reported (Van Den Berg, 2009). *Chilo partellus* can cause extreme stalk damage in maize and yield losses of up to 79% in sorghum grains (Van Den Berg, 2009). It resembles many other species of the genus *Chilo* morphologically, while the external genitalia of both sexes can be used for their identification (Chernoh, 2014). This spotted stem borer/maize stem borer severely damages grain of sorghum and maize, despite its potential to destroy millet and its varieties and many types of grass, including Sudan and Napier (Kfir *et al.*, 2002: Matama-kauma *et al.*, 2008).

9.2.1.1 Pest Taxonomy

The (*Chilo partellus*) was first described by Swinhoe in 1885. The most notable species frequently damage cereal crops and grasses in tropical areas. A thorough revision of its taxonomy was published by Bleszynski (1970), which clarified its taxonomic confusions. Before then, this pest species was often referred to as *C. zonellus* (Swinhoe), and later, Bhattacherjee (1971) described subspecies of *C. partellus*, which were *C. partellus* acutus and *C. partellus* kanpurensis.

Domain: Eukaryota
Kingdom: Animalia
Phylum: Arthropoda
Subphylum: Uniramia
Class: Insecta
Order: Lepidoptera
Family: Crambidae
Genus: *Chilo*
Species: *partellus*

(Wikipedia, 2023, downloaded at site https://en.wikipedia.org/wiki/Insect on 03–01–2023)

9.2.1.2 Pest Status and Distribution

The spotted stem borer is a widely distributed pest in the world. It is native to Asia, then it established itself in East Africa in the 1950s and spread toward Southern and Central Africa (Inter-African Phytosanitary Council, 1985). The dispersal of *C. partellus* is influenced by moisture gradient and altitude (Muhammad and Underwood, 2004; Ongamo *et al.*, 2006). Its population is more communal in dry mid-altitude and coastal areas, but it can also occur in moist mid-altitude of agroecological zones (EPPO, 2014). In many countries such as Sri Lanka, India, Afghanistan, Pakistan, Japan, Nepal, Indonesia, and Vietnam, it has been recognized as a key insect pest of maize crops (CABI, 2007). In Eastern African countries, including East Africa, South Africa, Sudan, Zambia, Eritrea, Uganda, Kenya, and Tanzania, it is also known as a notorious insect pest of sorghum and maize (Hutchison *et al.*, 2008). It has been spread in Israel on sorghum and maize crops (Yakir *et al.*, 2013).

9.2.1.3 Pest Identification

9.2.1.3.1 Eggs

Coinciding batches on the lower side of leaf surfaces nearly to midrib in rows of almost 3–5 in groups (50–100 eggs/group) are laid by adult females. They are flat, oval, and scale-like in shape. Eggs are whitish when laid fresh. Later, the eggs change to orange and then black before hatching.

9.2.1.3.2 Larva

The larvae of this pest have reddish-brown head capsules, while the remaining body is buttery pink to yellow brown, having dark spots and four longitudinal stripes along the back. Matured larvae are almost 25 mm in length in size with shiny prothoraxic shields. Abdominal prolegs and hooks are present in a broad circle (Kfir *et al.*, 2002).

9.2.1.3.3 Pupa

Pupae are slender, light yellow brown to dark red brown, and shiny in color, having a size of 15 mm long. Small spines can be seen on the dorsal anterior margins of the pupae's 5th and 7th abdominal segments, while the last abdominal segment has large spines. Larvae pupate in small chambers within the stem (Kfir *et al.*, 2002).

9.2.1.3.4 Adults

Pupae transform into adult moths with a wingspan of 20–30 mm. Female forewings are yellowish brown, with dark scales forming longitudinal strips and white hind wings. While in male moths, the forewings are pale brown, and the hind wings are pale straw color. The forewings have double black spots (Kfir *et al.*, 2002).

9.2.1.4 Biological Description and Life Cycle

Adult females oviposit on the midrib of leaves, and eggs hatch within 4–5 days, varying during winter and summer. The *C. partellus* have an average life span of 44–48 days during summer peak months (May–June), while its life cycle prolongs in winter months up to 60–64 days from egg to adult stage (Siddalingappa *et al.*, 2010). Adult moths are nocturnal as they remain active and mate at night just after eclosion from pupae (Lella and Srivastav, 2013). The optimum temperature for survival is 18–35°C (Khadioli *et al.*, 2014).

9.2.1.5 Mode and Symbol of the Damage

Maize stem borers are oligophagous in their feeding behavior. The larvae feed inside the leaf whorls. The larvae enter the stem by making mines in the early stages of the plant. The first instars (1st and 2nd) scratch the leaves' chlorophyll, leaving behind **yellow streaks** and **window/windowpane** appearance (Mwimali, 2014). Newly hatched caterpillars start feeding on leaves, leaving holes called **gunshot holes**. Lateral larva instars tunnel in the stem and kill the central shoot, which dries. This dried and dead central shoot of a plant is called **dead heart** (Mashwani *et al.*, 2015). Different damage symptoms can be seen during the attack, such as destruction of growing points, stunted growth of the plant, and improper nutrient translocation (Polaszek, 1998; Kfir *et al.*, 2002). It damages almost all growth stages, that is, the initial seedling stage, as well as the vegetative, earring, grain-filling, and silk, tassel, and cob-bearing stages.

9.2.1.6 Risk Period and Action Threshold

The action threshold for *C. partellus* is almost 10% dead hearts or infested plants per field (van den Berg *et al.*, 2015). All areas where its host plants are being grown and the hot climate are at risk of attack from permanent or seasonal populations (Yonow *et al.*, 2017).

9.2.1.7 Pest Scouting and Monitoring

Some maize cultivars are genetically modified and can withstand spotted borer infestation. Pest can be monitored by visual inspection by looking at the damage symptoms such as affected plant parts: internal feeding on growing points, dead hearts, frass visible, the abnormal color of inflorescence, empty grains, stunted stem growth, and complete dead plant (Chabi-Olaye *et al.*, 2008). Regular inspection is recommended to avoid yield losses due to the spotted borer.

9.2.1.8 Integrated Management

C. partellus is a major devastating insect pest and has been reported as a major entomological problem in the Poaceae family (Neupane and Subedi, 2019). Culminate yield losses due to this pest may destroy the entire crop (Iqbal *et al.*, 2017). Integrated management of this pest includes cultural and mechanical control, resistant varieties, biological control, sex pheromone control, and chemical control.

- Cultural and mechanical control, including agronomic practices such as mixed cropping, field sanitation, planting time (dates), and management of crop residues, are fundamental to minimizing borer infestation. These practices majorly contribute to the growth and development of pests by creating unfavorable conditions (Ghanashyam *et al.*, 2012). Thakur *et al.* (2018) illustrated that, comparatively, winter maize sowing has lower infestation than all-year sowing.

- Destruction of crop residues may destroy the diapause larvae inside the stubbles, stalks, and cobs; it is important to do this in the dry season (Kfir *et al.*, 2002).
- Tillage practices expose larvae and pupae to predators, that is, ants, spiders, reduviids, and birds (Olaye *et al.*, 2005). Mono-cropping reduces the host-finding ability of the pest (Hailu, 2018).
- The biological control method also plays an important role in its management, for example, organisms such as insects (as predators and parasitoids), fungi (entomopathogenic), viruses, and bacteria (as biopesticides) (Sufyan *et al.*, 2019).
- *Trichogramma chilonis* and *Apanteles flavipes* egg cards are also used to manage the pests (Farid *et al.*, 2007).
- Botanical pesticides such as neem extracts and general ash are the most eco-friendly management approach for *C. partellus* (Ogendo *et al.*, 2003).
- Botanical formulations are highly useful for *C. partellus* infestation control as they can enhance crop yield by 60% (Oben *et al.*, 2015).
- Contact and systemic insecticides like carbofuran (Furadon/Sunfuran/Curator 3G) @ 14 kg/acre, chlorentraniliprole (Ferterra 0.4 % G) @ 4 kg/acre, fipronil (Regent 80%WDG) @ 30 g/acre, imidacloprid + fipronil (Lesenta 80% WG) @ 100 g/acre, monomehypo 5G @ 14 kg/acre, and chlorantraniliprole + thiamethoxam (VIRTAKO 0.6% G) @ 4 kg/acre are some recommended insecticides against the spotted stem borer. Contact insecticides are effective at the initial stages of the crop before entering the pest into the stem, while systemic are effective at later stages (Rauf *et al.*, 2017). Seed dressing with imidacloprid and thiamethoxam can reduce infestation by up to 90%. Dust application of cypermethrin in whorls is effective at older stages as it can increase crop yield (Sharif, 2016).

9.2.2 Shoot Fly

Shoot fly, *Atherigona soccata* (Rondani) (Diptera: Muscidae), is a major pest that harms crop yields in fodder and grain. Maximum damage is done to the sorghum, maize, and millet crop during the humid and rainy seasons. It attacks when the seedlings are young (Bhupender *et al.*, 2017). Due to continuous cropping and the adoption of insect-resistant hybrids and enhanced sorghum cultivars, this pest has become a major problem for maize-, sorghum-, and millet-growing farmers.

9.2.2.1 Pest Taxonomy

Kingdom: Animalia
Phylum: Arthropoda
Class: Insecta
Order: Diptera
Family: Noctuidae
Genus: *Atherigona*
Species: *soccata*

(Wikipedia, 2023, downloaded at site https://en.wikipedia.org/wiki/Insect on 03–01–2023)

9.2.2.2 Pest Status and Distribution

It is distributed in the different growing areas of the world, that is, Asia, Africa, and Mediterranean Europe (Patel *et al.*, 2015). It is an invasive pest of Pakistan's sorghum, maize, and millet and can cause significant losses (Khaliq *et al.*, 2022). It has many host crops, including sorghum, wheat, barley, maize, soybean, millet, and grasses. The shoot fly attacks these crops between 7 and 30 days after seedling emergence; as a result, the crop can be destroyed by more than 50%. Shoot fly can results in 68% forage yield loss and 80–90% grain yield loss (Kahate *et al.*, 2014; Padmaja *et al.*, 2010). The shoot fly significantly influences the reduction of production in sorghum, maize, and millet (Kumar *et al.*, 2008). Under delayed planting in kharif, a 1% rise in shoot fly dead hearts led to a 143 kg grain yield/ha drop and an average loss of 90–100% (Dhaliwal *et al.*, 2004).

9.2.2.3 Pest Identification

The eggs are oval and white. They are laid singly underneath leaves. Their full-grown larvae are whitish yellow. Pupae are dark brown and barrel-shaped inside the dead stem or in soil. Adults are gray to brown and look like a housefly but are larger than house flies. The head and thorax of a female are pale yellow (Ogwaro, 1981; Kumari *et al.*, 2021).

9.2.2.4 Biological Description and Life Cycle

Under 25–30°C temperature, the life cycle is complete in 17–21 days and has 15–16 generations in a year. The female adult fly can lay up to 75 eggs underneath the leaves (Sharma *et al.*, 2003). Its incubation period varies between 1 and 3 days. After hatching from eggs, larvae move on the dorsal side of leaves, reach near the leaf sheath, and cut the emerging point, which causes wilt and dryness in central leaves. Its larval period is 8–10 days and 3–4 instars. The pupae take a week to change into adult form. An adult can survive 10–20 days (Kumari *et al.*, 2021).

9.2.2.5 Mode and Symbol of the Damage

All maggot stages cause damage in the form of dead hearts in all of its host crops. Damage of maggots at the seedling stage leads to typical dead hearts (Kamatar, 2003). It is observed that 5- to 30-day-old seedlings are susceptible to shoot fly attacks. Symptoms occurred after 3–4 days of pest attacks. Their larvae move towards the up side of the leaf after hatching, reach near the midrib, and cut the emerging point, resulting in wilting and drying of the central shoot, which causes a dead heart. Late infestation may damage the panicles and lead to the drying and rotting of panicles (Sajwan *et al.*, 2022). Their overall risk is medium, but crops are damaged badly at the seedling stage and when sown later. Cloudy weather is favorable for its fast growth and infestation (Keerthi *et al.*, 2017).

9.2.2.6 Risk Period and Action Threshold

The economic threshold for the shoot fly in maize and sorghum is 10% dead hearts or one egg per plant. Its population peaks in August and September. After the seedling emergence of 1–4 weeks is the peak risk period of crop fodder crops like maize and sorghum (CABI, 2007).

9.2.2.7 Pest Scouting and Monitoring

Monitor the crop by selecting at least five random plants from different spots in the field. Look for the dead hearts, which indicates a shoot fly attack in the crop (CABI, 2007).

9.2.2.8 Integrated Management

Many shoot fly management strategies have been implemented to reduce losses, including cultural/agronomic practices, host plant resistance, natural enemies, and synthetic insecticides (Kumar *et al.*, 2008).

Cultural control

- Early sowing of crops and intercropping, sorghum, maize, and millet with leguminous crops (Shekharappa and Bhuti, 2007)
- Use of resistant cultivars (Kalpande *et al.*, 2015; Mohammed *et al.*, 2016)
- Crop rotation and maintain the field hygiene conditions (Bhupender *et al.*, 2017)

Biological control

- *Trichogramma simmondsi* and *T. chilonis* are used as egg parasitoids of shoot flies (Singh and Sharma, 2002)
- *Opius* species and spiders are the predators of shoot fly larvae (Delobel and Lubega, 1984; Getu *et al.*, 2006)

Chemical control

- Treatment of seeds with imidacloprid with 6 mL/kg of seed reduces the incidence of shoot flies (Sandhu, 2016)
- Carbofuran 3G at 20 kg/hectare (Balikai, 1998)
- Thiomethoxam 5g/1 kg of seed (Daware *et al.*, 2011)

Physical control

- Yeast ammonium sulfide mixture in fishmeal traps captures adult shoot flies (Mohan, 1991)

9.2.3 Pyrilla

9.2.3.1 Pest Taxonomy

Kingdom: Animalia
Phylum: Arthopoda
Class: Insecta
Order: Hemiptera
Family: Lophopidae
Genus: *Pyrilla*
Species: *perpusilla*

(Wikipedia, 2023, downloaded at site https://en.wikipedia.org/wiki/Insect on 03–01–2023)

9.2.3.2 Pest Status and Distribution

Since 1903, *Pyrilla perpusilla* has been observed throughout much of Asia (Cotterell, 1954). Bangladesh and the eastern province of Afghanistan have localized severe pest outbreaks (Fennah, 1963; Dean, 1979; Miah *et al.*, 1986). This species has happened all over India, occasionally at pandemic levels (Stebbing, 1903; Rahman and Nath, 1940; Dhaliwal and Bains, 1985) Six districts in Nepal have reported the pest (Neupane, 1971). It has been reported in most of Pakistan's regions (Sheikh, 1968; Rahim, 1989). The species has been observed in Sri Lanka's eastern and south-central regions (Kumarasinghe and Ranasinghe, 1985, 1988). *Pyrilla perpusilla* is also found in other parts of Asia, including Burma, Indo-China, and Thailand (Fennah, 1963), and in Laos, Vietnam, Cambodia, and Indonesia (Varma, 1986).

9.2.3.3 Pest Identification

Eggs

Oval eggs were placed in a cluster underneath leaves, close to the midrib. The eggs are white to greenish yellow, with a size of about 0.9–1.0 mm long and 0.45–0.64 mm broad (Kumarasinghe and Wratten, 1996).

Nymphs

Nymphs that have emerged are between 0.8 and 1 mm long, are milky white in color, and develop through five instars to achieve adulthood. Every nymph instar has distinctive anal filaments slightly longer than the body.

Adults

The adult *P. perpusilla* is a soft-bodied, pale, tawny to yellow bug having a noticeably drawn-forward head. Male and female wingspan ranges are 16–18 mm and 19–21 mm, respectively (Kumarasinghe and Wratten, 1996).

9.2.3.4 Biological Description and Life Cycle

Females live significantly longer than males as adults, with an average life span of 14–200 days. The female experiences a pre-copulation period of around 11 days, and copulation occurs during the day and lasts about 2 hours. After a pre-oviposition phase of 3–47 days, the female typically needs 45–60 minutes to lay one egg, depending on the season and weather. There can be a delay of 2–25 days between subsequent oviposition. Females lay eggs on the abbatial surface of the leaf. They are shielded with a wax thread-like substance released by the female sex and laid out in 4–5 rows. Eggs are placed inside the base of the leaf case during the winter, providing a shield from unfavorable weather surroundings. The female typically chooses an inferior, hidden location shaded for oviposition.

The females lay eggs in winter between dried leaf sheaths and the stalks. A lifetime fecundity of 37–880 eggs are laid at a rate of 20–50. The incubation period spans 6–30 days, depending on the season. However, there have been reports of six instar nymphs in North India (Rajak *et al.*, 1987). *Pyrilla perpusilla* has five instars lasting 7–41 days, with the highest total nymph duration of 134. A temperature of around 30°C and a relative humidity (R.H.) of around 80% are suitable for nymphal development (Gupta and Ahmad, 1983). The insect has 3–4 generations in Pakistan and India.

9.2.3.5 Mode and Symbol of the Damage

Pyrilla perpusilla causes fungal infections by sucking plant sap and excreting honeydew onto leaves. Sugar yield and quality are impacted by this direct and indirect damage (Butani, 1964; Asre *et al.*, 1983). Since 1940, numerous articles have reported damages ranging from 2% to 34% in the (sucrose) content of the sugarcane and from 3 to 26% in the sugar concentration. Due to this pest, up to 28% of the potential yield is lost. Additionally, poor seed set growth and challenges in grinding cane from infected plants have been observed.

9.2.3.5.1 Risk Period and Action Threshold

By the middle of March, the overwintering nymphs mature into adults, and oviposition begins by April. The pest multiplied quickly throughout April and the first 2 weeks of May. In June, nymphs and adults moved to the recently cultivated crop. The action threshold for this pest is 3–5 nymphs or 5 adults or both 5 per plant.

9.2.3.6 Pest Scouting and Monitoring

Locate the pest underneath the leaf where adults and nymphs suck the cell fluid. Yellow and dry leaves are an indication of an infestation. Yellow patches appear on the leaves, which are easy to see.

9.2.3.7 Integrated Management

Agronomic control

Using the pest's phenology, different sowing and harvesting dates lessen its effects. Burning rubbish also had a positive impact on pest management. The pest was successfully suppressed by scattering *E. melanoleuca* cocoons and sugarcane waste in the harvested fields. The natural enemy of the crop for the following season was "planted" into the field (Joshi and Sharma, 1989). As the crop served as the pest's more or less constant host for 3–4 generations, the method of ratooning significantly boosted the pest populations. Crop rotation, however, had no discernible impact on pest management (Khan and Khan, 1966).

Biological control

P. perpusilla can be controlled by natural enemies, particularly the interactions between five parasitoid species and seven predator species. In the literature, 16 biological control agents have been reported to attack *P. perpusilla* (Butani, 1972). About 80% of parasitism of the *P. perpusilla* has been reported by egg parasitoids and the leftover 20% by a combination of nymphal–adult parasitoids and predators (Chaudhary and Sharma, 1988). The combination of parasitoids, predators, and diseases significantly reduces the *P. perpusilla* infestation. The parasitoid *Epiricania melanoleuca* (Fletcher) (Lepidoptera: Epipyropidae) seems the most effective against *P. perpusilla* (Fletcher, 1939), dropping pest populations by as much as 90–100%.

Cultural control

As eggs and larval instar stages of this pest are destroyed through this method, garbage burning or mulching has little effect on the pest's population. However, eliminating young branches from the ratoons' growth reduces pest populations.

Chemical control

Flooding chlorpyriphos @ 2 L acre^{-1} or Imidacloprid @ 750–1000 mL acre^{-1} proves effective in controlling pyrilla. A spray of carbosulfon @ 500 mL acre^{-1} and clothianadin @ 90 g acre^{-1} also proves very effective against pyrilla at the early stages of sugarcane crops.

9.2.4 Aphids

Aphids are the main group of plant virus vectors (Eastop, 1977; Chan *et al.*, 1991). They are small, soft-bodied insects with piercing-sucking mouth parts (Blackman and Eastop, 2000). Aphids (Homoptera: Insecta) contain roughly 4700 species, of which about 1000 are known to exist (Dohlen *et al.*, 2006). In the Adelgidae, Phylloxeridae, and Aphididae families, about 250 species of aphids nourish on horticultural or agronomic crops (Blackman and Eastop, 2000). They excrete honeydew through cornicles rich in plant sugars and other chemicals (Glen, 1973; Faria, 2005). Despite the broad range of aphids' biocenotic behavior, aphid species also significantly reduce the yield of horticultural or agronomic crops (Vereschagina and Gandrabur, 2014).

9.2.4.1 Pest Taxonomy

Kingdom: Animalia
- Phylum: Arthropoda
  - Class: Insecta
    - Order: Hemiptera
      - Family: Aphididae
        - Genus: *Aphis*
          - Species: *gossypii*

(Wikipedia, 2023, downloaded at site https://en.wikipedia.org/wiki/Insect on 03–01–2023)

9.2.4.2 Pest Status and Distribution

Aphids are both myrmecophilous (cared for by ants) and non-myrmecophilous (McPhee *et al.*, 2008). The stronger Dolichoderinae, Formicinae, and Myrmicinae are phylogenetically advanced subfamilies of ants that participate in these mutualistic interactions between ants and aphids (Delabie, 2001). According to a report, 19–26 aphid species from 37 host plants partnered with 16–23 ant species from different geographical regions worldwide (Idechiil *et al.*, 2007; Özdemir *et al.*, 2008; Kataria and Kumar, 2013; Mortazavi *et al.*, 2015).

9.2.4.3 Pest Identification

Eggs

Eggs laid by aphids can be either oval or chisel-shaped, held by supports or threads. Aphid eggs initially have a bright yellow-green tint, which darkens near hatching. Some aphids coat their eggs in wax to repel predators (Johannsen, 1912).

Nymph

Nymphs/larvae are orange, red, or yellowish maggots no longer than 1/8 in. (3 mm). The head is curled within the circular body (Auclair, 1967).

Adults

Aphid adults typically lack wings; however, most species can also be found with wings, especially during periods of high population or spring and fall. When the quality of the food source declines, the pest develops a winged population to spread to new plants (Braendle *et al.*, 2006).

9.2.4.4 Biological Description and Life Cycle

Aphids typically have different morphs throughout their life cycle of parthenogenesis (migrate in the spring months, alate and unwinged) and viviparous females (viviparity) in the summer (Hille Ris Lambers, 1966). Beginning in the summer, they fly from the hosts to hosts, whereas re-migrants return from the secondary hosts to the primary hosts in the fall. The sexual females and males larvae develop on the main host (gynoparae, only of females; andropara, only of males; Hand and Wratten, 1985). Some heteroecious aphid species have secondary hosts where the males can reproduce (Orantes *et al.*, 2012). During the few species, remigration occurs into the central summer months rather than the fall months, shortening the breeding season (Popova, 1967). Numerous aphid taxa contain odd morphs, such as pseudo-funda trices in the family Adelgidae, "soldier" larvae that cannot develop further or reproduce (Shaposhnikov, 1986). Various specialized individuals or morphs partially share certain functions due to polymorphism (Jonsson and Jonsson, 2001). Each morph has a distinct function and only occurs at certain points in the life cycle. The migratory and stationary apterous and alate females exhibit differences in morphological appearance, epigenesis, metabolism, behavior, and life cycle (Vereschagina and Gandrabur, 2014).

9.2.4.5 Mode and Symbol of the Damage

Aphids consume the new tender, succulent, and newly developing foliage, resulting in the shortening of internodes, rosetting, dwarfing, and abnormal root growth (Capinera, 1974). Plants that have been severely affected seem bushy or bonsai-like. Small plants or seedlings can quickly perish from an aphid infestation. In the year following an aphid infestation, deep-rooted plantings may experience injury and mortality, especially during a particularly harsh winter. Cold and aphid infestation reduce the life span and vigor of dormant asparagus crowns more than any of these factors acting alone would (Valenzuela and Bienz, 1989).

Aphids are reportedly one of the most significant cereal pests in the world (Vickerman and Wratten, 1979). During eating, they inject the plant with toxic saliva, which causes blight on buds, dimpling on fruits, curling on leaves, and appearance of splashes of color on the foliage (Hashmi, 1994). By excreting honeydew on the leaves of the plants, aphids create additional harm that eventually prevents photosynthesis by fostering the growth of sooty mold (Eastop and Blackman, 2000).

Aphid damage can be split into three groups under certain conditions. The first type of harm is characterized by the aphid colonies' substantial sap consumption as well as the chemical reaction on the plant (Yuan *et al.*, 2014). The second type has to do with the potential for quick intrapopulation change. The third type of harm is connected to aphids' capacity to spread viral infections to plants (Fereres and Moreno, 2009). The three types of harm are connected to the polymorphism of aphids (Vereschagina and Gandrabur, 2014).

Aphid infestations during the rabi season could increase from mid-February to early March (Dhanda *et al.*, 2020). The leaves with severe aphid infestations may curl, wilt, and exhibit yellow or even dead regions (Kadioglu *et al.*, 2012; Alam *et al.*, 2020). Shoots that have been severely attacked wither and deteriorate. Cereal aphid populations typically start on the lower leaves and leaf coverings crops (Mishra and Sharma, 2016). The inflorescence, particularly the branches of the panicles at the glume bases, is a food source for large populations (Arora, 2017).

9.2.4.6 Risk Period and Action Threshold

The economic threshold for aphids is 10–30% infested plants.

9.2.4.7 Pest Scouting and Monitoring

Consistent and systematic scouting and monitoring can help detect early aphid infestations and prevent further destructive epidemics. The most suitable tools/techniques for scouting/monitoring aphids in crops are the yellow sticky card (for flying/alate aphids) and physical inspection visually or with a hand lens (wingless aphids). Sticky-card installation at every 2000–4000 ft^2 is sufficient for monitoring aphids in crops (Stein *et al.*, 2015; Koralewski *et al.*, 2020).

9.2.4.8 Integrated Management

Nonchemical control (cultural, mechanical, physical, biological, reproductive control, etc.).

Aphid colonies are preyed on, parasitized, or pathogenized by several predators, parasitoids, and entomopathogens, respectively. Predators like coccinellid beetles and syrphid flies keep aphids in check on fodder crops. Excessive rainfall also washes away crop aphids (Arora, 2017). Aphids are consumed by larvae of lacewings (Neuroptera: Chrysopidae and Hemerobiidae) and both larvae and adults of ladybird beetles (Coleoptera: Coccinellidae) (Van Veen *et al.*, 2008). Entomopathogenic fungi are the most prevalent aphid pathogens, primarily of the Deuteromycotina and Zygomycotina families (Hajek and St. Leger, 1994). Several *Hymenoptera* species and some *Dipteran* species parasitize aphids. Those parasitoids are frequently used in biological control programs because they have the potential to significantly reduce aphid colony mortality (Boivin *et al.*, 2012). A mono-phyletic subfamily of the Braconidae family (Hymenoptera) with a focus on aphids is known as the Aphidiinae. The parasitoids are called koinobionts because the larvae grow inside the moving, feeding parasitoid host (Kambhampati *et al.*, 2000).

9.2.4.8.1 Chemical Control

Imidacloprid and acetamiprid are used for aphid control (Takahashi *et al.*, 1992). Chlorantraniliprole, flonicamid, azadirachtin, *Beauveria bassiana*, *Isaria fumosoroseus* thiamethoxam, tolfenpyrad, spirotetramat, cyantraniliprole, pyrifluquinazon, dinotefuran, bifenthrin, and others are some prominent aphidicides that are used to suppress aphid infestation at their dose rates recommended on different crops (Koralewski *et al.*, 2020).

9.2.5 Grasshoppers

Grasshoppers have large jumping hind legs and belong to the order Orthoptera in the Acrididae family. Grasshoppers attack crops, especially sorghum fodder (Latchininsky *et al.*, 2011). Grasshoppers also have migratory behavior (Cease *et al.*, 2015). Some pestiferous grasshoppers are short-horned, that is, *Melanolpus differentialis, Melanolpus bivitattus, Melanolpus femurrubrum*, and *Melanolpus sanguinipes* (Boone, 2019).

9.2.5.1 Pest Taxonomy

Kingdom: Animalia
Phylum: Arthropoda
Subphylum: Hexapoda
Class: Insecta
Order: Orthoptera
Family: Acrididae
Genus: *Melanolpus*
Species: *differentialis, bivitattus, femurrubrum, sanguinipes*

(Wikipedia, 2023, downloaded at site https://en.wikipedia.org/wiki/Insect on 03 01–2023)

9.2.5.2 Pest Status and Distribution

Grasshoppers are highly polyphagous insect pests with a wide range of host plants. They are distributed in Northern Mexico, the United States, Canada, and southern Ontario (Evans and Arthur, 2007). Pakistan's irrigated and rain-fed areas and geographical conditions are ideal for breeding grasshoppers. As a result, they pose a serious threat to crops and pastures (Soomro and Wagan, 2005).

9.2.5.3 Pest Identification

They are yellow to brown with a shiny appearance, having a size of 28–37 mm and 34–50 mm in polymorphic male and female adults, respectively. They have reddish-brown antennae, and their eyes are compound with brown to yellowish spots. They have a uniform and glossy forewing and pronotum. Both male and female adults have differently shaped cerci (Boone, 2019).

9.2.5.4 Biological Description and Life Cycle

They have one generation per year. The grasshopper's life cycle has three stages, that is, eggs, nymphs, and adults. Adult females lay eggs in a pod inside the soft soil layer, and a single pod has 8–30 eggs. The female grasshopper chooses to lay eggs in the soil bordered by grass roots. An average number of eggs is up to 100 eggs/female during her life span in the summer and fall seasons. Egg hatching takes place 10–15 days after laying. Hatched nymphs take 32 days to transform into adults (USDA, 2015; Kaufman *et al.*, 2007).

9.2.5.5 Mode and Symbol of the Damage

The grasshoppers are major voracious feeders of crops and grasslands that consume one-half of their body weight in all terrestrial habitats (Hewitt and Onsager, 2000). Nymphs and adults of grasshoppers chew stems and leaves of plants by defoliating entire fields and showing ragged to round-hole symptoms further extending from leaf margins to veins (Shishodia *et al.*, 2010: Cigliano *et al.*, 2000: Qureshi and Bhattim, 2006).

9.2.5.6 Risk Period and Action Threshold

Grasshoppers peak infestation period is hot summer from May to June. The action threshold for is 13–25 grasshoppers per square meter (Agri-Fact, 2002).

9.2.5.7 Pest Scouting and Monitoring

The scouting and monitoring method of grasshopper infestation differs from other insect pests. The person should walk the infestation area and count the number of jumping grasshoppers. 'T' (a 1 m stick) is the best sampling method for infestation estimation. The stick should be kept above the height of the crop (Agri-Fact, 2002).

9.2.5.8 Integrated Management

Cultural control, such as tillage operation, can be helpful by disrupting the eggs in the soil as they hinder egg hatching. The tillage practices expose the eggs to predators, parasites, and desiccation. Tillage practices are more successful in the summer season than fall. Carbaryl bait impregnated with a 2–5% formulation is the method of control in winters before the emergence of crops or when crops are just a few inches tall. A crop border treatment of 150 ft from the edges is suitable for reducing population. The seed treatment with insecticides such as imidacloprid and thiamethoxam effectively manages the grasshopper population (John and Gary, 2008). *Nosema* baits are also effective in the control of grasshoppers (Flint, 1998).

9.3 INSECT PESTS OF GUAR

Guar, also known as the cluster bean, is an annual summer legume that is drought hardy and mostly grown in Pakistan's arid and semiarid regions. Guar gum is a significant product from guar seeds used in several industries worldwide (Minhas *et al.*, 2021). The major countries cultivating guar are India, Pakistan, the United States, Italy, Morocco, Germany, and Spain (Punia *et al.*, 2009). A productive crop produces between 300 and 400 quintals of green fodder, 15 quintals of dried seeds, or 60 quintals of green pods per hectare under ideal environmental conditions (Sharma, 2018). However, various insect pests, including both sucking and chewing insect pests, severely reduce crop yield each year. These include hairy caterpillars, Bihar hairy caterpillars, stem flies, pod borers, aphids, jassids, and white flies.

9.3.1 Hairy Caterpillar

Two species of hairy caterpillars, hairy caterpillar (*Ascotis imparta*) and Bihar hairy caterpillar (*Spilosoma oblique*), generally damage the cluster bean (Gauaahar) crop (Dalwadi *et al.*, 2007). Bihar hairy caterpillar, *Spilosoma obliqua* (Walker) (Lepidoptera: Arctiidae), *Spilarctia oblique*, and *Emproctis Objecta* are polyphagous insect pests that feed on cluster beans, cowpea, Indian bean, brinjal, and potato. Many parasitoids and predators generally do not prefer it due to the dense hairs on the larval body (Jaydeep *et al.*, 2020).

9.3.1.1 Pest Taxonomy

Kingdom: Animalia
Phylum: Arthropoda
Class: Insecta
Order: Lepidoptera
Family: Erebidae
Genus: *Spilosoma*
Species: *obliqua*

(Wikipedia, 2023, downloaded at site https://en.wikipedia.org/wiki/Insect on 03-01-2023)

9.3.1.2 Pest Status and Distribution

It is an oriental and sporadic pest. It is widely distributed in Afghanistan (southeastern), Pakistan (northern areas), India, Bangladesh, Myanmar (Burma), and Bhutan (Sivakumar *et al.*, 2020). Its incidence is in Khyber Pakhunkhwa (KPK) and Punjab (Bhutto *et al.*, 2013).

9.3.1.3 Pest Identification

In full growth condition, it is covered with long grayish hairs (5 cm) and has a body length of 40–45 mm. The head and thorax are pale yellow, with black antennas (Dwivedi, 2020). The wing expansion of the moth is 50 mm.

9.3.1.4 Biological Description and Life Cycle

The eggs are pale yellow when near hatching and creamy white when laid. The female lays an average of 148–232 eggs. Egg, larval, and pupal duration range between 5 and 6, 20 and 25, and 8 and 9 days, respectively. Larvae have six instars. The length of the male varies from 51 to 55 mm, and the female moth varies from 55 to 59 mm. The total life duration of males is 37–42 days, while females range from 39–44 days (Dwivedi, 2020).

9.3.1.5 Mode and Symbol of the Damage

Larvae of this pest feed on chlorophyll on the underside of leaves, due to which the leaf color changes from brownish to yellow. In the later stages, the larvae eat the leaves from the border, and the leaves look like a net or web (Si-ming, 2007).

9.3.1.6 Risk Period and Action Threshold

Economic threshold level (ETL) is 20–25% of damaged leaves observed (Dwivedi, 2020). The risk period is from the end of April to the end of May (Bragard *et al.*, 2022).

9.3.1.7 Pest Scouting and Monitoring

Feeding injuries like cut leaves indicate caterpillar infestation in the crop. Crops should be monitored regularly with light traps and pheromone installation.

9.3.1.8 Integrated Management

Botanical practices

Two botanicals in 5% neem oil are effective alone and in mixture form. Combining 5% neem oil (16.49%, w/v) with Indoxacarb @ 5% gives sustainable and eco-friendly management of *S. obliqua* on the crop in the field conditions (Dwivedi, 2020). Other management practices are similar, like a red hairy caterpillar.

Cultural control

Growing trap crops like cowpea, castor, and jatropha on boundaries of the main crop to attract and trap female moths for oviposition and the caterpillars for feeding. Deep plowing 3–4 times before the onset of monsoon (pre-monsoon) causes the hibernating pupae to be exposed to the sun's rays and predatory birds. The alternate hosts should be removed and destroyed where caterpillars harbor.

Mechanical/physical control

Set up 3–4 light traps to trap the moths. The early-instar larvae on sieved leaves should be collected from the trap crops, and egg masses from the main crop and destroyed. The migration of the larvae can be prohibited by digging a 30-cm-deep and 25-cm-wide trench around the infested fields and filling it with water or any dust insecticide like carbaryl or sevin.

9.3.1.8.1 Chemical Control

Application of chlorpyriphos 20 EC at 1.5 L/ha and triazophos 40 EC at 0.8 L/ha or quinalphos 25 EC @ 1.5 L/ha.

- **Field sanitation:** After every 10 days
- **Light trap:** Use a 200W mercury vapor lamp (one trap/ha)

9.3.2 Jassids

The Cicadellidae family, which includes the jassids, an economically significant group of Auchenorrynchan Hemiptera, has roughly 2,445 described genera and 22,637 species worldwide, including 340 genera and 1,350 species in Asia (Viraktamath, 2006). Almost 17 host plant species belonging to eight families are infested by jassids (Rahman, 2014).

9.3.2.1 Pest Taxonomy

Kingdom: Animalia
Phylum: Arthropoda
Class: Insects
Order: Homoptera
Family: Cicadellidae
Genus: *Amrasca*
Species: *biguttula*

(Wikipedia, 2023, downloaded at site https://en.wikipedia.org/wiki/Insect on 03–01–2023)

9.3.2.2 Pest Status and Distribution

Among all the insect pests of guar, jassids have been reported as the major sucking pests (Bali *et al.*, 2022). *Amrasca biguttula* is a polyphagous insect in Argentina, Brazil, Greece, Spain, China, Pakistan, Iran, Syria, and the United States. It is widely dispersed in Australia and East, West, Southern, and Central Africa (Rahman, 2014).

9.3.2.3 Pest Identification

Eggs

The newly laid eggs appeared yellow, slightly oval, and translucent (Singh *et al.*, 2018).

Nymphs

Jassids can be found on the underside of the plant and between the veins of the leaves. The jassid nymphs are pale green, translucent, flattened, and wingless (Vennila *et al.*, 2007). The nymph of jassid experiences five instars (Singh *et al.*, 2018).

Adults

The adults' bodies are wedge-shaped, elongated, and pale green. Its forewings have two black spots on the vertex of the head and black specks on each apical edge. The insect moves diagonally over leaf surfaces (Jayasimha *et al.*, 2012). The adults move quickly and side to side while active (Jayarao *et al.*, 2015).

9.3.2.4 Biological Description and Life Cycle

Amrasca biguttula is a polyphagous insect that is present all year long on cotton, brinjal, and China roses, but it also migrates to other crops for survival, causing significant harm to many economically significant crops (Fuleiro and Rai, 1985). In the veins of leaves, leafhopper females lay their eggs (Agarwal *et al.*, 1978; Singh and Agarwal, 1988). The female deposits eggs into thin petioles, twigs, succulent leaves, and other plant parts. However, they favor laying their eggs in the leaf tissue's veins. The newly laid eggs are yellow, slightly oval, and translucent. A single female's fecundity ranges from 13 to 18 eggs. Eggs were incubated for an average of 3–7 days. The eggs are oval-shaped and taper slightly to the front. *A. biguttula* are hatch from eggs and pass through five instars. The newly hatched nymphs are translucent, yellowish, and without wings. The initial instar nymphal stage lasts 1 to 3 days. The antennae are setaceous, and the eyes are dark brownish black. Nymphs in their second instar live for 2–5 days. Jassids bear primitive wing pads in the second instar nymphs. The third instar nymphal stage completes development in 2–6 days. The fourth instar nymph lasts for 3–6 days. The nymphs in their fourth instar are yellowish green, and their wings reach their fourth abdominal segment. The nymph in its fifth instar is greenish yellow in appearance, and its wing pads extend to its ninth abdominal segment. It has long, setaceous antennae (Singh *et al.*, 2018). After that, nymphs go through metamorphosis and develop into adults, about 3.5 mm long, wedge-shaped, and pale green. The vertex and forewings have black dots (Vennila *et al.*, 2007). Adults move fast and side to side while active (Singh and Kumar, 2003). Male adult longevity ranges from 12 to 17 days, whereas female longevity ranges from 12 to 18 days. The leafhopper has a life span of 19–35 days (Thirumalaraju, 1984). Jassid is supposed to have 11 generations yearly throughout the summer (Iqbal *et al.*, 2008).

9.3.2.5 Mode and Symbol of the Damage

Jassids have been found to cause low to severe infestation from germination to maturity (Bindra and Singh, 1969; Puttaswamy *et al.*, 1977; Parihar, 1979; Satyavir, 1980; Pareek *et al.*, 1983). The nymph and adults inject their saliva into the tissues, causing toxemia, while they suck the sap from the cell from the ventral side of leaves. The damaged leaves initially turn a mild shade of green, then turn yellow, turn radish red, and eventually turn brick red or brown. Curling and a gradual drying out of the leaves accompany this transition. The plant's growth is stunted, negatively impacting the yield (Butani and Jotwani, 1983). A 40–56% yield loss has been reported due to *Amrasca biguttula* damage from young seedlings to mature crops (Krishnaiah and Kalode, 1984; Halder *et al.*, 2016). In addition to reducing yields through direct feeding activity, leafhoppers carry significant viral infections that result in severe infection and plant death. Jassids and whiteflies, in addition to causing immediate harm by dissipating, serve as the vectors for the yellow mosaic virus (Satyavir *et al.*, 1984). Jassids can reduce yield by up to 35–40% and increase by 60–70% under ideal conditions (Sultana *et al.*, 2016).

9.3.2.6 Risk Period and Action Threshold

The action threshold reached for jassids is 1–1.5 nymphs or adult/leaf (Afzal *et al.*, 2014).

9.3.2.7 Pest Scouting and Monitoring

Nymphs and adults of jassids can be seen on the undersides of the leaves. When disturbed, nymphs tend to migrate sideways. Adult jassids can fly easily, and both nymphs and adults follow a distribution pattern (Shivalingaswamy *et al.*, 2002). Examining the tip damage signs, curled leaves, and blackening of the terminal's tiny leaves, brick-red leaves is a basic criterion for monitoring and scouting jassid (Devi *et al.*, 2018). Underneath the leaves should be checked for adults and newborns. For every 50 ha of crop, 20–30 plants should be monitored.

9.3.2.8 Integrated Management

It is necessary to investigate the quantitative estimation of the population of pests and their natural enemies with the major abiotic environmental parameters to offer a solid foundation for the IPM approach against a pest of cluster beans (Bali *et al.*, 2022).

Botanical control

Neem oil, neem seed, neem leaf extract, neem seed kernel extract, and neem seed spraying against jassids are all very effective (Bali *et al.*, 2022).

Biological control

Jassid predators such as the spider (Krishnaiah and Kalode, 1984), *Chrysoperla* (Yadav and Patel, 1990), *C. septumpunctata* (Kaethner, 1991; Ravi Kumar *et al.*, 1999), and predatory mites *Amblysieus* spp. (Guddewar *et al.*, 1994) naturally suppress jassid populations (Rosaiah, 2001). *Bacillus thuringiensis*, *Beauvaria bassiana*, and *Metarrhizium anisopliae* prove the most effective among the environmentally friendly insecticides used to manage jassids after two applications in a field (Bali *et al.*, 2022).

Cultural control

Utilize the cultivars of insects that the Agriculture Department has recommended. Adopt a suitable crop rotation strategy and avoid growing cucurbits close to the primary crop. Install many yellow and blue sticky traps 30 cm off the ground to track the occurrence of sucking pests and estimate their intensity. Avoid excessive use of irrigation and fertilizer applications (Bali *et al.*, 2022).

Chemical control

Pre-sowing seed treatment with imidacloprid (70 WS) @ 5 g per kg **seed** or thiomethoxam (25 WG) at 3 g per kg seed proves effective against the initial infestation of jassid at the early growth stage of the crop. Nitenpyram 10% AS @ 200 mL, thiamethoxam 25 WG @ 24 g, thiacloprid 48 SC @ 125 g, acephate 97 DF @ 300 g, acephate 75 SP @ 250–375 g, chlorfenapyr 360 SP @ 75 g, etopfenprox 30 EC @ 200 mL, dinotefuron 20 SG @ 100 g, and pymetrazine 20% SC @ 300–400 mL per acre can be used when the infestation of jassids reaches to its ETL.

9.4 INSECT PESTS OF SUDAN GRASS AND NAPIER GRASS

9.4.1 Cutworms

Agrotis ipsilon (Hufnagel) (Lepidoptera: Noctuidae) is a black-colored cutworm that is particularly polyphagous and targets a variety of crops (Ram *et al.*, 2001; Binning *et al.*, 2015). Its ability to feed on almost 100 host plants, including maize, wheat crop, cotton crop, soybeans, vegetables, and various weeds, makes it one of the most destructive species of subterranean pests (Liu *et al.*, 2015). Its name comes from its propensity to bite through the stem to remove a seedling at the ground level. Due to their harm to several crops and widespread distribution, it is a sizable category of insect pests (Vendramim *et al.*, 1982).

This pest chews the seedling of the plant inside the ground due to its living habitat. In one night, only a single larva can eat and digest several plants from crops. Usually, they cut plants and drag them inside the soil and feed them there. Cutworms coil themselves upon disturbing. Studies show cutworms are found feeding during the day and are hidden at night (Showers, 1997).

9.4.1.1 Pest Taxonomy

Kingdom: Animalia
Phylum: Arthropoda
Class: Insecta
Order: Lepidoptera
Family: Noctuidae
Genus: Agrotis
Species: *ipsilon*

(Wikipedia, 2023, downloaded at site https://en.wikipedia.org/wiki/Insect on 03–01–2023)

9.4.1.2 Pest Status and Distribution

A. ipsilon is broadly scattered throughout the world and is recognized to injure the crops in North, Central, and South America; Asia; Europe; Oceania; Middle East; and Africa. *A. ipsilon* is a serious agricultural insect pest in the tropical and subtropical regions of the world, causing substantial losses in crop production in countries such as Chile, Brazil, Egypt, India, Myanmar, Poland, Spain, and the United States.

9.4.1.3 Pest Identification

Eggs

The eggs are spheres (Kumar, 2016b) with somewhat flattened edges and around 0.51–0.63 mm wide and 0.54–0.64 mm tall. The eggs are originally yellowish and pale and are distinguished by short rims that extend from the tip (Pathania, 2010).

Larvae

The larvae pass through six larval instars (Verma and Verma, 2002b). For instars I–VI, the body length ranges from roughly 1.9 to 3.7, 4.3 to 5.1, 7.0 to 9.0, 14.0 to 20.0, 23.0 to 27.0, and 38.0 to 45.0 mm, respectively. The larvae turn photo-negative after the fourth instar and occupy the day hiding in the earth. The larvae are active most of the night, peaking just after midnight and an hour before daylight (Lal and Rohilla, 2007). The more mature larvae tend to tear apart plants at the soil's surface and drag the plant tissues underneath. Most larvae are cannibalistic (Verma and Verma, 2001b).

Pupae

The full-grown larvae pupate in the soil inside clay cells (Verma, 2015). The period of the pupa ranges between 12 and 14 days (Verma and Verma, 2002b).

Adults

The adults' wingspan ranges from 40 to 5 mm, or rather big (Verma, 2015). It is simple to distinguish between male and female moths based on their antennae. Females have setaceous antennae, whereas males have bipectinate antennae (Pathania, 2010). After emergence, the females deposit eggs which hatch in 4–8 days.

9.4.1.4 Biological Description and Life Cycle

A. ipsilon's life span includes an egg, larva, pupa, and adult, just like other cutworms (Duportets *et al.*, 1998). The pests remain active from October to April and probably migrate to the mountains for further breeding during summer. The moth comes out at dusk and fly out until darkness sets in. The female moth deposits 200–350 creamy-white dome-shaped eggs at night in the cluster, each having 30–40 eggs on the underside of the leaves or in the soil. The eggs hatch out within 2–5 days in summer and 6–18 days in winter. The newly hatched larvae chew eggshells and move like a semilooper. The larval stage completes its development within 30–35 days. The larvae stage of *A. ipsilon* is active at night and feeds mostly on the stem and roots of different host plants. The host plant's leaves have small, irregular holes, and the later instar may cut and sever the stems below the soil's surface, killing the plant. Due to incomplete clipping, the host plant may begin to wilt (Fishel *et al.*, 2009). Larvae are damaging because they may swiftly travel from one plant to another after cutting them. Such eating behavior causes crops to be damaged disproportionately, and it may even destroy entire stands of crops. The later larval instars show cannibalistic behavior. The full-grown larvae develop an earthen chamber in soil and pupate there underground. The pupal stage completes its development within 10 days in summer and 30 days in winter (Hunt *et al.*, 2005).

9.4.1.5 Mode and Symbol of the Damage

Distinct kinds of cutworms may be distinguished based on how they eat. Surface cutworms consume enough food to topple the seedlings while feeding at or below the soil's surface. Cutworms that climb plants devour the buds, leaves, and reproductive parts of plants, whereas cutworms that live underground eat the underground stems and roots (Verma, 2015). The larvae cut the plants just above or a little distance below the soil's surface. Although most of the plant is still intact, the stem generally has enough tissues stripped for it to fall over (Ojha and Nath, 1987).

Cutworms can do enormous harm by cutting young plants after they reach the fourth instar. One larva can notch multiple plants in a single night (Verma, 2015). The younger larvae of *A. ipsilon* feed on the leaves, while adult cutworms feed towards the base of the stem (Kriti *et al.*, 2014). *A. segetum* is a determined insect pest of maize in the northwestern Himalayas and causes damage at the seedling stage. The issue is most severe during summer when circumstances are rain-fed (Sidhu, 2019). Low cutworm numbers can significantly lower the density of plants, particularly when bigger caterpillars of this cutworm are present at crop emergence (Showers *et al.*, 1983).

9.4.1.6 Risk Period and Action Threshold

The activity of moth *A. ipsilon, A. segetum*, and other cutworm species varies significantly throughout altered periods in various regions. March to August (Chandramohan, 1996) *and* November to April are considered the risk period (Yadav *et al.*, 1984; Nag and Nath, 1994). Moth activity peaked between March and April, with no activity between June and November (Vaishampayan and Singh, 1994). Almost three species of these cutworms, *A. ipsilon, A. segetum*, and *A. flammatra*, are the most common in Asia, accounting for 41.16–47.95%, 14.32–23.98%, and 15.08–23.77% prevalence, respectively (Bisht *et al.*, 2005).

The ETL level for *A. ipsilon* is 2–5% drooping or cutting of plants. The economic threshold level differs with cutworm size established on larvae feeding (Chandel *et al.*, 2013).

9.4.1.7 Pest Scouting and Monitoring

Light and pheromone traps monitor cutworms in cropping areas (Hendrix and Showers, 1990; Chandel *et al.*, 2013).

9.4.1.8 Integrated Management

Cutworms have been successfully managed using various strategies, including host-plant resistance and chemical and biological control (Walczak, 2002).

Cultural, mechanical, and physical control

The danger and severity of an *A. ipsilon* assault are decreased by autumn tillage, which buries crop remains, and spring tillage, which removes early spring weed growth before moth emergence (Ostlie and Potter, 2017). The larvae or pupae are also exposed by soil splitting or cultivating a month before spring seeding, making them easily accessible to generalist bird predators for eating (Chandel *et al.*, 2013). Always encourage predatory birds to visit the field by managing birdbaths and feeders near the field. Arrange weed piles at a different location in the field because larvae hide there, where the larvae can be collected and destroyed manually or chemically. Irrigation of the field regularly also helps to kill the larvae in the soil by suffocation. Drenching around the field, filling it with water, and dusting it with insecticide dust prove effective. Before sowing, the field should be chemigated with chlopyriphos (10 G @ 20 kg ha^{-1}).

Biological control

Periscepsia carbonaria (Panzer, 1798) and *Macrocentrus collaris* (Spinola, 1808), two parasitoids of *A. segetum*, are responsible for parasitization rates of 4.16–13.92% and 12.30–14.60%, respectively (Chandla, 1986; Kumari and Chandla, 2010). *Enicospilus merdarius* (Gravenhorst, 1829) is a significant parasitoid parasitizing *A. ipsilon*, *A. segetum*, and *A. spinifera* (Patel *et al.*, 1991). *Turanogonia chinensis* (Wiedemann, 1824), a tachinid larval-pupal parasitoid, parasitizes *A. segetum* up to 21.8% (Kalra, 1992). *Apanteles antipoda* Ashmead and *Cotesia ruficrus* (Haliday, 1834) parasitizes *A. ipsilon* larvae (Yousuf and Ray, 2009) up to 45% (Patil *et al.*, 2016).

Chemical control

Spreading of Bt mixed bran bait (2g Bt and 1 kg wheat bran) in piles or by broadcasting @ 10 kg bait ha^{-1} proves effective in attracting and killing larvae (Chandel and Chandla, 2003; Chandel *et al.*, 2008, 2013). Typically, it is advised to use chlorpyriphos 20EC @ 2 L to get rid of cutworms (Anonymous, 2018). The ETL for maize is 2–3% plant cut or wilted. The seedlings can also be treated with systemic insecticides like imidacloprid to give some defense against larvae damage (Bhagat, 2018). Emamectin benzoate, chlorpyriphos, clothianidin, cypermethrin, and triazophos are very effective (Chandel *et al.*, 2013; Anonymous, 2018; Chandel *et al.*, 2022).

9.4.2 Bugs (Chinch Bugs and Greenbugs)

9.4.2.1 Chinch Bugs

9.4.2.1.1 Pest Taxonomy

Kingdom: Animalia
Phylum: Arthropoda
Subphylum: Hexapoda
Class: Insecta
Order: Hemiptera
Family: Blissidae
Genus: *Blissus*
Species: *burmeister*

(Wikipedia, 2023, downloaded at site https://en.wikipedia.org/wiki/Insect on 03–01–2023)

9.4.2.1.2 Pest Status and Distribution

Blissidae (chinch bugs) belongs to the Lygaeoidea family with high species richness, and mostly resides in the Poaceae family of leaf sheaths. Lygaeoidea is Pentatomomorpha's second-biggest superfamily, with 14 families (Gao and Zhou, 2021). Blissidae comprises more than 420 species of 55 genera, extensively dispersed in tropical and subtropical regions in Southeast Asia, Australia, and China (Gao and Malipatil, 2019). Most are herbivores, but chinch bugs, *Blissus leucopterus*, are the most destructive pest of different grasses (Hussain *et al.*, 2014). Almost all chinch bugs are sap feeders and reside in leaf sheaths, with morphological specialization and severe flattening (Minghetti *et al.*, 2020).

9.4.2.1.3 Pest Identification

The chinch bug has three stages: egg, nymph, and adult. Nymphs are smaller, have different colors, and lack wings. Red or orange chinch bug nymphs have a white stripe on their abdomen. Adults are 0.2 in. (4.7 mm) long and have two types of wings: brachypterous and macropterous.

9.4.2.1.4 Biological Description and Life Cycle

A female can lay up to 300 eggs in her life span, averaging 4–5 each day, depending on the host plant and circumstances. One-mm-long oblong eggs are laid singly under thatch, leaf sheaths, or on leaf surfaces. They are white or cream at first and then become orange. Throughout four instars, nymphs darken and grow wings. As temperatures rise, insects' metabolism and development speed faster. Therefore, southern chinch bugs grow faster in the summer. After 8–25 days, a nymph emerges from the egg and takes 34–94 days to mature (4–13 weeks combined). Females live 25–55 days longer than males (42–100 days) (Kerr, 1966).

9.4.2.1.5 Mode and Symbol of the Damage

Chinch bugs may cause significant damage to grasses quickly and can spread to unlimited plants. The first signs of damage appear as patches of yellowing grass that swiftly become brown and disappear if the pest is not treated. The grass looks like it has undergone a dry spell after a while. These patchily damaged sections are often spherical and develop outward as the bugs migrate from unhealthy to healthy grass. The chinch bug will often target and cause the most severe damage to the turf in open radiant zones or under dry spell pressure initially (Vasquez and Buss, 2006; Nikpay *et al.*, 2019).

9.4.2.1.6 Risk Period and Action Threshold

The action threshold for chinch bugs is 20–25 bugs/ft^2 in the infested area of the crop. The peak infestation period is usually from July to early September.

9.4.1.7 Pest Scouting and Monitoring

To properly manage any pest, monitor plant material for pest activity, abundance, and management susceptibility. Chinch bugs aggregate, which helps identify infestations. Adults can fly, but they prefer to stroll, making them easier to spot. When populations are high, individuals may rest on leaf blades or crawl across the field (Addesso *et al.*, 2012). Chinch bugs are most active in the middle of the afternoon on warm, sunny days. The easiest way to find them is to cut the grass near where it is dead and look at the moss, the undersides of grass stolons, and the soil's surface.

9.4.2.1.8 Integrated Management

Cultural control

- Proper irrigation, fertilization, and mowing increase the healthiness of turfgrass.
- Outbreaks of chinch bugs frequently coincide with periods of drought stress.
- Pest-resistant plants are an effective integrated pest management (IPM) strategy. Pest-resistant cultivars can withstand pest feeding, repel pests physically or chemically, or both (Buss *et al.*, 2018).

Biological control

- Big-eyed predator bugs, *Geocoris* spp. (Hemiptera: Geocoridae), *Xylocoris vicarius* (Hemiptera: Anthocoridae) and *Lasiochilus pallidulus* (Hemiptera: Lasiochilidae) are frequent.
- Parasite *Eumicrosoma benefica* Gahan (Hymenoptera: Scelionidae) (Cherry, 2011) is a potential parasitoid of chinch bug.

Chemical control

Various insecticides like clothianidin, imidacloprid or thiamethoxamm deltamethrin, cyfluthrin, bifenthrin, permethrin, and lambda-cyhalothrin can be used for chinch bug control (El Aalaoui *et al.*, 2022).

9.4.2.2 Greenbugs

9.4.2.2.1 Pest Taxonomy

Kingdom: Animalia
Phylum: Arthropoda
Subphylum: Hexapoda
Class: Insecta
Order: Homoptera
Family: Aphididae
Genus: *Schizaphis*
Species: *graminum*

(Wikipedia, 2023, downloaded at site https://en.wikipedia.org/wiki/Insect on 03–01–2023)

9.4.2.2.2 Pest Status and Distribution

The greenbug, *Schizaphis graminum* (Rondani) (Aphididae: Homoptera), is a pest of nearby 70 graminaceous (Vakhide, 2014). The greenbug is found in North, South, and Central America, Europe, Africa, and Middle East Asia (Macharia *et al.*, 2016). The greenbug was initially identified as a species that feeds and causes damage to bluegrass (Street *et al.*, 1978). The pest transmits the yellow dwarf mosaic virus in many cereal crops (Chapin *et al.*, 2001). Their attack was also found in Pakistan at different locations. Medium to high infestations were found in Charsadda, Multan, and Malkot during a preliminary assessment of Punjab and the North Western Frontier Province of Pakistan (Inayatullah, 1993). They attacked the different Poaceae family crops, including oats, barley, wheat, millet, maize, and meadow plants (brome, cocksfoot, fescue, or wild grass species) (Tofangsazi *et al.*, 2012).

9.4.2.2.3 Pest Identification

The adults are small, elongated, and without wings. The head is yellow- and green- straw-colored, while the thorax and abdomen are yellowish. The antenna of adults is uniformly dusky.

9.4.2.2.4 Biological Description and Life Cycle

The females lay eggs with parthenogenesis in warm and mild climates. At the same time, females mated with a winged male in the winter season. Nymphs are produced directly from females. Nymphs have three instars completed in 7 to 9 days. Adult green bugs lay 5–7 nymphs per day. These bugs are minor, 1.3–2.1 mm elongated and oval-shaped, with a head and first part of thorax straw to pale green and with light-medium to green-colored abdomen. Their cornicles are pale in color with dark tips. Their total life span is 35–40 days (Moraes *et al.*, 2004).

9.4.2.2.5 Mode and Symbol of the Damage

The greenbug has enzymes in its saliva that break down the plant's cell walls (Al-Mousawi *et al.*, 1983). Their nymphs and adults both nourish on the lower and upper side of leaves by sucking the phloem fluid from the leaves. Their initial feeding causes yellow and then red spots on the leaves. Their midrib becomes narcotic as the spots cover most of the leaf. Intense feeding causes broad leaf and root loss and, eventually, plant death. They produce honeydew, which produces sooty mold and leads to the chlorosis of leaves (Pendleton *et al.*, 2009). They also cause viral diseases such as barley dwarf virus in wheat and sorghum (Nuessly, 2005).

9.4.2.2.6 Risk Period and Action Threshold

The decision to treat plants with insecticides is 25–50 bugs per plant losses, which is an indication to control this pest.

9.4.2.2.7 Pest Scouting and Monitoring

The field should be checked daily from the seedling to the tillering stage. If discoloration occurs in the field, it should be checked for bugs.

9.4.2.2.8 Integrated Management

Biological control

- *Lysiphlebus testaceipes* (Jones *et al.*, 2014) and *Diaeretiella rapae* (Jokar *et al.*, 2012) are two natural parasitoids of greenbugs.
- *Coccinella septempunctata*, *Hippodamia convergence* (Royer *et al.*, 2008), Syrphid flies, and green lacewing (Jessie *et al.*, 2019) are predators of greenbugs and play an important role in overcoming the natural field population of greenbugs.

Cultural control

- One or more resistant varieties are used to overcome greenbugs (Mohammadi *et al.*, 2018).
- Intercropping with an alternate host may reduce its damage.
- Greenbugs are less abundant in late-sowing crops than the early ones.

Chemical control

Spinosad and nicotinoid (Emam *et al.*, 2013), Thiomethoxam (Moscardini *et al.*, 2014), and many other insecticides like clothianidin, imidacloprid or thiamethoxamm deltamethrin, cyfluthrin, bifenthrin, permethrin, and lambda-cyhalothrin can be used for its control

(El Aalaoui and Sbaghi, 2022)

9.4.3 Grasshoppers

All aspects regarding pest taxonomy, pest status and distribution, pest identification, biological description and life cycle, mode and symbol of the damage, risk period and action threshold, pest scouting and monitoring, and integrated management of grasshoppers are the same as mentioned above in Section 9.2.5.

9.5 INSECT PESTS OF COWPEA

9.5.1 Hoppers

Leaf hoppers (Homoptera: Cicadellidae) are the most common sap-feeding insect pests of cowpea crops, causing pod and foliage damage (Swaminathan *et al.*, 2007). There are three most common species of leaf hoppers in cowpea crops, that is, *Empoasca kerri*, *Empoasca signata*, and *Empoasca* spp. The leaf hopper (*Empoasca kerri* Pruthi) is the major sucking hopper among all three species in the early stage of the crop (Patel *et al.*, 2012).

9.5.1.1 Pest Taxonomy

Kingdom: Animalia
 Phylum: Arthropoda
  Subphylum: Hexapoda
   Class: Insecta
    Order: Homoptera
     Family: Cicadellidae
      Genus: *Empoasca*
       Species: *kerri*, *signata*

(Wikipedia, 2023, downloaded at site https://en.wikipedia.org/wiki/Insect on 03–01–2023)

9.5.1.2 Pest Status and Distribution

Leafhoppers were minor pests of their host, but recently, they changed their status and appeared to spread larger crop areas in severe patches, causing economic damage to crops (Rachappa *et al.*, 2016). These leaf hoppers' damage yield losses of 26–54% and in fodder up to 42.11% (Dabhade *et al.*, 2012; Nasruddin *et al.*, 2014). These leaf hoppers are widely distributed in Burma, China, India, Myanmar, Nepal, Sri Lanka, and Pakistan (Lu *et al.*, 2014).

9.5.1.3 Pest Identification

Leafhopper adults are yellowish green with two black spots on each front wing at the margins of the wings (apical). They also have two prominent black spots on the vertex of their head. Nymphs are also pale yellow or green yellow, having a wedge shape both in adults and nymphs (Tazerouni *et al.*, 2019). Adults are pale green and wedge-shaped, with a 1/8 to 1/4 in. size. Adults have spines on their hind legs. Nymphs are also wedged (wing pads are extended up to the fifth abdominal segment) in shape, and pale green eggs are elongated in shape and white to yellow (Tnau, 2016).

9.5.1.4 Biological Description and Life Cycle

Adult females deposit eggs in the leaf veins (Tnau, 2016). Females can lay 15–37 eggs, and their oviposition period is 5–7 days. Eggs hatch within 6–13 days depending on the temperature. The hatching period prolongs in summer compared to winter. They have five nymphal instars and become adults after 8–22 days. Their total life span is 29–42 days. Adult longevity of males and females is 9 and 17 days, respectively. In special environmental conditions, adults can live up to 102 days. They remain active around the year, but their peak duration is November to January.

9.5.1.5 Mode and Symbol of the Damage

Nymphs and adults of *Empoasca* species draw cell sap from vascular tissues of the leaves (most preferred are the first three-terminal leaves), and during sap sucking, they inject saliva, which disrupts the photosynthesis movements of the plant (Singh *et al.*, 2008). Feeding injury produces symptoms like yellowing foliage by damaging the cells, and leaves turn upward, which leads to stunted growth of the infested plants in the crop. 'Hopper Burn' is a common symptom on leaves due to hopper injuries (Nielsen *et al.*, 1990).

9.5.1.6 Risk Period and Action Threshold

They remain active around the year, but their peak duration is November to January. Their action threshold is 5–10 nymphs, adults, or both plants (Nasruddin *et al.*, 2016).

9.5.1.7 Pest Scouting and Monitoring

Cowpea crops should be monitored closely in the early vegetative stage of the crop. Twenty different plants from five different spots within the field should be checked to estimate the leafhopper's population/plant. The underside of leaves should be checked for nymphs and adults (Chasen *et al.*, 2014). A sweep net should be used to estimate the population of nymphs and adults/sweep net in the field (Krupke, 2016).

9.5.1.8 Integrated Management

- Intercropping of cowpea with maize, sorghum, finger millet, cassava, and green gram is the major contributor to reducing the infestation of sap-sucking hoppers in many regions worldwide (Karungi *et al.*, 2000).
- Manipulating the planting dates (early or late sowing) is also a strategy to reduce pest attacks, 'host evasion,' and plant density of cowpea crops (Karungi *et al.*, 2000).
- Intercropping with seed dressing with chemicals is also helpful in pest infestation reduction. Handpicking nymphs and adults is impossible due to their minute size (Karungi *et al.*, 2000).
- Chemical control is sap-sucking insects' most effective management method (Sutaria *et al.*, 2010). Seed dressing and foliar application of some insecticides, such as neonicotinoids (imidacloprid, thiamethoxam, acetamiprid) alone and in the mixture have great efficacy against sap-sucking leafhoppers (Antu *et al.*, 2016).
- Pre-sowing seed treatment with imidacloprid (70 WS) @ 5 g per kg **seed** or thiomethoxam (25 WG) at 3 g per kg seed proves effective against the initial infestation of leafhoppers at the early growth stage of the crop.
- Nitenpyram 10% AS @ 200 mL, thiamethoxam 25 WG@ 24 g, thiacloprid 48 SC @ 125 g, acephate 97 DF @ 300 g, acephate 75 SP @ 250–375 g, chlorfenapyr 360 SP @ 75 g, etopfenprox 30 EC @ 200 mL, dinotefuron 20 SG @ 100 g, and pymetrazine 20% SC @ 300–400 mL per acre can be used when the infestation of leafhoppers reaches to its ETL.

9.5.2 Aphids

The most common and well-known cowpea pest is the aphid, *Aphis craccivora* Koch. (Hemiptera: Aphididae) around the world (Sadeghi *et al.*, 2009). Aphids are major sap-sucking and phloem feeders. Aphids conceal honeydew, increasing the survival chance of fungus spores present on the leaves of the plants. Aphids are important in transmitting plant diseases as vectors (Klingler *et al.*, 2001).

9.5.2.1 Pest Taxonomy

Kingdom: Animalia
Phylum: Arthropoda
Subphylum: Hexapoda
Class: Insecta
Order: Hemiptera
Family: Aphididae
Genus: *Aphis*
Species: *craccivora*

(Wikipedia, 2023, downloaded at site https://en.wikipedia.org/wiki/Insect on 03–01–2023)

9.5.2.2 Pest Status and Distribution

Aphis craccivora is a cosmopolitan insect pest and Palearctic origin. This aphid species is abundantly found in the Mediterranean area of subtropics and tropics. They heavily infest India, the Philippines, Thailand, the southern United States, Latin America, and tropical Africa (Jackai and Daoust, 1986). It has been expanded to Siberia, Canada, Chile, and Argentina (Berim, 2015).

9.5.2.3 Pest Identification

These cowpea aphids are minute dark brown–bodied insects with black-and-white legs. Nymphs are gray with a waxy appearance on their body. The adult's size is almost 2 mm in length. They feed on the tender shoots of immature leaves and pods of the plants. There are three life stages. that is, eggs, nymphs, and adults (Obopile and Ositile, 2010). The slightly dusted and waxed immature nymphs are shiny black with white to yellow antenna tips. They have six segmented antennae and are known as winged and unwinged aphids from 1.2 to 2.2 mm in size with cross-bars on their abdomen in females (Blackman and Eastop, 2000).

9.5.2.4 Biological Description and Life Cycle

They are ovoviviparous, in which females give birth to young ones. Females can live up to 9–25 days and molt four times in a life span (Berim, 2015). The optimal temperature and relative humidity for their survival are from 24 to 28.5°C and 65%, respectively (Mayeux, 1984).

9.5.2.5 Mode and Symbol of the Damage

It is a polyphagous pest like other species of aphids with a range of 50 host plants from the Leguminosae family (Mehrparvar *et al.*, 2012). Adults and nymphs attack seedlings later on flowers, buds, and pods in cowpea in hot and dry areas. Losses due to these aphids have been reported from 20–100% in cowpea crops (Kataria and Kumar, 2013). These aphids damage plants in two ways, that is, directly and indirectly. Aphids suck the cell sap and inject saliva into the plant parts where they feed, which favors the growth of fungal spores, dis-translocate the flux of phloem, and alters or hinder the photosynthetic activities of the plants; due to this, plants become sick and stunted (Kataria and Kumar, 2013). Light absorption is reduced in plant-effective areas due to sticky honeydew (Smith and Boyko, 2007). They also affect the nutrient absorption of plants by disrupting the physiological mechanism of mineral elements (Shegro *et al.*, 2012; Gerrano *et al.*, 2017).

9.5.2.6 Risk Period and Action Threshold

The peak risk period of this aphid is from April to June and then September to November. The action threshold of this *A. craccivora* is not established in cowpea crops yet (Hertel *et al.*, 2013).

9.5.2.7 Pest Scouting and Monitoring

See Section 9.2.4.

9.5.2.8 Integrated Management

See Section 9.2.4.

9.5.3 Foliage Beetles

Cowpea crop is prone to pest infestation in both field and storage stages. During storage, quantitative and qualitative losses arise from physical, chemical, or biological factors such as insects as their pests (Emeasor *et al*., 2007). Foliage beetles such as *Callosobruchus chinensis*, *Callosobruchus maculatus*, *Callosobruchus phaseoli*, and *Callosobruchus pulcher* are important pests of cowpea and legume crops (Prevett, 1961). The *C. maculatus* is a cosmopolitan pest of cowpea and has economic importance by infesting crops, which results in dry weight loss, nutritional value reduction, and reduced seed germination ability that ultimately reduces the commercial value of peas (Ofuya, 2001). Infestation initially starts in the field and increases during the storage of seeds (Adedire *et al*., 2004). This *C. maculatus* s infestation is up to 100% after 6 months of storage (Maina, 2011), resulting in 60% weight loss of seeds (Umeozor, 2005).

9.5.3.1 Pest Taxonomy

Kingdom: Animalia
Phylum: Arthropoda
Subphylum: Hexapoda
Class: Insecta
Order: Coleoptera
Family: Chrysomelidae
Genus: *Callosobruchus*
Species: *maculatus*, *phaseoli*, *pulcher*, *chinensis*

(Wikipedia, 2023, downloaded at site https://en.wikipedia.org/wiki/Insect on 03–01–2023)

9.5.3.2 Pest Status and Distribution

The cosmopolitan pest *C. maculatus* originates in Africa. Later, it spread to tropical and subtropical areas where its hosts are available (Beck and Blumer, 2014). It is a highly serious, economically important insect pest of legumes in the field and storage. *C. maculatus* population can grow exponentially by reducing the crop's market value (Beck and Blumer, 2014; Southgate, 1979; Vásquez *et al*., 2005). *C. maculatus* and other species, that is, *phaseoli*, *pulcher*, and *chinensis*, are beetles with their common names cowpea weevil (not true weevils) or cowpea seed beetle. These are the leaf beetle family Chrysomelidae members with cosmopolitan distribution on every continent except Antarctica (Tran and Credland, 1995; Cope and Fox, 2003).

9.5.3.3 Pest Identification

The cowpea weevil, unlike true weevil, completely lacks the 'snout'. It is elongated in shape and reddish brown. It bears black/gray very hard forewings (elytra) with two central black markings/spots. The last abdominal segment also bears two spots of black color. The female and male individuals of this sexually dimorphic beetle are darker and brown, respectively, with anal plates large and dark in females and smaller and lighter in males (Beck and Blumer, 2014; Utida, 1972; Fatima *et al*., 2016).

9.5.3.4 Biological Description and Life Cycle

C. maculatus is a holometabolic (life stages; egg, larva, pupa, adult) insect. It has two types of morphs, one is flightless or sedentary, and the other is a flight or dispersal morph. Both morphs are physiologically and morphologically different from one another. Sedentary morphs have a short life span and high fecundity than dispersal. Sedentary morphs distinguish male and female characters, but dispersal morph has no distinguishing characteristics (Beck and Blumer, 2014; Utida, 1972). These beetles are temperature-sensitive as they influence the generation time from the egg to the adult stage (Howe and Currie, 1964). One generation completion duration is 3–4 weeks at 30°C and 30% temperature and relative humidity, respectively (Beck and Blumer, 2014).

Eggs

A single female deposits eggs on seeds at a rate of more than 100 at a time per seed during its lifetime. Eggs are oval or spindle-shaped, with transparent color and gummy attachment to seeds (Beck and Blumer, 2014). The size of the eggs is 0.74 mm and 0.38 mm in length and width, respectively. Mitchell (1975) declared that the foliage beetle's life cycle could be increased when they lay eggs multiple times on each seed.

Larva

Larvae are whitish and opaque in color; they bore inside the seed upon hatching. Foliage beetles have four larval instars before pupation (Devereau *et al.*, 2003).

Pupae

The pupa of these beetles is white, having sizes of 3.87 mm and 1.76 mm in length and width, respectively. The seed's shell turns thinner when larvae pupate inside the shell (Beck and Blumer, 2014).

Adults

Adults of *C. maculatus* are metallic in color having pale spots on its body. Both male and female head color is black, and wings are brown with black patches. A distinguishing character at the end of the abdomen (plate's color and size) is visible in both females (plate's size large with longitudinal line) and males (small plate size and stripes absent). After 24 to 48 hours of emergence, they become sexually mature and live for 2–3 weeks (Edvardsson and Tregenza, 2005). Females prefer to mate multiple times with virgin males. Those females that mate multiple times lay larger eggs than those that mate only once in life (Messina and Slade, 1999).

9.5.3.5 Mode and Symbol of the Damage

After hatching, larvae feed on seeds and shells filled with frass. The larvae remain inside and pupate there. Adults chew the seed coat from the inside and come out of the seed by destroying it completely (Devereau *et al.*, 2003). Its larvae cause seed injury by feeding, which reduces the amount present in grains of carbohydrates and proteins that ultimately degrade the grains' nutritional quality, forcing the growers to sell their commodities at lower rates (Murdock *et al.*, 2003; Allotey and Oyewo, 1993). Hu *et al.*, 2009 estimated that *C. maculatus* caused 35%, 73%, and 7–13% yield/grain losses in Central America, Kenya, and South America, respectively.

9.5.3.6 Risk Period and Action Threshold

Its ETL is unknown yet under storage conditions (Hamdi *et al.*, 2017).

9.5.3.7 Pest Scouting and Monitoring

Grains are harvested promptly to avoid infestation and transfer to storage. After 2–3 weeks of storage, commodities are monitored to see the infestation; if infestation is present, then infested grains should be removed directly (Maina, 2011). Visual inspection is recommended under field conditions.

9.5.3.8 Integrated Management

Managing stored grain pests like *C. maculatus* has been inconvenient due to synthetic insecticides, which have environmental risks (Odeyemi *et al.*, 2006). Many botanical extracts show insecticidal properties against this pest (Sanon *et al.*, 2010). The extract of *Vitex negundo* and eucalyptus in powder was mixed with a black gram in the ratio of 3% (w/w mixture), which reduced the infestation rate and oviposition in their hosts (Rahman and Talukder, 2006). The essential oils (extracts) such as *Ocimum amricanum*, *Hyptis suaveolens*, and *Hyptis spicigera* have high mortality effects against *C. maculatus* (Ilboudo *et al.*, 2010). The jute bags, treated with plant leaf extracts (*A. indica, C. collinus*, and *J. curcas*), reduced the infestation and emergence rate (Raja *et al.*, 2001). Black pepper and clove oils also have insecticidal properties when seeds are treated with them (Mahdi and Rahman, 2008). Neem extracts act as repellent for this pest (Elhag, 2000).

Some biocontrol agents act as ectoparasitoids of *C. maculatus*. An ectoparasitoid (*D. basalis*) can parasitoid its population up to 80% with some essential oils (Ketoh *et al.*, 2005).

An autocidal control method (sterile insect technique) is highly effective for managing stored grain insect pests (Fetoh, 2011; Bakri *et al.*, 2005). *Beauveria bassiana* can also reduce the *C. maculatus* population during storage (Shams *et al.*, 2011). Its larval and pupal stages can be controlled up to 100% by applying *Bacillus thuringiensis* (Malaikozhundan and vinodhini, 2018). Using gamma rays and cobalt-60 sources can prevent the proper development of all its growth stages (Bhalla *et al.*, 2008). These rays may induce semi-sterility in adults (Ibrahim *et al.*, 2017).

Temperature and relative humidity strongly influence the growth of *C. maculatus*. Higher temperatures, such as sunlight heating grains, can effectively control the *C. maculatus* (Adebayo and Anjorin, 2018). Reducing the temperature of godowns and freezing the whole area for a long period can completely control this pest (Johnson and Valero, 2000).

Bhalla *et al.* (2008) reported that all growth stages (egg, larva, pupa, adult) could effectively be controlled by low temperatures up to −14ºC. Microwave radiation up to 240 W with 2450 MHz for 2–3 minutes can destroy the pest (Barbosa *et al.*, 2017). Light spectrums such as white and red light play an important source (as a lethal factor) in reducing damage in stored grains (Kehinde *et al.*, 2019).

The use of high gaseous concentrations of ozone and exposure time up to 500–1500 ppm cause mortality in all growth stages (Pandiselvam *et al.*, 2019). Modifying the environment inside the storage houses, like increasing CO_2 and decreasing O_2 levels can significantly reduce egg and adult mortality (Ofuya and Reichmuth, 2002).

9.5.4 Pod Borer

Cowpea, *Vigna unguiculata* Walp, is the main legume crop in subtropical and tropical regions of southeastern Asia and Africa (Srinives *et al.*, 2007). Insect pests are the major constraints in the production of this crop (Mahalakshmi *et al.*, 2016). Pod borers are important cowpea pests causing damage from germination to maturity (Dhakal *et al.*, 2018; Choudhary *et al.*, 2017). Two important species (*Maruca vitrata* Fabricious and *Maruca testulalis* Geyer) of pod borer are prevalent in legumes attacking cowpea crops (Mia, 1998). The larvae of this borer make webs (feed inside the web) on the leaves, flowers, pods, and inflorescence. The highest infestation can be seen at the development stage of flowering and pods (Mahalakshmi *et al.*, 2016). Higher temperatures and relative humidity favor the infestation rate of this borer in cowpea crops (Akhilesh and Paras, 2005).

9.5.4.1 Pest Taxonomy

Kingdom: Animalia
- Phylum: Arthropoda
  - Subphylum: Hexapoda
    - Class: Insecta
    - Order: Lepidoptera
      - Family: Crambidae
        - Genus: *Maruca*
          - Species: *vitrata, testulalis*

(Wikipedia, 2023, downloaded at site https://en.wikipedia.org/wiki/Insect on 03-01-2023)

9.5.4.2 Pest Status and Distribution

The losses caused by *M. vitrata* or spotted borer are difficult to estimate, but according to an estimate, US$30 million in losses occur annually due to this pest (ICRISAT, 1992). Grain yield losses of 20–60%, 54.5%, 72%, and more than 50% are estimated in India, Bangladesh, Nigeria, and Nepal, respectively (Regmi *et al.*, 2012). Giraddi *et al.* (2000) reported 100% flower bud losses in black gram. Yield losses due to *M. vitrata* vary (25–85%) from host to host and depend on the variety and the country (Singh *et al.*, 1990). *M. vitrata* (Lepidoptera: Crambidae) originated in Indo-Malaysia and Australia (Rose and Singh, 1989; Margam *et al.*, 2011a). The most probable origin of the genus *Maruca* is Indo-Malaysia, but now it also supports an origin from Southeast Asia (Periasamy *et al.*, 2015). This pest has been widely distributed from Cape Verde Island (West Africa) to Fiji and Samoa. It also has been reported from Hawaii, Australia, and Southeast Asia (Caldwell, 1945). Central and South American areas, the Caribbean Islands, and southern Florida have also reported this pest (Heppner, 1995). As its host range is wide, it shows migratory behavior from coastal to dry savanna parts (south-to-north gradient; Arodokoun *et al.*, 2006). Spotted pod borer or *M. vitrata* (Lepidoptera: Crambidae) is a pan-tropical pest of legume crops, mainly cowpea. It has been known by different common names, that is, *Maruca* pod borer, bean pod borer, soybean pod borer, legume pod borer, and mung moth. It can destroy crops from 20–80% upon harvesting. Its favorite feeding sites are flower buds, young stems, flower peduncles, and young pods (FAO, 2015).

9.5.4.3 Pest Identification

The larvae are greenish white and bear brown heads. The back of each segment bears two pairs of dark spots. The adults bear forewings of light brown color that have white markings, while the hind wings are white, and their lateral edges bear brown markings (Srinivasan *et al.*, 2021)

9.5.4.4 Biological Description and Life Cycle

M. vitrata eggs are translucent and oval, having faint sculpturing on the chorion with a size of 0.65 mm and 0.45 mm in length and width, respectively. There are five larval instars with different head capsule sizes (1st instar: 0.16–0.19 mm, 2nd instar: 0.31–0.38 mm, 3rd instar: 0.5–0.63 mm, 4th instar: 0.75–1 mm, and 5th instar: 1.25–1.38 mm). The larvae's color varies from green to brown (Adati *et al.*, 2004). Theh pupa changes color with the development days, from light brown to red brown and dark brown before emergence. Male (with a distinct ring at the last abdominal segment) and female (the ring is absent) pupae are used to distinguish the character (Hassan, 2007). Adults of *M. vitrata* are moths with 13–25 mm wingspan and a dark brown body. Female body weight is more than males, and they have nine abdominal segments. Both males and females have specific abdominal characteristics, such as external genitalia, which look like a forked hairy tip, and in females, the abdomen is blunt-tipped (Ochieng, 1981).

The life cycle of this moth requires almost 22–25 days, depending on favorable environmental conditions. Adult moths are nocturnal and remain inactive during the day and rest under the host plant leaves (Lu *et al.*, 2007). Adult life span may increase or decrease (minimum 3–10 and maximum 29) days depending on food quality and temperature (Naveen *et al.*, 2009; Chi *et al.*, 2005). Pupae usually emerge at night (Huang and Peng, 2001). Adult moths mate after 3 days of their emergence from pupa (Lu *et al.*, 2007). They mate between the night's 4th and 12th hour of darkness when the temperature (20–25°C) and relative humidity (80%) is optimum. Females mate once, and males multiple times during their life span (Hassan, 2007). Gravid females oviposit singly or in batches (2–6) on vegetative parts of host plants just after the 3–4 days of pairing. A female can lay 200–800 eggs (Naveen *et al.*, 2009). Eggs start hatching after 3–5 days, and larvae emerge in the evening. Larval development time is completed in 8–14 days (usually requires 24 days) depending upon abiotic factors (Adati *et al.*, 2004). The pupal duration is almost 3–14 days, and larvae stick to the host plant within a silken cocoon (Adati *et al.*, 2004). Its population does not undergo diapause as it is maintained on various host plants (Tamò *et al.*, 2002).

9.5.4.5 Mode and Symbol of the Damage

The larvae of *M. vitrata* are the most important growth stage, which infests cowpea plants. The larvae feed on tender portions of plants like stems, flowers, young pods, and flower buds but feed on flowering parts preferably (Liao and Lin, 2000; Chi *et al.*, 2003). Stamens and pistils are the targets of flower sites (Traore *et al.*, 2013). Initially, first instars were found feeding together inside the flowers, but after some time, they started dispersing or moving slowly from flower to flower, remaining alone per flower and destroying floral parts (Traore *et al.*, 2013). All five instars may consume almost 4 to 6 flowers before pupation to complete their life cycle (Traore *et al.*, 2013). Different instars larvae prefer to feed on flowers and pods. First- to second-instar larvae prefer to feed on flowers as they are less mobile, while the fourth and fifth larval instars prefer to feed on pods due to mobility (Liao and Lin, 2000). First-instar larvae can also be seen in a newly formed pod, while third- to fifth-instar larvae bore into the old pod and feed inside (Traore *et al.*, 2013).

9.5.4.6 Risk Period and Action Threshold

The peak risk period of *M. vitrata* has been observed in an early sown crop of cowpea in June and July. Depending on the temperature, they complete their one generation in July and the second between July and September. Their fecundity is low during off seasons as they prefer to migrate during November and December (Sharma *et al.*, 1999). The economic threshold for *M. vitrata* is one larva per 2 m row or 3 larvae plant^{-1} (Patel *et al.*, 2020; Srinivasan *et al.*, 2021).

9.5.4.7 Pest Scouting and Monitoring

Visual inspection: Damaged pods, prematurely dropped pods, and deformed shape pods on or off the vine can be seen clearly, which tells about the infestation. The naked eye or hand lens can be used to see eggs near the flower. Larvae entrance holes on the pods and seed pods indicate the infestation (Paddock, 1976). Sometimes, pupation occurs inside the pods, and wet frass exuded by larvae is visible. Monitoring can also be done by installing light traps or pheromone traps in the field (ARN, 2000).

9.5.4.8 Integrated Management

Farmers rely on many management methods based on financial status and knowledge (Rosendahl *et al.*, 2009). More than 90% of growers depend on chemical control (Bhattarai *et al.*, 2011). Insecticide application at the flower and pod stage may enhance yield by up to 100% (Kamara *et al.*, 2007). Some insecticides include methomyl, cypermethrin, alphacypermethrin, carbaryl, lambda-cyhalothrin, betacyfluthrin, emmamectine benzoate, indoxacarb, and spinosad are highly effective against spotted pod borer (Mohapatra and Srivastava, 2008). A mixture of deltamethrin, lambda-cyhalothrin, and dimethoate showed high toxicity at the pods infestation stage against *M. vitrata* (Kamara *et al.*, 2007). Cypermethrin and dimethoate in the mixture formed give great results in cowpea fields (Dzemo *et al.*, 2010).

Parasitoids, that is, *Phanerotoma lecobasis, Pristomerus* species, *Testudobracon* species, and *Apanteles taragamae*, are the egg and larval parasitoids of spotted pod borers (Srinivasan *et al.*, 2007; Arodokoun *et al.*, 2006). Nucleopolyhedrovirus and *M. anisoplae*, and *Beauveria bassiana* are entomopathogenic microbes that are highly effective against *M. vitrata* (Tamò *et al.*, 2011; Ekesi *et al.*, 2011; Lee *et al.*, 2007). *Bacillus thuringiensis* (endotoxins) showed high susceptibility against this insect pest (Yule and Srinivasan, 2013; Srinivasan, 2008).

Potentialities of neem concentration @ 50,000 ppm showed 90% larval mortality (Yule and Srinivasan, 2013). Ekesi (2000) reported that garlic bulb, *Piper guineese*, and neem seed extracts effectively reduced the egg hatching of spotted pod borers. Five percent neem seed kernel extract and 0.5 mL/L H_2O dichlorvos showed excellent efficacy against spotted pod borers (Gopali *et al.*, 2010).

Cultural practices like weeding in the crop, planting time and density, and intercropping are vital in reducing this pest infestation in cowpea crops (Asiwe *et al.*, 2005). The range of plant spacing 1–1.5 m and planting distance at 30×20 cm^2 or 60×20 cm^2 is highly recommended to reduce the spotted pod borer infestation (Asiwe *et al.*, 2005). Sandeep *et al.* (2013) evaluated that legume crop productivity can be increased by bird perching and summer plowing, which reduces borer damage by up to 85% in the field. Handpicking of eggs and larvae daily is also helpful in reducing the pest population, ultimately reducing the infestation.

Use of sex pheromone (E, E)—10, 12-Hexadecadienol and (E)—10-Hexadecenal) traps have high efficiency in Nigeria and Kano with light traps against *M. vitrata* moth trapping at night during the cowpea crop season (Downham *et al.*, 2003; Bottenberg *et al.*, 1997).

9.5.5 Sucking Bugs

The sucking bugs are the species complex of Hemiptera that attack pods and the seeds of cowpeas, *Vigna unguiculata* (L.) (Olufemi and Odebiyi, 2001). Among them, some are the most known species, i.e., *Clavigralla tomentosicollis* (Stal.), *Nezara viridula* L., *Mirperus jaculus* W., *Anoplocnemis curvipes* (Fab.), *Halymorpha halys* (Stal.), and *Riptortus dentipes* (Fab.) (Hoebeke and Carter, 2003). These sucking bugs feed primarily on fresh pods, tender shoots, and seeds by inserting their stylet and sucking the sap from plant parts (Pedigo, 2002; Lucini and Panizzi, 2018; Grozea *et al.*, 2016).

9.5.5.1 Pest Taxonomy

Kingdom: Animalia
Phylum: Arthropoda
Subphylum: Hexapoda
Class: Insecta
Order: Hemiptera
Family: Pentatomidae and Coreidae
Genus: *Halymorpha, Clavigralla, Riptortus*
Species: *Halymorpha halys, Clavigralla tomentosicollis, Riptortus dentipes*

(Hoebeke and Carter, 2003)

9.5.5.2 Pest Status and Distribution

The brown marmorated stink bug (*Halyomorpha halys* Stal.) is a highly polyphagous bug, basically native to China, Korea, Taiwan, and Japan, but now has been introduced in the United States (originating from China through a single introduction (Lee *et al.*, 2013a; Xu *et al.*, 2014). After its first introduction to the United States, it spread to Pennsylvania, Philipsburg, and then to other states (Hamilton, 2009; Nielsen *et al.*, 2013). Its population abundance exploded in the Mid-Atlantic states in 2010 (Leskey *et al.*, 2012b). *H. halys* has been spread in 41 states and the District of Columbia (Leskey *et al.*, 2014). Based on its nature, it has been known to different levels of categories; in West Virginia, New Jersey, Maryland, Pennsylvania, Virginia, and Delaware, it is a severe nuisance and agriculture pest, while in Ohio, New York, North Hampshire, and Tennessee, it is a nuisance and agriculture pest. It is only a nuisance insect pest in California, Michigan, New Hampshire, and Rhode Island (Leskey and Hamilton, 2012). In two states, Washington and Oregon, it has been shifted from a nuisance to only a pest of agriculture

(Wiman *et al.*, 2014). Despite its wide host range, this pest has established itself in other countries (New Zealand, North America, France, Germany, Italy, Canada, and the continents Australia and Asia) of the world due to its bioclimatic variables (Zhu *et al.*, 2012; Callot and Brua, 2013; Walker, 2009; Harris, 2010; Fogain and Graff, 2011).

9.5.5.3 Pest Identification

These bugs belong to the order Hemiptera and are hemimetabolous, having three developmental stages, that is, egg, nymph, and adult. These adult bugs have abdominal edges and white-and-black banding on the antenna that discriminates the character of this species. They have a 'shield' on their body in almost all stink bugs. Adults are almost 17 mm in length, having a brown-gray color. Male adults have claspers on their ventral abdominal segment, while these claspers are absent in females. Eggs are laid in the form of mass or group on the underside of the leaves; their color is light green upon laying later it changes into white before hatching (Taylor *et al.*, 2014).

9.5.5.4 Biological Description and Life Cycle

H. halys species is multivoltine, having 4–6 generations, 1–2 around the year (Nielson *et al.*, 2008a). Neonates emerge from eggs and feed on eggshells after 3–6 days of oviposition. A total of five nymphal instars have a blackish head, red color eyes, and orange-red abdomen along with markings and different longevities: second (3–5 days after first instar emergence), third (12–13 days of hatching, dark brown), fourth (19–20 days of hatching), and fifth (26–27 days of hatching). Thirty-two to 35 days are required to complete their life cycle at an optimum temperature of 30°C (Nielsen *et al.*, 2008). Some adults are nonreproductive and prefer to hibernate in natural or artificial shelters. Females of this pest are known to be polyandrous (Ingels and Varela, 2014).

9.5.5.5 Mode and Symbol of the Damage

Cowpea is an economically important crop throughout the tropics and subtropics in Asia, South America, and Africa (Dugje *et al.*, 2009). These pod-sucking bugs have over 100 host plants (Bergmann *et al.*, 2013). Feeding on a wide range of vegetable and legume crops causes significant injury to the marketable part of the crop. Scarred, deformed pods and faded sunken areas are some symptoms of the feeding of these bugs on beans. Oparaeke *et al.* (2000) reported that pod-sucking bugs are a serious pest of this crop. Their cylindrical nymphs and adults rest on foliage and feed over young shoots and pods (Oparaeke *et al.*, 2000). Nymphs and adults suck the cell sap from the crop's young, developing shoots and green pods (Kamara *et al.*, 2007: Dzemo *et al.*, 2010). This feeding causes deformed seeds, drying of premature seeds, and abortion of seeds and pods, ultimately resulting in the crop seed's economic loss (Egho, 2010; Degri *et al.*, 2012).

9.5.5.6 Risk Period and Action Threshold

Pod-sucking bugs can damage crops from the flowering stage to the final mature pods. The highest risk period is usually from early pod filling to late pod ripening; moreover, summer legumes remain at risk until pods do not mature properly. An action threshold of 2–3 bugs/10 plants is considered for taking action, or 2–3 bugs per meter row (CABI, 2016).

9.5.5.7 Pest Scouting and Monitoring

For the monitoring of cowpea-sucking bugs, they should be located after 45 days of planting crops. Monitoring in the morning is better as bugs remain active at this time. Shriveling pods at the early stage is an indication of damage. Feeding puncture wounds can be seen clearly on host plants. During monitoring, nymphs and adults can be found under cowpea leaves (CABI, 2016).

9.5.5.8 Integrated Management

Biological control

Natural enemies such as predators and parasitoids of egg, nymphal, and adult parasitoids of cowpea-sucking bugs have been reported in Asia and other regions of the world. The genus *Trissolcus* has a high level of parasitism ranging from 63–85%. Pathogens such as *Ophiocordyceps nutans* and intestinal viruses of *P. stali* from the order Hemiptera also greatly contribute to the mortality of these pod-sucking bugs (Qiu *et al*., 2007; Hou *et al*., 2009; Yang *et al*., 2009; Sasaki *et al*., 2012; Leskey *et al*., 2013; Talamas *et al*., 2013).

Cultural control

Intercropping of cowpea with maize, sorghum, finger millet, cassava, and green gram significantly reduces the infestation of pod-sucking bugs in many regions worldwide. Intercropping with seed dressing with chemicals is also helpful in pest infestation reduction (Isubikalu *et al*., 1999; Karungi *et al*., 2000). Manipulating the planting dates (early or late sowing) is also a strategy to reduce pest attacks, 'host evasion,' and plant density of cowpea crops.

Chemical control

Chemical control is the cheapest and most effective for insect pests. A wide range of insecticides includes Bifenthrin, permethrin, and fenpropathrin; neonictotinoids like thiamethoxam, clothianidin, and dinotefuron; carbamates like oxamyl and methomyl are highly effective against cowpea-sucking bugs (Leskey *et al*., 2012a, c; Kuhar *et al*., 2012a–d).

REFERENCES

Adati, T., S. Nakamura, M. Tamò, and K. Kawazu. 2004. Effect of Temperature on Development and Survival of the Legume Pod Borer, *Maruca vitrata* (Fabricius) (Lepidoptera: Pyralidae) Reared on a Semisynthetic Diet. *Applied Entomology & Zoology* 39, no. 1: 139–45.

Addesso, K. M., H. J. McAuslane, and R. Cherry. 2012. Aggregation Behavior of the Southern Chinch Bug (Hemiptera: Blissidae). *Environmental Entomology* 41, no. 4: 887–95.

Adebayo, R. A., and O. O. Anjorin. 2018. Assessment of Entomocidal Effects of Solar Radiation for the Management of Cowpea Seed Beetle, *Callosobruchus maculatus* (F.) (Coleoptera: Chrysomelidae) in Stored Cowpea. *Global Journal of Science Frontier Research* 18: 21–6.

Adedire, C.O., and J.O. Akinneye. 2004. Biological Activity of Tree Marigold, *Tithonia diversifolia* on Cowpea Seed Bruchid, *Callosobruchus maculatus* (Coleoptera: Bruchidae). *Annals of Applied Biology* 144, no. 2: 185–9.

Afzal, M., S.M. Rana, M.H. Babar, I.U. Haq, Z. Iqbal, and H.M. Saleem. 2014. Comparative Efficacy of New Insecticides against Whitefly, *Bemisia tabaci* (Genn.) and Jassid, *Amrasca devastans* (Dist.) on Cotton, Bt-121. *Biologia (Pakistan)* 60, no. 1: 117–21.

Agarwal, R. A., S.K. Banerjee, and K.N. Katiyar. 1978. Resistance to Insects in Cotton to *Amrasca devastans* (Distant). *Cotton Fiber Tropical* 33: 404–14.

Agri-Fact Sheet. 2002. *Grasshoppers: Life Cycle, Damage Asessment and Management Strategy*. Agdex 622–24, Alberta Agriculture, Food and Rural Development, Canada. .

Ahmed, S., A. Shahzad, M. Naeem, and M. Y. Ashraf. 2003. Determination of Efficacy of Cypermethrin, Regent and Carbofuran against *Chilo partellus* Swin. and Biochemical Changes Following Their Application in Maize Plants. *International Journal of Agriculture and Biology* 5: 30–5.

Akhilesh, K., and N. Paras. 2005. Study of the Effect of Meteorological Factors on the Population of Insect Pests Infesting UPAS 120 Cultivar of Pigeon Pea. *Journal of Maharashtra Agricultural University* 30: 190–2.

Alam, M.J., M.K. Nahar, M. Khatun, M. Rashid, and K. Ahmed. 2020. Impact Assessment of Different Sowing Dates on Maize Aphid, *Rhopalosiphum* madis infestation in Bangladesh. *Sustainability in Food & Agriculture* 1: 87–94.

Allotey, J., and A. Oyewo. 1993. Some Aspects of the Biology and Control of *Callosobruchus maculatus* (F.) on Some Stored Soybean, *Glycine max* (L.) Merr., Varieties under Tropical Conditions. *Proceedings of Ghana Science Association* 26: 14–18.

Al-Mousawi, A. H., P. E. Richardson, and R. L. Burton. 1983. Ultra Structural Studies on Green Bug (Hemiptera: Aphididae) Feeding Damage to Susceptible and Resistant Wheat Cultivars. *Annals of the Entomological Society of America* 71: 964–71.

Anonymous. 2018. Annual Report. *All India Network Project on White Grubs and Other Soil Arthropods*, Department of Entomology, Palampur, 2004, 23.

Antu, M., D. M. Korat, and P. Sreelakshmi. 2016. Management of Leaf Hopper Infesting Cowpea by Seed Treatment and Foliar Spray with Neonicotinoid Insecticides. *International Journal of Farm Sciences* 6, no. 4: 95–103.

Arif, M. J., M. D. Gogi, M. Sufian, A. Nawaz, and R. M. Sarfraz. 2017. Principles of Insect Pests' Management. In *Sustainable Insect Pest Management. MAS Computers, Press Market, Aminpur Bazar, Faisalabad, Pakistan.* © University of Agriculture, edited by M. J. Arif, J. E. Foster, and J. Molina-Ochoa. Pakistan: Faisalabad: 17–47.

Arn, H. 2000. The Pherolist. http://www-pherolist.slu.se/index.html.

Arodokoun, D. Y., M. Tamò, C. Cloutier, and J. Brodeur. 2006. Larval Parasitoids Occurring on *Maruca vitrata* Fabricius (Lepidoptera: Pyralidae) in Benin, West Africa. *Agriculture, Ecosystems & Environment* 113, no. 1–4: 320–5.

Arora, R. 2017. Emerging Technologies for Integrated Pest Management in Forage Crops. In *Theory and Practice of Integrated Pest Management*, edited by A. Arora, B. Singh, and A. K. Dhawan. Jodhpur: Scientific Publishers: 321–37.

Asiwe, J. A. N., S. Nokoe, L. E. N. Jackai, and F. K. Ewete. 2005. Does Varying Cowpea Spacing Provide Better Protection against Cowpea Pests? *Crop Protection* 24, no. 5: 465–71.

Asre, R., P. Gupta, and A. Pawar. 1983. Control of Sugarcane *Pyrilla* by Its Natural Enemies. *Indian Farming* 33: 37–8.

Atwal, A. S. 1976. *A Book of Agriculture Pests of India and South East Asia*. 2nd ed., 159. New Delhi, India: Kalyani Publisher: 159.

Auclair, J. L. 1967. Effects of Light and Sugars on Rearing the Cotton Aphid, *Aphis gossypii*, on a Germ-free and Holidic Diet. *Journal of Insect Physiology* 13, no. 8: 1247–68.

Bakri, A., K. Mehta, and D. R. Lance. 2005. Sterilizing Insects with Ionizing Radiation. In *Sterile Insect Technique*. Dordrecht: Springer: 233–68.

Bali, R., A. K. Singh, P. K. Kumawat, T. Hussain, M. Singh, R. N. Sharma, S. Bajiya, and S. S. Kakraliya. 2022. Seasonal Incidence and Eco-friendly Management of Jassids in Cluster Bean [*Cyamopsis tetragonoloba* (L.) Taub]. *Biological Forum—An International Journal*, 14, no. 1: 1316–20.

Balikai, R. A. 1998. Chemical Control of Shoot Fly, *Atherigona soccata* Rondani in *Rabi sorghum*. *Karnataka Journal of Agricultural Sciences* 11: 1082–4.

Barbosa, D. R. E. S., L. F. da Silva Fontes, P. R. R. Silva, J. A. Neves, A. F. de Melo, and A. B. Esteves Filho. 2017. Microwave Radiation to Control *Callosobruchus maculatus* (Coleoptera: Chrysomelidae) Larvae in Cowpea Cultivars. *Austral Entomology* 56, no. 1: 70–4.

Beck, C. W., and L. S. Blumer. 2014. *A Handbook on Bean Beetles, Callosobruchus Maculatus*. National Science Foundation, Emory University, or Morehouse College, USA.

Bergmann, E., K. M. Bernhard, G. Bernon, M. Bickerton, S. Gill, C. Gonzales, G. C. Hamilton, C. Hedstrom, K. Kamminga, and C. Koplinka-Loehr. 2013. Host Plants of the Brown Marmorated Stink Bug in the U.S. Accessed September 15, 2013. http://www.stopbmsb.org/where-is-bmsb/host-plants/#host_plants_table.

Berim, M. N. 2015. 'Pests: *Aphis craccivora* Koch. —Groundnut Aphid'. Interactive Agricultural Ecological Atlas of Russia and Neighboring Countries. *Agro Atlas*. Accessed February 12, 2015.

Bhagat, R. C. 2018. Diversity of Noctuid Caterpillars Belonging to Subfamily Amphipyrinae, Catocalinae, Heliothinae, Noctuinae and Plusiinae on Host Crops/Plants of Jammu and Kashmir State (India). *International Journal of Current Research in Biosciences & Plant Biology* 5, no. 1: 52–9.

Bhalla, S., K. Gupta, B. Lal, M. L. Kapur, and R. K. Khetarpal. 2008. Efficacy of Various Nonchemical Methods against Pulse Beetle, *Callosobruchus maculatus* (Fab.). In *Endure International Conference (12–15 October) on Diversifying Crop Protection*. France: La Grande Motte: 1–4.

Bhattacherjee, N. S. 1971. Studies on the Maize and Jowar Stem Borer, *Chilo zonellus* (Swinhoe). *Entomologist* 104, no. 1302–1303: 298–304.

Bhattarai, M., R. Patricio, S. Yule, M. H. Wu, and R. Srinivasan. 2011, 7th ASAE Conference Meeting the Challenges Facing Asian Agriculture and Agricultural Economics Toward a Sustainable Future, October 13–15, 2011, Hanoi, Vietnam.

Bhupender, S., N. Kumar, and H. Kumar. 2017. Seasonal Incidence and Management of Sorghum Shoot Fly, *Atherigona soccata* Rondani – A Review. *Forage Research* 42, no. 4: 218–24.

Bhutto, A. Akbar, N. M. Soomro, and M. F. Khan. 1971 28.2. *Increased Attack of Rice Stem Bore Complex Coupled with Enhanced Yield in Response of Nitrogen Application on Paddy Crop* vol. 2013. Pakistan: Entomological Society of Karachi: 175–80.

Bindra, O. S., and H. Singh. 1969. Pea Stem Borer, *Melanogromyza (Agromyza) phaseoli* Tryon. (Diptera, Agromyzidae). *Pesticides* 3, no. 7: 19–21.

Binning, R. R., J. Coats, X. Kong, R. L. Hellmich. 2015. Susceptibility to Bt Proteins Is Not Required for *Agrotis Ipsilon* Aversion to Bt Maize. *Pest Management Science* 71: 601–6.

Bisht, R. S., D. M. Tripathi, and P. N. Mishra. 2005. Seasonal Abundance of Cutworms (*Agrotis* spp: Noctuidae: Lepidoptera) Caught at Light Trap in Garhwal Himalaya. *Indian Journal of Entomology* 67, no. 3: 247–51.

Blackman, R. L., and V. F. Eastop. 2000. *Aphids on the World's Crops: An Identification and Information Guide*. 2nd ed. Chichester: Wiley: 414.

Bleszynski, S. 1970. A Revision of the World Species of *Chilo zincken* (Lepidoptera: Pyralidac). *Bulletin of the British Museum* 25: 101–95.

Boivin, G., T. Hance, and J. Brodeur. 2012. Aphid Parasitoids in Biological Control. *Canadian Journal of Plant Science* 92, no. 1: 1–12.

Boone, M. 2019. Species *Melanoplus* Differentials-Differential Grasshoppers. [Online]. Bug Guide. Accessed July 03, 2020.

Bottenberg, H., M. Tamò, D. Arodokoun, L. E. N. Jackai, B. B. Singh, and O. Youm. 1997. *Advances in Cowpea Research* vol. 1997. Ibadan, Nigeria: International Institute of Tropical Agriculture (IITA) and Japan International Center for Agricultural Sciences (JIRCAS), IITA: 271–84.

Braendle, C., G. K. Davis, J. A. Brisson, and D. L. Stern. 2006. Wing Dimorphism in Aphids. *Heredity* 97, no. 3: 192–9.

Bragard, C., P. Baptista, E. Chatzivassiliou, P. Gonthier, J. Jaques Miret, A. F. Justesen, A. MacLeod, C. S. Magnusson, P. Milonas, S. Parnell, R. Potting, P. L. Reignault, E. Stefani, H. Thulke, W. V. der Werf, A. V. Civera, L. Zappalà, A. Lucchi, P. Gómez, and J. Yuen. 2022. Commodity Risk Assessment of *Malus domestica* Plants from Turkey. *EFSA Journal. European Food Safety Authority* 20, no. 5: e07301.

Buss, E. A., B. Whitman, and A. G. Dale. 2018. *Managing Southern Chinch Bug in Warm Season Turfgrasses*. University of Florida-IFAS Extension. ENY-325/LH036.

Butani, D. 1972. Parasites and Predators Recorded on Insect Pests of Sugarcane in India. *Indian Sugar* 22: 17–31.

Butani, D. K., and M. G. Jotwani. 1983. Insects as a Limiting Factor in Vegetable Production. *Pesticides* 17: no. 9: 6–8.

Butani, D. K., and M. G. Jotwani. 1964. Sugarcane Leaf Hopper *Pyrilla perpusilla* walker-A Review. Indian Sugarcane [Journal] 9.

CABI. 2007. Crop protection compendium. *Selected Texts for Chilo partellus*. 2007 ed. CA B International. Wallingford, UK: CA B International.

CABI. 2016. https://doi.org/10.1079/PWKB.20167801446.

Caldwell, N. E. H. 1945. Bean Pests in Queensland. *Queensland Agricultural Journal* 60: 156–71.

Callot, H., and C. Brua. 2013. *Halyomorpha halys* (Stål, 1855), la Punaise Diabolique, Nouvelle Espèce pour la Faune de France (Heteroptera Pentatomidae). L'Entomologiste 69: 69–71.

Capinera, J. L. 1974. Damage to Asparagus Seedlings by *Brachycolus asparagi*. *Journal of Economic Entomology* 67, no. 3: 447–8.

Cease, A. J., J. J. Elser, E. P. Fenichel, J. C. Hadrich, J. F. Harrison, and B. E. Robinson. 2015. Living with Locusts: Connecting Soil Nitrogen, Locust Outbreaks, Livelihoods, and Livestock Markets. *BioScience* 65, no. 6: 551–8.

Chabi-Olaye, Adenirin, Fritz Schulthess, and Christian Borgemeister. 2008. Effects of Nitrogen and Potassium Combinations on Yields and Infestations of Maize by *Busseola fusca* (Lepidoptera: Noctuidae) in the Humid Forest of Cameroon. *Journal of Economic Entomology* 101, no. 1: 90–8.

Chan, C., A. Forbes, and D. Raworth. 1991. *Aphid-Transmitted Viruses and Their Vectors of the World. Research Branch. Agriculture Canada* vol. 216.

Chandel, R. S., and V. K. Chandla. 2003. Managing Tuber Damaging Pests of Potato. *Indian Horticulture* 48: 15–7.

Chandel, R. S., V. K. Chandla, K. S. Verma, and M. Pathania. 2013. Insect Pests of Potato in India: Biology and Management. In *Insect Pests of Potato: Global Perspectives on Biology and Management*, edited by P. Giordanengo, C. Vincent, and A. Alyokhin. Waltham, MA: Academic Press Elsevier, Inc: 227–70.

Chandel, R. S., K. R. Dhiman, V. K. Chandla, and R. Desh. 2008. Insect Pests of Potato-I: Root and Tuber Eating Pests. *Pestology* 32: 39–46.

Chandel, R. S., K. S. Verma, A. Rana, S. Sanjta, A. Badiyala, S. Vashisth, R. Kumar, and A. S. Baloda. 2022. The Ecology and Management of Cutworms in India. *Oriental Insects* 56, no. 2: 245–70.

Chandla, V. K. 1986. *Insect pest complex of potato crop in Shimla Hills and their management* [PhD thesis]. Solan (H.P.) (India): Department of entomology–Apiculture. Dr. Y.S. Parmar University of Horticulture and Forestry.

Chandramohan, N. 1996. Light Trap Catches of Cutworms in the Nilgiris. Madras. *Agricultural Journal* 83, no. 1: 503–5.

Chapin, J. W., J. S. Thomas, S. M. Gray, D. M. Smith, and S. E. Halbert. 2001. Seasonal Abundance of Aphids (Homoptera: Aphididae) in Wheat and Their Role as Barley Yellow Dwarf Virus Vectors in the South Carolina Coastal Plain. *Journal of Economic Entomology* 94, no. 2: 410–21.

Chasen, Dietrich, Backus, and Cullen. 2014. *Potato Leafhopper Ecology and IPM Focused on Alfalfa. JIPM.*

Chaudhary, J., and S. Sharma. 1988. Biological Control of *Pyrilla* Using Parasites and Predators. *Biological Technology for Sugarcane Pest Management*: 186–206.

Chernoh, E. 2014. Maize Stalk Borers. *Pliantwise Knowledge Bank www.plantwise.org/knowledgebank and BIONET-EAFRINET*, Accessed June 20, 2016. http://keys.lucidcentral.org/keys/v3/eafrinet/index.htm.

Cherry, R. 2011. Distribution of *Eumicrosoma benefica* (Hymenoptera: Scelionidae) in Southern Chinch Bug (Hemiptera: Blissidae) Populations. *Florida. Entomologist* 94: 352–3.

Chi, Y., Y. Sakamaki, K. Tsuda, and K. Kusigemachi. 2003. The Seasonal Abundance of the Legume Pod Borer, *Maruca vitrata* in Kagoshima, Japan. *Memoirs of the Faculty of Agriculture Kagoshima University* 38: 41–4.

Chi, Y., Y. Sakamaki, K. Tsuda, and K. Kusigemati. 2005. Effect of Temperature on Oviposition and Adult Longevity of the Legume Pod Borer, *Maruca vitrata* (Fabricius) (Lepidoptera: Crambidae). *Japanese Journal of Applied Entomology & Zoology* 49, no. 1: 29–32.

Choudhary, A. L., A. Hussain, and M. D. Choudhary. 2017. Bioefficacy of Newer Insecticides against Aphid. *Aphis Craccivora Koch on Cowpea* 6, no. 4: 1788–92.

Cigliano, M. M., D. M. L. Wysiecki, and C. Lange. 2000. Grasshopper (Orthoptera, Acrididae) Species Diversity in the Pampas, Argentina. *Diversity & Distributions* 6: 81–91.

Cope, J. M., and C. W. Fox. 2003. Oviposition Decisions in the Seed Beetle, *Callosobruchus maculatus* (Coleoptera: Bruchidae): Effects of Seed Size on Superparasitism. *Journal of Stored Products Research* 39, no. 4: 355–65.

Cotterell, G. 1954. Notes on Insects Injuries to Crops in Afganistan. *Plant Protection Bulletin.* Food and Agriculture Organization 2: 53–5.

Dabhade, P. L., J. G. Bapodra, R. T. Jethva, R. T. Rathod, and M. V. Dhabi. 2012. Estimation of Yield Losses Due to Major Insect Pests of Groundnut in Gujarat. *Legume Research* 35, no. 4: 354–6.

Dalwadi, M. M., D. M. Korat, and B. D. Tank. 2007. Population Dynamics of Major Insect-Pests of Indian Bean in Relation to Weather Parameters. *Research on Crops* 8, no. 3: 672–7.

"Datasheet—*Aphis craccivora*". CAB International. 6 December 2013. Accessed July 16, 2014.

Daware, D. G., P. P. Ambilwade, R. J. Kamble, and B. B. Bhosle. 2011. Bio-efficacy of Insecticides Against Sorghum Shoot Fly *Atherigona soccata* (Rondani). *Indian Journal of Entomology* 73, no. 3: 227–9.

Dean, G. J. 1979. The Major Insect Pests of Rice, Sugarcane and Jute in Bangladesh. *Pans* 25, no. 4: 378–85.

Degri, M. M., Y. T. Maina, and B. I. Richard. 2012. Effects of Plant Extracts on Post Flowering Insect Pests and Grain Yield of Cowpea (*Vigna unguiculata* (L.) Walp.) in Maiduguri, Semi Arid Zone of Nigeria. *Journal of Biology, Agriculture & Healthcare* 2, no. 3: 46–51.

Delabie, J. H. C. 2001. Trophobiosis between Formicidae and Hemiptera (Sternorrhyncha and Auchenorrhyncha): An Overview. *Neotropical Entomology* 30, no. 4: 501–16.

Delobel, A. G. L., and M. C. Lubega. 1984. Rainfall as a Mortality Factor in the Sorghum Shootfly, *Atherigona soccata* Rondani (Diptera: Muscidae) in Kenya. *Insect Science and Its Application* 2: 67–71.

De-Prins, J., and W. De-Prins. 2016. *Afromoths, Online Database of Afrotropical Moth Species (Lepidoptera). World Wide Web Electronic Publication*, Accessed June 22, 2016. http://www.afromoths.net.

Devereau, A. D., I. Gudrups, J. H. Appleby, and P. F. Credland. 2003. Automatic, Rapid Screening of Seed Resistance in Cowpea, *Vigna unguiculata* (L.) Walpers, to the Seed Beetle Callosobruchus maculatus (f.) (Coleoptera: Bruchidae) Using Acoustic Monitoring. *Journal of Stored Products Research* 39, no. 1: 117–29.

Devi, Y. K., S. Pal, and D. Seram. 2018. Okra Jassid, *Amrasca biguttula biguttula* (Ishida) (Hemiptera: Cicadellidae) Biology, Ecology and Management in Okra Cultivation. *Journal of Emerging Technologies & Innovative Research* 5: 333–43.

Dhakal, R., R. Ghimire, M. Sapkota, S. Thapa, A. K. Bhatta. 2018. Effects of Different Insecticides on Cowpea Aphid (*Aphis craccivora* Koch). *International Journal of Global Science Research* 5, no. 2: 819–28.

Dhaliwal, G., R. Arora, and A. Dhawan. 2004. Crop Losses Due to Insect Pests in Indian Agriculture: An Update. *Indian Journal of Ecology* 31, no. 1: 1–7.

Dhaliwal, Z., and S. Bains. 1985. Numerical Response of Some Important Parasitoids of *Pyrilla* perpusilla (Walker) to the Variable Host Population. *Indian Journal of Ecology* 12: 126–32.

Dhanda, S., S. Singh Yadav, S. Yadav, S. Kumari, and C. 2020. The population dynamics of insect-pest infesting tomato and there relation with different abiotic factor. ~ 1158 ~ J. Author. *Journal of Entomology & Zoology Studies* 8: 1158–61.

Dohlen, C. D., C. A. Rowe, and O. E. Heie. 2006. A Test of Morphological Hypotheses for Tribal and Subtribal Relationships of *Aphidinae* (Insecta: Hemiptera: Aphididae) Using DNA Sequences. *Molecular Phylogenetics and Evolution* 38: 316–29.

Downham, M. C. A., D. R. Hall, D. J. Chamberlain, A. Cork, D. I. Farman, M. Tamò, D. Dahounto, B. Datinon, and S. Adetonah. 2003. *Journal of Chemical Ecology* 29, no. 4: 989–1011.

Dugje, I. Y., L. O. Omoigui, F. Ekeleme, A. Y. Kamara, and H. Aleigbe. 2009. *Farmers' Guide to Cowpea Production in West Africa. IITA*. Ibadan: Nigeria: 19pp.

Duportets, L., C. Gadenne, M. C. Dufour, and F. Couillaud. 1998. The Pheromone Biosynthesis Activating Neuropeptide (PBAN) of the Black Cutworm Moth, *Agrotis ipsilon*: Immunohistochemistry, Molecular Characterization and Bioassay of Its Peptide Sequence. *Insect Biochemistry & Molecular Biology* 28, no. 8: 591–9.

Dwivedi, S. A. 2020. Sustainable Ecofriendly Management Polyphagous Defoliator Pests of Groundnut: A Review. *Plant Archives* 20: 282–7.

Dzemo, W. D., A. S. Niba, and J. A. N. Asiwe. 2010. *African Journal of Biotechnology* 9, no. 11: 1673–9.

Eastop, V. 1977. Worldwide Importance of Aphids as Virus Vectors. In *Aphids as Virus Vectors*, edited by K. F. Harris and K. Maramorosch. Academic Press, Elsevier Inc., USA. 3–62.

Eastop, V. F., and R. Blackman. 2000. *Aphids on the World's Crops. An Identification Guide*. 2nd ed. John Wiley & Sons Ltd, UK.

Edvardsson, M., and T. Tregenza. 2005. Why Do Male *Callosobruchus maculatus* Harm Their Mates? *Behavioral Ecology* 16, no. 4: 788–93.

Egho, E. O. 2010. Studies on the Control of Major Insect Pests and Yield of Cowpea (Vigna unguiculata (L.) Walp.) Under Calendar and Monitored Application of Synthetic Chemical in Abraka, Southern Nigeria. *Archieves of Applied Science Research* 2, no. 4: 224–34.

Ekesi, S. 2000. Effect of Volatiles and Crude Extracts of Different Plant Materials on Egg Viability of *Maruca Vitrata* and *Clavigralla romentosicollis*. *Phytoparasitica* 28, no. 4: 305–10.

Ekesi, S., R. S. Adamu, and N. K. Maniania. 2011. *Crop Protection* 21: 589–95.

El Aalaoui, M. E., and M. Sbaghi. 2022. Field and Laboratory Evaluation of d-Limonene, Mineral Oil, and Potassium Salts of Fatty Acid against *Nysius raphanus* (Hemiptera: Lygaeidae). *Journal of Agricultural & Urban Entomology* 38, no. 1: 1–15.

Elhag, E. A. 2000. Deterrent Effects of Some Botanical Products on Oviposition of the Cowpea Bruchid *Callosobruchus maculatus* (f.)(Coleoptera: Bruchidae). *International Journal of Pest Management* 46, no. 2: 109–13.

Emam, R. A., A. E. R. M. Ali, and S. H. El-Ghareeb. Mannaa. and S. M. *Abdel-Aal*. 2013. Effect of some insecticides in reducing the population of two aphid species, *Rhopalosiphum maidis* and *Schizaphis graminum* on sorghum varieties, Horus, and Dorado. Assiut. *J. Agricul. Sci* 44, no. 1: 38–51.

Emeasor, K. C., S. O. Emosairue, and R. O. Ogbuji. 2007. Preliminary Evaluation of The efficacy of Mixed Powders of *Piper guineense* (Schum and Thonn) and *Thevetia peruviana* (Persoon) against *Callosobruchus maculatus* (f.) (Coleoptera: Bruchidae). *Nigerian Journal of Entomology* 24: 114–8.

European Plant Protection Organisation (EPPO). 2014. Data Sheets on *Sesamia Calamistis*. http://gd.eppo.int/taxon/SESACA.

Evans, V. Arthur. 2007. Grasshopper, Crickets, and Katydids: Order Orthoptera. *Field Guide to Insects and Spiders of North America*. Sterling Publishers, Inc.: 94.

FAO. 2018. *Integrated Management of the Fall Armyworm on Maize a Guide for Farmer Field Schools in Africa*. Rome: Food and Agriculture Organization of the United Nations.

FAO. 2020. *World Food and Agriculture – Statistical Yearbook 2020*. Rome: Food and Agriculture Organization of the United Nations. https://doi.org/10.4060/cb1329en.

Fao, Org. *Archieved from the Original on 22 December 2015*. Retrieved June 9 2012.

Faria, C. A. d. 2005. *The Nutritional Value of Aphid Honeydew for Parasitoids of Lepidopteran Pests*. Université de Neuchâtel.

Farid, A. K., A. K. Amin, S. U. K. Khattak, and A. Sattar. 2007. Studies on Maize Stem Borer, *Chilo partellus* in Peshawar Valley. *Pakistan Journal of Zoology* 39, no. 2: 127–31.

Fatima, M. Shah, A. Usman, K. Sohail, M. Afzaal, B. Shah, M. Adnan, N. Ahmed, K. Junaid, S. R. A. Shah, and Inayat-Ur-Rahman. 2016. Rearing and Identification of *Callosobruchus maculatus* (Bruchidae: Coleoptera) in Chickpea. *Journal of Entomology & Zoology Studies* 4: 264–6.

Fennah, R. G. 1963. The Species of *Pyrilla* (Fulgoroidea: Lophopidae) in Ceylon and India. *Bulletin of Entomological Research* 53, no. 4: 715–35.

Fereres, Alberto, and Aranzazu Moreno. 2009. Behavioural Aspects Influencing Plant Virus Transmission by Homopteran Insects. *Virus Research* 141, no. 2: 158–68.

Fetoh, B. E. A. 2011. Latent Effects of Gamma Radiation on Certain Biological Aspects of the Red Palm Weevil (*Rhynchophorus ferrugineus* Olivier) as a New Control Technology. *J. Agric. Technol.* 7, no. 4: 1169–75.

Fishel, F., W. C. Bailey, M. L. Boyd, W. G. Johnson, M. H. O'Day, L. Sweets, and W. J. Wiebold. 2009. *Introduction to Crop Scouting*. Extension Publications (MU).

Fletcher, T. B. 1939. A New Epipyrops from India (Lep., epipyropidae). *Bulletin of Entomological Research* 30, no. 3: 293–4.

Flint, M. L. 1998. *Pests of the Garden and Small Farm*. 2nd ed. Oakland: University Calif. Agric. Nat. Res. Publ.: 3332.

Fogain, R., and S. Graff. 2011. First Records of the Invasive Pest, *Halyomorpha halys* (Hemiptera: Pentatomidae), in Ontario and Quebec. *Journal of the Entomological Society of Ontario* 142: 45–8.

Fuglie, K., M. Peters, and S. Burkart. 2021. The Extent and Economic Significance of Cultivated Forage Crops in Developing Countries. *Frontiers in Sustainable Food Systems* 5: 712136.

Fuleiro, J. R., and S. Rai. 1985. Determination of Vulnerable Stage of Crop Growth to Leafhopper Attack in Okra. *Indian Journal of Entomology* 47, no. 2: 238–9.

Gao, C., and Y. Zhou. 2021. Review of the Genus *Cavelerius* (Heteroptera: Blissidae) with Descriptions of Three New Species from China and Southeast Asia. *Acta Entomologica Musei Nationalis Pragae* 61, no. 1: 113–32.

Gao, C., and M. B. Malipatil. 2019. Revision of the Genus Sadoletus Distant, with Description of New Species from China and Australia (Hemiptera: Heteroptera: Heterogastridae). *Zootaxa* 4613, no. 2: 251–89.

Gerrano, A. S., W. S. J. Rensburg, and P. O. Adebola. 2017. Nutritional Composition of Immature Pods in Selected Cowpea [*Vigna unguiculata* (L.) Walp.] Genotypes in South Africa. *Australian Journal of Crop Science* 11, no. 2: 134–41.

Getu, E., A. Tadesse, M. Negeri, T. Tefera, H. Tsaheye, and A. Dejene. 2006. Review of Entomological Research on Maize, Sorghum and Millet Maize, Sorghum and Millet. Increasing Crop Production through Improved Plant Protection–Volume I: 167.

Ghanashyam, B., B. P. Achhami, P. N. Sharma Thakur, and R. P. Mainali. 2012. Review and Finding, Existing Problems and Future Strategies of Maize Entomological Research of Nepal. In *Proceedings of Workshop on Review and Strategy Development, of Entomological Research Works in Nepal*. Lalitpur: Khumaltar.

Giraddi, R. S., K. Amaranath, K. B. Chandrashekar, and R. S. Patil. 2000. Late Sowing and Dry Spell Cause Pest Outbreak in Kharif Pulses. *Insect & Environment* 6: 24.

Glen, D. M. 1973. The Food Requirements of *Blepharidopterus angulatus* (Heteroptera: Miridae) as a Predator of the Lime Aphid, *Eucallipterus tiliae*. *Entomologia Experimentalis & Applicata* 16, no. 2: 255–67.

Gogi, M. D., A. Nawaz, M. Safian, R. M. Sarfraz, and O. E. Liburd. 2017. Biorational Approaches in Pest Management. In *Sustainable Insect Pest Management. MAS Computers, Press Market, Aminpur Bazar, Faisalabad, Pakistan.* © University of Agriculture, edited by M. J. Arif, J. E. Foster, and J. Molina-Ochoa. Faisalabad, Pakistan: 231–83.

Gopali, J. B., T. Raju, D. M. Mannur, andY. Suhas. 2010. *Karnataka Journal ofAgricultural Sciences* 23: 35–8.

Gravenhorst, J. L. C. 1829. *Ichneumonologia Europaea.* Pars II. Vratislaviae, Wrocław, Western Poland, pp. 1–989.

Grozea, I., A. M. Virteiu, R. Stef, A. Cabaret, L. Molnar, V. Marcu, and D. Draga. 2016. The Spread of *Nezara viridula* (Hemiptera: Pentatomidae) Species from Its First Occurrence in Romania. *Bulletin of University of Agricultural Sciences & Veterinary Medicine Cluj-Napoca. Horticulture* 73, no. 2.

Guddewar, M. B., A. Shukla, R. Chander, S. Pandey, and M. L. Saini. 1994. *Tabenaemomama coronaria* R. Br. (Apocynaceae) A Potential Source of Botanical Insecticides. *Plant Protection Bulletin* 46: 1–5.

Gupta, M., and I. Ahmad. 1983. Morphology of the Indian Sugarcane Leafhopper, *Pyrilla perpusilla* Walker. *Folia Morphologica* 31, no. 4: 325–30.

Gupta, S., K. Handore, and P. Ip. 2010. Effect of Insecticides against *Chilo partellus* (Swinhoe) Damaging *Zea mays* (Maize). *International Journal of Parasitology Research* 2, no. 2: 4–7.

Hailu, G., S. Niassy, K. R. Zeyaur, N. Ochatum, and S. Subramanian. 2018. Maize–Legume Intercropping and Push–Pull for Management of Fall Armyworm, Stemborers, and Striga in Uganda. *Agronomy Journal* 110, no. 6: 2513–22.

Hajek, A. E., and R. St. Leger. 1994. Interactions between Fungal Pathogens and Insect Hosts. *Annual Review of Entomology* 39, no. 1: 293–322.

Halder, J., S. K. Sanwal, D. Deb, A. B. Rai, and B. Singh. 2016. Mechanisms of Physical and Biochemical Basis of Resistance against Leaf-Hopper (*Amrasca biguttula biguttula*) in Different Okra (*Abelmoschus esculentus*) Genotypes. *Indian Journal of Agricultural Sciences* 86, no. 4: 481–4.

Haliday, A. H. 1834. Essay on the Classification of Parasitic Hymenoptera. *Entomological Magazine* 2, no. ii: 93–106.

Hamdi, S. H., S. Abidi, D. Sfayhi, M. Z. Dhraief, M. Amri, E. Boushih, M. Hedjal-Chebheb, K. M. Larbi, and J. Mediouni Ben Jemâa. 2017. Nutritional Alterations and Damages to Stored Chickpea in Relation with the Pest Status of *Callosobruchus maculatus* (Chrysomelidae). *Journal of Asia-Pacific Entomology* 20, no. 4: 1067–76.

Hamilton, G. C. 2009. Brown Marmorated Stink Bug. *American Entomologist* 55, no. 1: 19–20.

Hand, S. C., and S. D. Wratten. 1985. Production of Sexual Morphs by the Monoecious Cereal Aphid *Sitobion avenae*. *Entomologia Experimentalis & Applicata* 38, no. 3: 239–47.

Harris, A. C. 2010. *Halyomorpha halys* (Hemiptera: Pentatomidae) and *Protaetia brevitarsis* (Coleoptera: Scarabeidae: Cetoniinanae) Intercepted in Dunedin. *Weta* 40: 42–4.

Hashmi, A. 1994. *Insect Pest Management.* Islamabad: Cereal and Cash Crops of Pakistan Agriculture Research Council.

Hassan, M. N. 2007. *Reinvestigation of the female sex pheromone of the legume podborer* Maruca vitrata *(Lepidoptera: Crambidae)* [PhD thesis]. UK: University of Greenwich.

Hendrix, W. H., and W. B. Showers. 1990. Evaluation of Differently Colored Bucket Traps for Black Cutworm and Armyworm (Lepidoptera: Noctuidae). *Journal of Economic Entomology* 83, no. 2: 596–8.

Heppner, J. B. 1995. Atlas of Neotropical Lepidoptera. *Checklist: Part 2* Hyblaeoidea-Pyraloidea—Tortricoidea. Miscellaneous atlas of Neotropical Lepidoptera.

Hertel, K., and K. Roberts. and P. *Bowden P.* 2013. Insect and Mite control in field crops. New South Wales DPI. ISSN 1441–1773.

Hewitt, J. A., and G. B. Onsager. 2000. A Method for Forecasting Potential Losses from Grasshopper Feeding on Northern Mixed Prairie Forages. *Journal of Range Management* 35: 53–7.

Hille Ris Lambers, D. 1966. Polymorphism in Aphididae. *Annual Review of Entomology* 11, no. 1: 47–78.

Hoebeke, E. R., and M. E. Carter. 2003. *Halyomorpha halys* (Stål) (Heteroptera: Pentatomidae): A Polyphagous Plant Pest from Asia Newly Detected in North America. *Proceedings of the Entomological Society of Washington* 105: 225–37.

Hou, Z., H. Liang, Q. Chen, Y. Hu, and H. Tian. 2009. Application of *Anastatus* sp. against *Halyomorpha halys*. *Forest Pest & Disease* 4, no. 39: 40: 43.

Howe, R. W., and J. E. Currie. 1964. Some Laboratory Observations on the Rates of Development, Mortality and Oviposition of Several Species of Bruchidae Breeding in Stored Pulses. *Bulletin of Entomological Research* 55, no. 3: 437–77.

Hu, Fei, Guo-Na Zhang, and Jin-Jun Wang. 2009. Scanning Electron Microscopy Studies of Antennal Sensilla of Bruchid Beetles, *Callosobruchus chinensis* (L.) and *Callosobruchus maculatus* (F.) (Coleoptera: Bruchidae). *Micron* 40, no. 3: 320–6.

Huang, C. C., and W. K. Peng. 2001. Emergence, Mating and Oviposition of the Bean Pod Borer, *Maruca vitrata* (F.) (Lepidoptera: Pyralidae). *Formosan Entomology* 21: 37–45.

Hunt, D. J., M. Luc, and R. H. Manzanilla-López. 2005. 2 Identification, Morphology and Biology. *Plant Parasitic Nematodes in Subtropical and Tropical Agriculture* 11.

Hussain, R., R. Perveen, M. Ali, and M. Kazim. 2014. New Record of Lygaeinae (Hemiptera: Heteroptera: Lygaeidae) from Pakistan (Parachinar), Khyber Pakhtunkhwa. International *J. Fauna. Biological. Studies* 1, no. 6: 50–3.

Hutchison, W. D., R. C. Venette, and D. Bergvinson. 2008. *Chilo partellus*: Pest Distribution Profile. *Harvet Choice*: 1–5pp.

Ibrahim, H. A., S. Fawki, M. M. A. Abd El-Bar, M. A. Abdou, D. M. Mahmoud, and E. E. El-Gohary. 2017. Inherited Influence of Low Dose Gamma Radiation on the Reproductive Potential and Spermiogenesis of the Cowpea Weevil, *Callosobruchus maculatus* (F.) (Coleoptera: Chrysomelidae). *Journal of Radiation Research & Applied Sciences* 10, no. 4: 338–47.

ICRISAT. 1992. *The Medium Term Plan*. International Crops Research Institute for the Semi-Arid Tropics. Patancheru, Andhra Pradesh, India.

Idechiil, O., R. Miller, K. Pike, and L. Hansen. 2007. Aphids (Hemiptera: Aphididae), Ants (Hymenoptera: Formicidae) and Associated Flora of Palau with Comparisons to Other Pacific Islands. *Micronesica* 39: 141–70.

Ilboudo, Z., L. C. B. Dabiré, R. C. H. Nébié, I. O. Dicko, S. Dugravot, A. M. Cortesero, and A. Sanon. 2010. Biological Activity and Persistence of Four Essential Oils towards the Main Pest of Stored Cowpeas, *Callosobruchus maculatus* (F.) (Coleoptera: Bruchidae). *Journal of Stored Products Research* 46, no. 2: 124–8.

Inayatullah, C., M. N. Nahid Ehsan-Ul-Haq, and M. F. Chaudhry. 1993. Incidence of Greenbug, *Schizaphis graminum* (Rondani) (Homoptera: Aphididae) in Pakistan and Resistance in Wheat against It. *International Journal of Tropical Insect Science* 14, no. 2: 247–54.

Ingels, M., and L. Varela. 2014. Pest Notes: Brown Marmorated Stink Bug. UC ANR Publication 74169. http://ipm.ucanr.edu/PMG/PESTNOTES/pn74169.html. University of California State-wide IPM Program.

Inter-African Phytosanitary Council. 1985. *Chilo Partellus (Swinhoe). Distribution Maps of Major Crop Pests and Diseases in Africa Map No. 199.* Yaoundé, Cameroon: Inter-African Phytosanitary Council.

Iqbal, J., H. Mansoor, A. Muhammad, T. Shahbaz, and A. Amjad. 2008. Screening of Okra Genotypes Against Jassid, *Amrasca biguttula biguttula* (Ishida)(Homoptera: Cicadellidae). *Pakistan Journal of Agricultural Sciences* 45, no. 4: 448–51.

Iqbal, J., S. U. Farooq, A. S. Alqarni, H. Ali, A. Zeshan, and M. J. Ansari. 2017. Management of Maize Stem Borer (*Chilo partellus*) with Insecticides of Three Formulations under Field Conditions. *Indian Journal of Agricultural Sciences* 87, no. 12: 1720–4.

Isubikalu, P., J. M. Erbaugh, A. R. Semana, and E. Adipalu. 1999. Influence of Farmer Production Goals on Cowpea Pest Management in Eastern Uganda: Implication for Developing IPM Programmes. *African Crop Science Journal* 7, no. 4: 539–48.

Jackai, L. E. N., and R. A. Daoust. 1986. Insect Pests of Cowpeas. Annual Review of Entomology. *Annual Reviews* 31, no. 1: 95–119.

Jayarao, B., S. B. Abulkhader, L. K. Naik, and M. M. Vinaykumar. 2015. Assessment of Biology and Morphometric Characteristics of Different Stages of Leafhopper, *Amrasca biguttula biguttula* (Ishida) on Okra. *Bisscan* 10, no. 2: 671–4.

Jayasimha, G. T., R. R. Rachana, M. J. Manjunatha, and V. B. Rajkumar. 2012. Biology and Seasonal Incidence of Leafhopper, *Amrasca biguttula biguttula* (Ishida) (Hemitpera: Cicadellidae) on Okra. *Pest Management in Horticultural Ecosystems* 18: 149–53.

Jaydeep, H., D. Kushwaha, and A. B. Rai. 2020. Biology and Feeding Potential of *Eocanthecona Furcellata* (Wolff) on Its Lesser Known Prey, *Spilosoma obliqua* (Walker). *Journal of Biological Control* 34, no. 2: 109–12.

Jessie, C. N., K. L. Giles, T. A. Royer, M. E. Payton, N. C. Elliott, and W. P. Jessie. 2019. Suitability of Schizaphis Graminum Parasitized by *Lysiphlebus testaceipes* as Intraguild Prey for *Chrysoperla rufilabris*. *Southwestern Entomologist* 44, no. 1: 21–33.

Johannsen, O. A. 1912. *Insect Notes for 1912*. Maine Agricultural Experiment Station.

John, B. C., and L. H. Gary. 2008. *A Guide to Grasshopper Control in Cropland*. NebGuide G1627, University of Nebraska-Lincoln Extension, Institute of Agriculture and Natural Resources. https://extension-publications.unl.edu/assets/pdf/g1627.pdf.

Johnson, J. A., and K. A. Valero. 2000. *Control of Cowpea Weevil, Callosobruchus maculatus, Using Freezing Temperatures*.

Jokar, M., M. Zarabi, S. Shahrokhi, and M. Rezapanah. 2012. Host-stage preference and functional response of aphid parasitoid *Diaeretiella rapae* (McIntosh) (Hym.: Braconidae) on greenbug. *Schizaphis graminum* (Rondani)(Hem: Aphididae). *Archi. Phytopathology. Plant Protection* 45, no. 18: 2223–35.

Jones, Douglas B., Kristopher L. Giles, N. C. Elliott, and M. E. Payton. 2007. Parasitism of Greenbug, *Schizaphis graminum*, by the Parasitoid *Lysiphlebus testaceipes* at Winter Temperatures. *Environmental Entomology* 36, no. 1: 1–8.

Jonsson, B., and N. Jonsson. 2001. Polymorphism and Speciation in Arctic Charr. *Journal of Fish Biology* 58, no. 3: 605–38.

Joshi, R., and S. Sharma. 1989. Augmentation and Conservation of *Epiricania melanoleuca* Fletcher, for the Population Management of Sugarcane Leafhopper, *Pyrilla perpusilla* walker, Under Arid Conditions of rajasthan. *Indian Sugar* 39: 625–8.

Kadioglu, Asim, Rabiye Terzi, Neslihan Saruhan, and Aykut Saglam. 2012. Current Advances in the Investigation of Leaf Rolling Caused by Biotic and Abiotic Stress Factors. *Plant Science* 182: 42–8.

Kaethner, M. 1991. No Side Effects of Neem Extracts on *Chrysoperla carnea* (Steph.) and *Coccinella septempunctata* L. [*Anzeiger fuer Schaedlingskunde Pflanzenschutz Umweltschutz* (*Germany, FR*] 64: 97–9.

Kahate, N. S., S. M. Raut, P. H. Ulemale, and A. F. Bhogave. 2014. Management of Sorghum Shoot Fly. *Popular Kheti* 2: 72–4.

Kalpande, V. V., B. A. Sakhare, R. B. Ghorade, and A. P. Lad. 2015. Potential Parents for Yield and Its Component Traits in Rabi Sorghum. *Bioscan* 10, no. 2: 837–39.

Kalra, V. K. 1992. Parasitism of *Agrotis segetum* Denis and Schiff. by *Turanogonia chinensis* Wied. on *Brassica* Cultivars in Haryana. *Journal of Insect Science* 5, no. 1: 86–7.

Kamara, A. Y., D. Chikeye, L. O. Omoigui, and I. Dugje. 2007. Influence of Insecticide Spraying Regimes and Cultivars on Insect Pests and Yield of Cowpea in the Dry Savanna of North Eastern Nigeria. *Journal of Food, Agriculture & Environment* 5, no. 1: 154–8.

Kamara, A. Y., D. Chikoye, L. O. Omoigui, and I. Y. Dugje. 2007. *Journal of Food Agriculture & Enviroment* 5: 154–8.

Kamatar, M. Y., and P. M. Salimath. 2003. To Shootfly, *Atherigona soccata* Rondani. *Indian Journal of Plant Protection* 31, no. 1: 73–7.

Kambhampati, S., W. Völkl, and M. Mackauer. 2000. Phylogenetic Relationships among Genera of Aphidiinae (Hymenoptera: Braconidae) Based on DNA Sequence of the Mitochondrial 16S RRNA Gene. *Systematic Entomology* 25, no. 4: 437–45.

Karungi, J., E. Adipala, S. Kyamanywa, M. W. Ogenga-Latigo, N. Oyobo, and L. E. N. Jackai. 2000. Pest Management in Cowpea. Part 2. Integrating Planting Time, Plant Density and Insecticide Application for the Management of Cowpea Field Pests in Eastern Uganda. *Crop Protection* 19, no. 4: 343–7.

Kataria, R., and D. Kumar. 2013. On the Aphid-Ant Association and Its Relationship with Various Host Plants in the Agroecosystems of Vadodara, Gujarat, India. *Halteres* 4: 25–32.

Kaufman, K., Eaton, and R. Eric. 2007. Grasshoppers. *Kaufman Field Guide to Insects of North America*. Houghton Mifflin Books: 74–5.

Keerthi, M. C., P. A. Somasekhar, M. Bheemanna, and D. Krishnamurthy. 2017. Effect of Weather Parameter on Seasonal Incidence of Sorghum Shoot Fly, *Atherigona soccata* Rondani (Diptera: Muscidae). *Journal of Entomology & Zoology Studies* 5, no. 5: 1684–7.

Kehinde, F. O., G. A. Dedeke, O. I. Popoola, and P. O. Isibor. 2019. Potential of Light Spectra as a Control of Cowpea Weevil, *Callosobruchus maculatus*, Activity. *IOP Conference Series: Earth & Environmental Science*, 8th International Biotechnology Conference, Exhibition and WorkshopMarch 11–13 2018, Nigeria 210.

Kerr, S. H. 1966. Biology of the Lawn Chinch Bug, Blissus insularis. *Florida Entomologist* 49, no. 1: 9–18.

Ketoh, G. K., H. K. Koumaglo, and I. A. Glitho. 2005. Inhibition of *Callosobruchus maculatus* (f.) (Coleoptera: Bruchidae) Development with Essential Oil Extracted from *Cymbopogon schoenanthus* L. Spreng. (Poaceae), and the Wasp *Dinarmus basalis* (Rondani) (Hymenoptera: Pteromalidae). *Journal of Stored Products Research* 41, no. 4: 363–71.

Kfir, Rami, W. A. Overholt, Z. R. Khan, and A. Polaszek. 2002. Biology and Management of Economically Important Lepidopteran Cereal Stem Borers in Africa. *Annual Review of Entomology* 47: 701–31.

Khadioli, N., Z. E. H. Tonnang, E. Muchugu, G. Ong'amo, T. Achia, I. Kipchirchir, J. Kroschel, and B. R. Le Ru. 2014. Effect of Temperature on the Phenology of *Chilo partellus* (Swinhoe) (Lepidoptera, Crambidae); Simulation of Life-Table Parameters and Visualization of Spatial Pest's Risk in Africa. *Bulletin of Entomological Research* 104, no. 6: 809–22.

Khaliq, A., S. Masood, H. Rauf, U. Faheem, K. Mahmood, A. Basit, and T. Muhammad. 2022. Efficacy of Seed Dressing Insecticides at Different Doses for the Control of Sorghum Shoot Fly *Atherigona soccata* (Rond.) (Diptera: Muscidae). *Biological. Clinical. Science & Research J.*1: 1–8.

Khan, M. R., and S. Khan. 1966. Bionomics and Control of *Pyrilla Pusana* Distant in Peshawar Region. *Agriculture Pakistan* 17: 385–414.

Klingler, J., I. Kovalski, L. Silberstein, G. A. Thompson, and R. Perl-Treves. 2001. Mapping of Cotton-Melon Aphid Resistance in Melon. *Journal of the American Society for Horticultural Science* 126, no. 1: 56–63.

Koralewski, T. E., H.-H. Wang, W. E. Grant, M. J. Brewer, N. C. Elliott, J. K. Westbrook, A. Szczepaniec, A. Knutson, K. L. Giles, and J. P. Michaud. 2020. Integrating Models of Atmospheric Dispersion and Crop-Pest Dynamics: Linking Detection of Local Aphid Infestations to Forecasts of Region-Wide Invasion of Cereal Crops. *Annals of the Entomological Society of America* 113, no. 2: 79–87.

Krishnaiah, N. V., and M. B. Kalode. 1984. Evaluation of Neem Oil, Neem Cake and Other Nonedible Oil Cakes against Rice Pests. *Indian Journal of Plant Protection* 12, no. 2: 101–7.

Kriti, J. S., M. A. Dar, and H. Z. Khan. 2014. Biological and Taxonomic Study of Agriculturally Important Noctuid Pests of Kashmir. *World Journal of Agricultural Research* 2, no. 2: 82–7.

Krupke. 2016. *Soybean Insect Control Recommendations*. USDA National Institute of Food and Agriculture, Purdue University, West Lafayette, IN 47907, United States. www.the-education-store.com.

Kuhar, T. P., H. Doughty, K. Kamminga, A. Wallingford, C. Philips, and J. Aigner. 2012a. Evaluation of Insecticides for the Control of Brown Marmorated Stink Bug in Bell Peppers in Virginia Experiment 1, 2011. *Arthropod Management Tests* 37: E37.

Kuhar, T. P., H. Doughty, K. Kamminga, A. Wallingford, C. Philips, and J. Aigner. 2012b. Evaluation of Insecticides for the Control of Brown Marmorated Stink Bug in Bell Peppers in Virginia Experiment 2, 2011. *Arthropod Management Tests* 37: E38.

Kuhar, T. P., H. Doughty, K. Kamminga, A. Wallingford, C. Philips, and J. Aigner. 2012c. Evaluation of Insecticides for the Control of Brown Marmorated Stink Bug in Bell Peppers in Virginia Experiment 3, 2011. *Arthropod Management Tests* 37: E39.

Kuhar, T. P., H. Doughty, K. Kamminga, A. Wallingford, C. Philips, and J. Aigner. 2012d. Evaluation of Insecticides for the Control of Brown Marmorated Stink Bug in Bell Peppers in Virginia Experiment 4, 2011. *Arthropod Management Tests* 37: E40.

Kuhar, T. P., H. Doughty, K. Kamminga, A. Wallingford, C. Philips, and J. Aigner. 2016b. *Pest management in potato: Farmer's knowledge, perception and problems in Himachal Pradesh* [MSc Thesis]. Palampur, India: Department of Entomology, CSKHPKV. *137 p.p.*

Kumar, A. A., B. V. S. Reddy, H. C. Sharma, and B. Ramaiah. 2008. Shoot Fly (*Atherigona soccata*) Resistance in Improved Grain Sorghum Hybrids. *Journal of SAT Agricultural Research* 6: 1–4.

Kumar, H. 2002. Resistance in Maize to Larger Grain Borer, *Prosphanus truncates* (Horn) (Coleoptera: Bostrichidae). *Journal of Stored Products Research* 38, no. 3: 267–80.

Kumarasinghe, N., and M. Ranasinghe. 1985. Life History and Monthly Incidence of Sugarcane Leaf Hopper *Pyrilla perpusilla* Singhalensis (Homoptera: Lophopidae) in Kantale. *Proceedings of the 41st Annual Session of the Sri Lanka Association of Advancement of Science* 35.

Kumarasinghe, N., and M. Ranasinghe. 1988. Effect of Environmental Factors on Populations of the Sugarcane Leaf Hopper *Pyrilla perpusilla* Singhalensis. *Proceedings of the 44th Annual Sessions of the Sri Lanka Association of Advancement of Science* 70.

Kumarasinghe, N. C., and S. D. Wratten. 1996. The Sugarcane Lophopid Planthopper *Pyrilla perpusilla* (Homoptera: Lophopidae): A Review of Its Biology, Pest Status and Control. *Bulletin of Entomological Research* 86, no. 5: 485–98.

Kumari, A., M. Goyal, R. Kumar, and R. S. Sohu. 2021. Morphophysiological and Biochemical Attributes Influence Intra-Genotypic Preference of Shoot Fly [*Atherigona soccata* (Rondani)] among Sorghum Genotypes. *Protoplasma* 258: 87–102.

Kumari, N., and V. K. Chandla. 2010. Save Parasites and Microbes of Cutworms, *Agrotis segetum* (Schiff) Pest of Potato in Shimla Hills. *Uttar Pradesh Journal of Zoology* 30, no. 1: 21–5.

Lal, R., and H. R. Rohilla. 2007. Cutworms, Their Biology and Management. In *Recent Trends in Biology and Management of Polyphagous Pests of Agricultural Importance*, edited by J. P. Bhanot, H. R. Rohilla, and V. K. Kalra. Hisar: Centre of Advance Studies, Department of Entomology, CCS HAU: 190–7.

Latchininsky, A., G. Sword, M. Sergeev, M. M. Cigliano, and M. Lecoq. 2011. Locusts and Grasshoppers: Behavior, Ecology, and Biogeography. *Psyche* 2011: 1–4.

Lee, Doo-Hyung, Brent D. Short, Shimat V. Joseph, J. Christopher Bergh, and Tracy C. Leskey. 2013a. Review of the Biology, Ecology, and Management of *Halyomorpha halys* (Hemiptera: Pentatomidae) in China, Japan, and the Republic of Korea. *Environmental Entomology* 42, no. 4: 627–41.

Lee, S.-T., R. Srinivasan, Y. J. Wu, and N. S. Talekar. 2007. Occurrence and Characterization of a Nucleopolyhedrovirus from *Maruca vitrata* (Lepidoptera, Pyralidae) Isolated in Taiwan. *BioControl* 52, no. 6: 801–19.

Lella, R., and C. P. Srivastav. 2013. *Screening of Maize Genotypes Against Stem Borer Chilo partellus* L. in Kharif Season. *International Journal of Applied Biology and Pharmaceutical Technology* 4, no. 4: 394–403.

Leskey, T. C., and G. C. Hamilton. 2012. Brown Marmorated Stink Bug Working Group Meeting, Accessed August 12 2013. http://www.northeastipm.org/neipm/assets/File/BMSB-Working-Group-Meeting-Report-Nov-2012.pdf.

Leskey, T. C., B. D. Short, B. R. Butler, and S. E. Wright. 2012a. Impact of the Invasive Brown Marmorated Stink Bug, *Halyomorpha halys* (Stål), in mid-Atlantic Tree Fruit Orchards in the United States: Case Studies of Commercial Management. *Psyche* 5, no. 3: 50–62.

Leskey, Tracy C., Brent D. Short, and Doo-Hyung Lee. 2014. Efficacy of Insecticide Residues on Adult *Halyomorpha halys* (Stål) (Hemiptera: Pentatomidae) Mortality and Injury in Apple and Peach Orchards. *Pest Management Science* 70, no. 7: 1097–104.

Leskey, Tracy C., Doo-Hyung Lee, Brent D. Short, and Starker E. Wright. 2012c. Impact of Insecticides on the Invasive *Halyomorpha halys* (Stål) (Hemiptera: Pentatomidae): Analysis on the Insecticide Lethality. *Journal of Economic Entomology* 105, no. 5: 1726–35.

Leskey, T. C., G. C. Hamilton, A. L. Nielsen, D. F. Polk, C. Rodriguez-Saona, J. C. Bergh, D. A. Herbert, T. P. Kuhar, D. Pfeiffer, G. P. Dively, C. R. R. Hooks, M. J. Raupp, P. M. Shrewsbury, G. Krawczyk, P. W. Shearer, J. Whalen, C. Koplinka-Loehr, E. Myers, D. Inkley, K. A. Hoelmer, D. Lee, and S. E. Wright. 2012b. Pest Status of the Brown Marmorated Stink Bug, *Halyomorpha halys* in the USA. *Outlooks on Pest Management* 23, no. 5: 218–26.

Leskey, T. C., G. C. Hamilton, D. J. Biddinger, M. L. Buffington, C. Dieckhoff, G. P. Dively, H. Fraser, T. Gariepy, C. Hedstrom, and D. A. Herbert. 2014. Datasheet for *Halyomorpha halys* (Stål), (Hemiptera: Pentatomidae). In *Crop Protection Compendium*. Wallingford, UK: CA B International.

Liao, C. T., and C. L. Lin. 2000. Occurrence of the Legume Pod Borer, Maruca Testulalis Geyer (Lepidoptera: Pyralidae) on Cowpea (*Vigna unguiculata* Walp.) and Its Insecticides Application Trial. *Plant Protection Bulletin* 42: 213–22.

Liu, Y. Q., X. W. Fu, H. Q. Feng, Z. F. Liu, and K. M. Wu. 2015. Transregional Migration of *Agrotis ipsilon* (Lepidoptera: Noctuidae) in North-East Asia. Annals of Entomology Society of America 108: 519–27.

Lu, P. F., H. L. Qiao, X. P. Wang, X. Wang, and C. Lei. 2007. The Emergence and Mating Rhythms of the Legume Pod Borer, *Maruca vitrata* (Fabricius, 1787) (Lepidoptera: Pyralidae). *Pan-Pacific Entomologist* 83, no. 3: 226–34.

Lu, S. H., and D. Z. Qin. 2014. Alafrasca Sticta a New Genus and Species of the Tribe Empoascini (Hempitera: Cicadellidae: Typhlocybinae) with a Checklist of the Tribe from China. *Zootaxa* 3779, no. 1: 9–19.

Lucini, T., and A. R. Panizzi. 2018. Electropenetrography (EPG): A Breakthrough Tool Unveiling Stink Bug (Pentatomidae) Feeding on Plants. *Neotropical Entomology* 47, no. 1: 6–18.

Macharia, M., D. Tebkew, W. Agum, and M. Njuguna. 2016. Incidence and Distribution of Insect Pests in Rain-Fed Wheat in Eastern Africa. *African Crop Science Journal* 24, no. 1: 149–55.

Mahalakshmi, M. S., M. Sreekanth, M. Adinarayana, Y. P. Reni, Y. Rao, Y., and E. Narayana. 2016. Incidence, Bionomics and Management of Spotted Pod Borer in Major Pulse Crops in India—A Review. *Agricultural Reviews* 37, no. 1: 19–26.

Mahdi, S. H. A., and M. K. Rahman. 2008. Insecticidal Effect of Some Spices on Callosobruchus maculatus (Fabricius) in Black Gram Seeds. *Uni. J. Zool. Rajsh. Uni.* 27: 47–50.

Maina, Y. T. 2011. Effects of Different Levels of Infestation and Storage Durations on the Development of *Callosobruchus maculatus* (Fab.) in Stored Cowpea *Vigna unguiculata* (L.) Walpers. *Production Agriculture & Technology* 7, no. 1: 49–54.

Malaikozhundan, Balasubramanian, and Jayaraj Vinodhini. 2018. Biological Control of the Pulse Beetle, *Callosobruchus maculatus* in Stored Grains Using the Entomopathogenic Bacteria, Bacillus thuringiensis. *Microbial Pathogenesis* 114: 139–46.

Margam, Venu M., Brad S. Coates, Malick N. Ba, Weilin Sun, Clementine L. Binso-Dabire, Ibrahim Baoua, Mohammad F. Ishiyaku, John T. Shukle, Richard L. Hellmich, Fernando G. Covas, Srinivasan Ramasamy, Joel Armstrong, Barry R. Pittendrigh, and Larry L. Murdock. 2011a. Geographic Distribution of Phylogenetically Distinct Legume Pod Borer, *Maruca vitrata* (Lepidoptera: Pyraloidea: Crambidae). *Molecular Biology Reports* 38, no. 2: 893–903.

Mashwani, M. A., F. Ullah, S. Ahmad, K. Sohail, M. Usman, and S. F. Shah. 2015. Infestation of Maize Stem Borer, *Chilo partellus* Swinhoe in Maize Stubbles and Stalks. *Acad. Journal of Agricultural Research* 3, no. 7: 094–8.

Matama-Kauma, T., F. Schulthess, B. P. Le Rü, J. Mueke, J. A. Ogwang, and C. O. Omwega. 2008. Abundance and Diversity of Lepidopteran Stem Borers and Their Parasitoids on Selected Wild Grasses in Uganda. *Crop Protection* 27, no. 3–5: 505–13.

Mayeux, A. 1984. The Groundnut Aphid. Biology and Control. *Oleagineux* 39: 425–34.

McPhee, K. E., E. Groden, and F. A. Drummond. 2008. *Ant-Homopteran Relationships: Relevance to an Ant Invasion in Maine*. School of Biology and Ecology.

Mehmood, H. Z., H. Afzal, A. Abbas, S., Hassan, and A. Ali. 2022. Forecasts About Livestock Production in Punjab-Pakistan: Implications for Food Security and Climate Change. *Journal of Animal & Plant Sciences* 35, no. 5: 1347–55.

Mehrparvar, M., S. M. Madjdzadeh, A. N. Mahdavi, M. Esmaeilbeygi, and E. Ebrahimpour. 2012. Morphometric Discrimination of Black Legume Aphid, *Aphis craccivora* Koch (Hemiptera: Aphididae), Populations Associated with Different Host Plants. *North-West J Zool.* 8, no. 1: 172–80.

Messina, F. J., and A. F. Slade. 1999. Expression of a Life-History Trade-Off in a Seed Beetle Depends on Environmental Context. *Physiological Entomology* 24, no. 4: 358–63.

Mia, M. D. 1998. Current Status of Insect Research of Pulse Crops and Future Research Needs. Proceeding of Workshop on Disease Resistance Breeding in Pulses, Dhaka: 87–93.

Miah, M., A. Qudrat-e-Khuda, and M. Shahjahan. 1986. The Problems of *Pyrilla perpusilla* and the Impact of Its Natural Enemies [of Sugarcane in Bangladesh]. *Bangladesh Journal of Zoology (Bangladesh).*

Minghetti, E., H. Braun, F. Matt, and P. M. Dellapé. 2020. *Tympanoblissus ecuatorianus* Dellapé and Minghetti, gen. et sp. nov, a New Apterous Genus of Chinch Bug (Hemiptera: Blissidae) from Ecuador: The Only Known Blissid with an Abdominal Mechanism for Sound Production. *Austral Entomology* 59, no. 3: 535–40.

Minhas, R., L. H. Akhtar, M. Zubair, M. S. Bukhari, M. Akram, R. Ullah, M. S. Ali Buzmi, A. Jabbar, M. S. Akhtar, and A. W. Chughtai. 2021. Evolution of New Guar Variety BR-21 with High Yield Potential through Pure-line Selection. *Egyptian Journal of Agronomy* 43, no. 2: 207–14.

Mishra, M., and H. L. Sharma. 2016. Population Dynamics of Major Insect Pests of Soyabean. *Indian Journal of Entomology* 78, no. 1: 92.

Mitchell, R. 1975. The Evolution of Oviposition Tactics in the Bean Weevil, *Callosobruchus maculatus* (F.). *Ecology* 56, no. 3: 696–702.

Mohammadi Anaii, M., M. Pahlavan Yali, and M. Bozorg Amir-Kalaee. 2018. Resistance of Nine Wheat Cultivars and Lines to Greenbug, *Schizaphis graminum* (Rondani) in Iran. *J. Agricul. Sci. Technol.* 20, no. 6: 1173–85.

Mohammed, R., S. M. Rajendra, A. A. Kumar, B. P. K. Kishor, B. V. S. Reddy, and H. C. Sharma. 2016. Components of Resistance to Sorghum Shoot Fly, *Atherigona soccata. Euphytica* 207: 419–38.

Mohan, S. 1991. Only Female Sorghum Shoot Flies in Fish Meal Traps. *Indian Journal of Plant Protection* 19: 77–8.

Mohapatra, S. D., and C. P. Srivastava. 2008. Toxicity of Biorational Insecticides Against Spotted Pod Borer, *Maruca vitrata* (Geyer) in Short Duration Pigeon Pea. *Indian Journal of Entomology* 70, no. 1: 61–3.

Moraes, J. C., M. M. Goussain, M. A. Basagli, G. A. Carvalho, C. C., and École, M. V. 2004. Sampaio. *Silicon Influence on the Tritrophic Interaction: Wheat Plants, the Greenbug Schizaphis Graminum* (Rondani) (Hemiptera: Aphididae), and its natural enemies, *Chrysoperla externa* (Hagen)(Neuroptera: Chrysopidae) and *Aphidius colemani* Viereck (Hymenoptera: Aphidiidae). Neotropical. Entomology 33: 619–24.

Mortazavi, Z. S., H. Sadeghi, N. Aktac, Ł. Depa, and L. Fekrat. 2015. Ants (Hymenoptera: Formicidae) and Their Aphid Partners (Homoptera: Aphididae) in Mashhad Region, Razavi Khorasan Province, with New Records of Aphids and Ant Species for Fauna of Iran. *Halteres* 6: 4–12.

Moscardini, V. F., P. C. Gontijo, J. P. Michaud, and G. A. Carvalho. 2014. Sublethal Effects of Chlorantraniliprole and Thiamethoxam Seed Treatments When *Lysiphlebus testaceipes* Feed on Sunflower Extrafloral Nectar. *BioControl* 59, no. 5: 503–11.

Muhammad, L., and E. Underwood. 2004. The Maize Agricultural Context in Kenya. In *Environmental Risk Assessment of Genetically Modified Organisms: A Case Study of Bt Maize in Kenya* vol. 1, edited by A. Hilbeck, and D. A. Andow. Wallingford: CABI Publishing, Oxford shire, UK: 1–2.

Murdock, L. L., D. Seck, G. Ntoukam, L. Kitch, and R. E. Shade. 2003. Preservation of Cowpea Grain in Sub-Saharan Africa-Bean/Cowpea CRSP Contributions. *Field Crops Research* 82, no. 2–3: 169–78.

Mwimali, M. G. 2014. *Genetic analysis and response to selection for resistance to two stem borers,* Busseola fusca *and* Chilo partellus, *in tropical maize germplasm* [PhD Diss.] African Centre for Crop Improvement, and School of Agriculture. *Earth and Environmental Science College of Agriculture, Engineering and Sciences.* University of KwaZulu-Natal Republic of South Africa.

Nag, A., and P. Nath. 1994. Abundance and Frequency of Emergence of Adult Cutworm, *Agrotis ipsilon* (Hufn.) Moth on Light Trap. *Bulletin of Entomological Research* 35, no. 1–2: 143–9.

Nasruddin, A., F. Abdul, S. B. Muhammad, and E. S. Ahwiyah 2014. Potential Damages, Seasonal Abundance and Distribution of *Empoasca terminalis* Distant (Homoptera: Cicadellidae) on Soybean in South Sulawesi, Indonesian. *Journal of Entomology* 11, no. 2: 93–102.

Nasruddin, A., M. Jayadi, and E. Syam'un. 2016. A. E. Said. and Jumardi. Action Thresholds of *Empoasca terminalis* (Homoptera: Cicadellidae) on Soybean in South Sulawesi. *Indonesia. J. Entomol.* 13: 33–9.

Naveen, V., M. I. Naik, M. Manjunatha, B. K. Shivanna, and S. Sridhar. 2009. Biology of Legume Pod Borer, *Maruca testulalis* Geyer on Cowpea. *Karnataka Journal of Agricultural Sciences* 22: 668–9.

Neupane, F. 1971. Studies on the Effectiveness of Different Insecticides against the Nymphs of Sugarcane Leaf Hopper, *Pyrilla perpusilla* walker. *Nepalese Journal of Agriculture*: 131–7.

Neupane, S., and S. Subedi. 2019. Life Cycle Study of Maize Stem Borer (*Chilo partellus* Swinhoe) under Laboratory Condition at National Maize Research Program, Rampur, Chitwan, Nepal. *Journal of Agriculture & Natural Resources* 2, no. 1: 338–46.

Nielsen, A. L., G. C. Hamilton, and D. Matadha. 2008. Developmental Rate Estimation and Life Table Analysis for *Halyomorpha halys* (Hemiptera: Pentatomidae). *Environmenta Entomology* 37: 348–55.

Nielsen, Anne L., Kristian Holmstrom, George C. Hamilton, John Cambridge, and Joseph Ingerson-Mahar. 2013. Use of Black Light Traps to Monitor the Abundance and Spread of the Brown Marmorated Stink Bug. *Journal of Economic Entomology* 106, no. 3: 1495–502.

Nielsen, G. R., W. O. Lamp, and G. W. Stutte. 1990. Potato Leafhopper (Homoptera: Cicadellidae) Feeding Disruption of Phloem Translocation in Alfalfa. *Journal of Economic Entomology* 83, no. 3: 807–13.

Nuessly, G. S., and R. T. Nagata. 2005. *Greenbug,* Schizaphis graminum *(Rondani) (Insecta: Hemiptera: Aphididae).* Institute of Food and Agricultural Sciences, University of Florida.

Oben, Esther Obi, Nelson Neba Ntonifor, Sevilor Kekeunou, and Martin Nkwa Abbeytakor. 2015. Farmer's Knowledge and Perception on Maize Stem Borers and Their Indigenous Control Methods in South Western Region of Cameroon. *Journal of Ethnobiology & Ethnomedicine* 11, no. 1: 77.

Obopile, M., and B. Ositile. 2010. Life Table and Population Parameters of Cowpea Aphid, *Aphis craccivora* Koch (Homoptera: Aphididae) on Five Cowpea *Vigna unguiculata* (L. Walp.) Varieties. *Journal of Pest Science* 83, no. 1: 9–14.

Ochieng, R. S., J. B. Okeyo-Owuor, and Z.T. Dabrowski. 1981. Studies on the Legume Pod Borer, *Maruca testulalis* (Geyer): Mass Rearing on Natural Food. *Insect Science & its Application* 1: 269–72.

Odeyemi, O. O., O. A. Gbaye, and O. Akeju. 2006. Resistance of *Callosobruchus maculatus* (Fab.) to Pirimiphos Methyl in Three Zones in Nigeria. In 9th International Working Conference on Stored Product Protection: 15–8.

Ofuya, T. I., and C. Reichmuth. 2002. Effect of Relative Humidity on the Susceptibility of *Callosobruchus maculatus* (Fabricius)(Coleoptera: Bruchidae) to Two Modified Atmospheres. *Journal of Stored Products Research* 38, no. 2: 139–46.

Ofuya, T.I. 2001. Biology, Ecology and Control of Insect Pests of Stored Legumes in Nigeria. In *Pests of Stored Cereals and Pulses in Nigeria*, edited by T.I. Ofuya, and N.E.S. Lale. Nigeria: Dave Collins Publication: 24–58.

Ogendo, J.O., S.R. Belmain, A. L. Deng, and D.J. Walker. 2003. *Comparison of Toxic and Repellent Effects of* Lantana camara *L. with* Tephrosia vogelii *Hook and a Synthetic Pesticide against* Sitophilus zeamais *motschulsky (Coleoptera: Curculionidae) in Stored Maize Grain. Ins. Sci* [App] vol. 23, no. 2: 127–35.

Ogwaro, K., and E. D. Kokwaro. 1981. Development and Morphology of the Immature Stages of the Sorghum Shootfly, *Atherigona soccata* Rondani. *International Journal of Tropical Insect Science* 1, no. 4: 365–72.

Ojha, K.N., and P. Nath. 1987. Studies on Relative Preference of Different Maize Cultivars to *Agrotis ipsilon* in Varanasi. *Indian Journal of Entomology* 1: 17–20.

Chabi-Olaye, A., C. Nolte, F. Schulthess, and C. Borgemeister. 2005. Relationships of Intercropped Maize, Stem Borer Damage to Maize Yield and Land-Use Efficiency in the Humid Forest of Cameron. *Bulletin of Entomological Research* 95, no. 5: 417–27.

Olufemi, O., R. Pitan, and J. A. Odebiyi. 2001. Crop Losses in Cowpea Due to the Pod-Sucking Bugs *Riptortus dentipes, Mirperus jaculus, Anoplocnemis curvipes* and *Nezara viridula. Insect Science & Its Application* 21, no. 3: 237–41.

Ongamo, G.O., B. P., S. LeRu, P. Dupas, P. A. Moyal. and J. F. 2006. Distribution, Pest Status and Agro-Climatic Preferences of Lepidopteran Stem Borers of Maize in Kenya. *Annales de la Société Entomologique de France* 42, no. 2: 171–7.

Oparaeke, A. M., M. C. Dike, and C. I. Amatobi. 2000. Insecticide Potential of Extracts of Garlic Bulb and African Nutmeg Dunal Seed for Insect Pest Control on Cowpea. ESN Occasional Publication 32: 169–74.

Orantes, L.C., W. Zhang, M.A.R. Mian, and A.P. Michel. 2012. Maintaining Genetic Diversity and Population Panmixia through Dispersal and Not Gene Flow in a Holocyclic Heteroecious Aphid Species. *Heredity* 109, no. 2: 127–34.

Ostlie, K., and B. Potter. 2017. *Insect-Pests of Corn–Stand Reducers*. University of Minnesota.

Özdemir, I., N. Aktaç, S. Toros, N. Kılınçer, and M.O. Gürkan. 2008. Investigations of the Associated Between Aphids and Ants on Wild Plants in Ankara Province (Turkey). *Munis Entomology & Zoology* 3: 606–13.

Paddock, E. L. 1976. Bean Pod Borer. Exclusion and Detection Manual. California Department of Food and Agriculture.

Padmaja, P.G., R. Madhusudhana, and N. Seetharama. 2010. *Sorghum Shoot Fly*. Hyderabad: Directorate of Sorghum Research.

Pandiselvam, R., V. Thirupathi, S. Mohan, P. Vennila, D. Uma, S. Shahir, and S. Anandakumar. 2019. Gaseous Ozone: A Potent Pest Management Strategy to Control *Callosobruchus maculatus* (Coleoptera: Bruchidae) Infesting Green Gram. *Journal of Applied Entomology* 143, no. 4: 451–9.

Panzer. 1798. Periscepsia Carbonaria. Fauna Europaea project, Fauna Europaea Secretariat, Berlin, Germany.

Pareek, B.L., R.C. Sharma, and C.P.S. Yadav. 1983. Record of Insect Faunal Complex on Moth Bean, *Vigna aconitifolia* (Jacq.) Marechal in Semi-arid Zone of Rajasthan. *Bulletin of Entomology* 24, no. 1: 44–5.

Parihar, D.R. 1979. Out Break of *Katra, Amrascta moorei* in the Rajasthan Desert. *Annals of Arid Zone* 18, no. 1/2: 140–1.

Patel, A.G., D.N. Yadav, and R.C. Patel. 1991. Natural Enemies of *Agrotis* spp. from Gujarat. Gujarat Agricultural University Research [Journal] 17, no. 1: 130–1.

Patel, H. C., P. K. Borad, and N. B. Patel. 2020. Determination of Economic Injury and Threshold Level of *Maruca vitrata* (Geyer) in Green Gram. *International Journal of Current Microbiology & Applied Sciences* 9, no. 9: 3211–5.

Patel, H. V., R. K. Kalaria, R. M. Patel, and G. R. Bhanderi. 2015. Biochemical Changes Associated in Different Sorghum Genotypes against Shoot Fly, *Atherigona soccata* (Rondani) Resistant. *Trends in Biosciences* 8, no. 11: 2867–71.

Patel, P. S., I. S. Patel, B. Panickar, and Y. Ravindrababu. 2012. Management of Sucking Pests of Cowpea Through Seed Treatment. *Trends in Biosciences* 5, no. 2: 138–9.

Patil, S. S., C. Kamble, and T. V. Sathe. 2016. Biocontrol Potential of *Cotesia ruficrus* Hal. (Hymenoptera: Braconidae) Against Different Lepidopterous Pests. *Biolife* 4, no. 2: 343–46.

Pathania, V. 2010. *Distribution and biology of cutworms infesting potato in Himachal Pradesh* [PhD thesis]. Palampur, India: Department of Entomology, CSKHPKV. *87 p.p.*

Pedigo, L. P. 2002. *Entomology and Pest Management.* 4th ed. NJ: Prentice Hall.

Pendleton, Bonnie B., Anastasia L Palousek L. P. Copeland, and G. J. Michels. 2009. Effect of Biotype and Temperature on Fitness of Greenbug (Hemiptera: Aphididae) on Sorghum. *Journal of Economic Entomology* 102, no. 4: 1624–7.

Periasamy, Malini, Roland Schafleitner, Krishnan Muthukalingan, and Srinivasan Ramasamy. 2015. Phylogeographical Structure in Mitochondrial DNA of Legume Pod Borer (*Maruca vitrata*) Population in Tropical Asia and Sub-Saharan Africa. *PLOS One* 10, no. 4: e0124057.

Polaszek, A. 1998. *African Cereal Stem Borers: Economic Importance, Taxonomy, Natural Enemies and Control.* Wallingford, UK: CABI Publishing: 530, pp.

Popova, A. A. 1967. *Adaptations of Aphids to Feeding on Their Host Plants.* Leningrad: Nauka. 292 pp.

Prevett, P. F. 1961. Field Infestation of Cowpea (*Vigna unguiculata*) Pods by Beetles of the Families—Bruchidae and Curculionidae in Northern Nigeria. *Bulletin of Entomological Research* 52, no. 4: 635–45.

Punia, A., R. Yadav, P. Arora, and A. Chaudhury. 2009. Molecular and Morphophysiological Characterization of Superior Cluster Bean (*Cyamopsis tetragonoloba*) Varieties. *Journal of Crop Science & Biotechnology* 12, no. 3: 143–8.

Puttaswamy, B. L. V. 1977. Record of pests infesting moth bean, *Phaseolus aconitifolia* (Jacq.) a potential pulse crop. *Curr. Res.* Gowda, and T. M. M. Ali 6, no. 4: 69–71.

Qiu, L. F., Z. Q. Yang, and W. Q. Tao. 2007. Biology and Population Dynamics of *Trissolcus halyomorphae. Scientia Silvae Sinica* 43: 62–5.

Qureshi, R., and G. R. Bhattim. 2006. Ethnobotinanical Observation of *Achyransthes aspera* Linn and *Aeriva* spp, with Special Reference to the People of Nara Desert. *Hamdard Medicus* XLIX(1): 43–8.

Rachappa, V., Y. Suhas, S. Vennila, O. P. Sharma, and P. Shivayogeppa. 2016. Influence of Climate Change on Occurrence of Green Leafhopper *Empoasca kerri* Pruthi on Pigeon Pea. *Journal of Experimental Zoology India* 19, no. 2: 1163–6.

Rahim, A. 1989. Abiotic Factors Influencing Development and Longevity of *Tetrastichus pyrillae* Craw. (Hymenoptera: Eulophidae), an egg parasite of *Pyrilla perpusilla* walk. *Pakistan Journal of Scientific & Industrial Research* 32: 820–2.

Rahman, A., and F. A. Talukder. 2006. Bioefficacy of Some Plant Derivatives That Protect Grain Against the Pulse Beetle, *Callosobruchus maculatus. Journal of Insect Science* 6, no. 1: 1–10.

Rahman, A. 2014. *Development of management practices against Jassid* (Amrasca devastans) *in okra* ([Doctoral dissertation]. Dhaka: Department of Entomology, Sher-E-Bangla Agricultural University.).

Rahman, K. A., and R. Nath. 1940. Bionomics and Control of the Indian Sugar-Cane Leaf-Hopper, *Pyrilla perpusilla*, wlk. *Bulletin of Entomological Research* 31, no. 2: 179–90.

Raja, N., A. Babu, S. Dorn, and S. Ignacimuthu. 2001. Potential of Plants for Protecting Stored Pulses from *Callosobruchus maculatus* (Coleoptera: Bruchidae) Infestation. *Biological Agriculture & Horticulture* 19, no. 1: 19–27.

Rajak, R., A. Pawar, M. Misra, J. Prasad, A. Verma, and G. Singh. 1987. Sugarcane *Pyrilla* Epidemic 1985—A Case Study. *Plant Protection Bulletin, (India)* 39: 1–9.

Ram, G., S. S. Misra, and K. P. M. Dhamayanthi. 2001. Relative Susceptibility of Advanced Hybrids and Promising Cultivars of Potato, *Solanum tuberosum* L. to Greasy Cutworm, *Agrotis ipsilon* (Hufnagel) in North-Eastern Plains. *Journal of Entomological Research* 25, no. 3: 183–7.

Rauf, A., M. Ayyaz, F. Baig, M. N. Naqqash, and M.J. Arif. 2017. Response of *Chilo partellus* (Swinhoe) and Entomophagous Arthropods to Some Granular and New Chemistry Formulations in *Zea mays* L. 5, no. 3: 1351–6.

Ravi Kumar, S.S., G.G. Kulkarni, G.K. Basavana, and C.P. Mallapur. 1999. Insecticidal Property of Some Indigenous Plant Extracts against Lepidopterous Pests of Safflower. *Annals of Agri Bio Research* 4: 49–52.

Regmi, R., R. Tiwari, and R. Thapa. 2012. Ecofriendly Management of Spotted Pod Borer (Maruca Vitrata Fabricious) on Yardlong Bean in Chitwan, Nepal. *International Journal of Research* 1, no. 6: 386–94.

Rosaiah, R. 2001. Performance of Different Botanicals against the Pests Complex in Bhendi. *Pestology* 25: 17–9.

Rose, H.S., and A. P. Singh. 1989. Use of Internal Reproductive Organs in the Identification of Indian Species of the Genus *Maruca walker* (Pyraustinae: Pyralidae: Lepidoptera). *Journal of Advanced Zoology* 10: 99–103.

Rosendahl, Ingrid, Volker Laabs, Cyrien Atcha-Ahowé, Braima James, and Wulf Amelung. 2009. Insecticide Dissipation from Soil and Plant Surfaces in Tropical Horticulture of Southern Benin, West Africa. *Journal of Environmental Monitoring* 11, no. 6: 1157–64.

Royer, T. A., K. L. Giles, M. M. Lebusa, and M. E. Payton. 2008. Preference and Suitability of Greenbug, *Schizaphis graminum* (Hemiptera: Aphididae) Mummies Parasitized by *Lysiphlebus testaceipes* (Hymenoptera: Aphidiidae) as Food for *Coccinella septempunctata* and *Hippodamia convergens* (Coleoptera: Coccinellidae). *Biological Control* 47, no. 1: 82–8.

Sadeghi, Amin, Els J M J. M. Van Damme, and Guy Smagghe. 2009. Evaluation of the Susceptibility of the Pea Aphid, *Acyrthosiphon pisum*, to a Selection of Novel Biorational Insecticides Using an Artificial Diet. *Journal of Insect Science* 9: 1–8.

Sajjad, M. 2019. Constrains in Fodder Production in Pakistan. Weekly Technology Times. Downloaded on March 7, 2023 from site. http://www.technologytimes.pk/2019/05/02/constrains-fodder-production-pakistan.

Sajwan, R., L. Rawat, M. Joshi, P. Kumar, A. Mamgain, and G. Kharola. 2022. Importance, Taxonomy, Bionomics and Damaging Symptoms of Various Insect Pests Associated with Millets. *Vigyan Varta* 3, no 10: 50–63.

Sandeep, K., S. K. A. V. Kannaujia, and V. K. Singh. 2013. *Plant Archives* 13, no. 1: 171–2.

Sandhu, G. S. 2016. Evaluation of Management Components Against Shoot Fly in Sorghum. *Ann. Pl. Protec. Sci* 24: 67–70.

Sanon, Antoine, Niango M. Ba, Clementine L. Binso-Dabire, and Barry R. Pittendrigh. 2010. Effectiveness of Spinosad (Naturalytes) in Controlling the Cowpea Storage Pest, *Callosobruchus maculatus* (Coleoptera: Bruchidae). *Journal of Economic Entomology* 103, no. 1: 203–10.

Sasaki, F., T. Miyamoto, A. Yamamoto, Y. Tamai, and T. Yajima. 2012. Relationship Between Intraspecific Variations and Host Insects of *Ophiocordyceps nutans* Collected in Japan. *Mycoscience* 53, no. 2: 85–91.

Satyavir. 1980. *Seasonal Incidence of Insect Pests of Mothbean and Cowpea Crops*. Annual Progress Report vol. 1980. Jodhpur: Central Arid Zone Research Institute: 53.

Satyavir, S.K. Jindal, and S. Lodha. 1984. Screening of Moth Bean Cultivars against Jassid, White Fly and Yellow Mosaic Virus. *Annals of Arid Zone* 23, no. 2: 99–103.

Shams, G., H. S. Mohammad, I. Sohrab, S. Mahmoud, and A. Shahram. 2011. A Laboratory Assessment of the Potential of the Entomopathogenic Fungi *Beauveria bassiana* (Beauvarin) to Control *Callosobruchus maculatus* (f.) (Coleoptera: Bruchidae) and *Sitophilus granarius* (L.) (Coleoptera: Curculionidae). *African Journal of Microbiology Research* 5, no. 10: 1192–6.

Shaposhnikov, G.C. 1986. New Species of the Genus *Dysaphis* Börner (Homoptera, Aphidinea) and Peculiarities of the Taxonomic Work with Aphids. *Entomologicheskoe Obozrenie* 65: 535–50.

Sharif, M.Z. 2016. *Maize Stem Borer* (Chilo partellus)*; Destructive Insect Pest of Maize Crop in Pakistan.*

Sharma, H.C., K.B. Saxena, V. R. Bhagwat. 1999. *The Legum e Pod Borer,* Maruca vitrata*: Bionomics and Management.* Information Bulletin no. 55. International Crops Research Institute for the Semi-Arid Tropics (ICRISAT), Patancheru, India. 42 pp. ISBN 92-9066-406-1.

Sharma, H. C., S. Taneja, N. K. Rao, and P. K. E. Rao. 2003. Evaluation of Sorghum Germplasm for Resistance to Insect Pests. Information Bulletin No. 63. *International Crops Research Institute for the Semi-Arid Tropics, Patancheru, A.P. (INDIA)*: pp. 17–27.

Sharma, P. N., and P. Gautam. 2010. Assessment of Yield Loss in Maize Due to Attack by the Maize Borer, *Chilo partellus* (Swinhoe). *Nepal Journal of Science & Technology* 11: 25–30.

Sharma, S. 2018. Cluster Bean (Guar). In *Forage Crops of the World* vol. I: Major Forage Crops. Apple Academic Press: 205–24.

Shegro, A., N. G. Shargie, A. van Biljon, and M. T. Labuschagne. 2012. Diversity in Starch, Protein and Mineral Composition of Sorghum Landrace Accessions from Ethiopia. *Journal of Crop Science & Biotechnology* 15, no. 4: 275–80.

Sheikh, F. 1968. Control of Sugarcane Pests by Air in Mardan District during Kharif-1964. *Agriculture Pakistan* 19: 673–84.

Shekharappa, S., and S. G. Bhuti. 2007. Integrated Management of Sorghum Shoot Fly, *Atherigona soccata* Rondani. *Karnataka Journal of Agricultural Sciences* 20: 535–6.

Shishodia, M. S., K. Chandra, and S. K. Gupta. 2010. An Annotated Checklist of Orthoptera (Insecta) from India. *Rec. Zool. Surv.*14: 1–366.

Shivalingaswamy, T. M., S. Satpathy, B. Singh, and A. Kumar. 2002. Predator–Prey Interaction Between Jassid (*Amrasca biguttula biguttula*, Ishida) and a Staphylinid in Okra. *Vegetable Science* 29, no. 2: 167–9.

Showers, W. B., L. Von Kaster, and P. G. Mulder. 1983. Corn Seedling Growth Stage and Black Cutworm (Lepidoptera: Noctuidae) Damage. *Environmental Entomology* 12, no. 1: 241–4.

Showers, W. B. 1997. Migratory Ecology of the Black Cutworm. *Annual Review of Entomology* 42: 393–425.

Siddalingappa, C. T., V. Hosamani, and Y. Shivasharanappa. 2010. Biology of Maize Stem Borer, *Chilo partellus* (Swinhoe) Crambidae: Lepidoptera. *International Journal of Plant Protection* 3, no. 1: 91–3.

Sidhu, K. S. 2019. *Management of cutworms* Agrotis *spp. in maize* [MSc thesis]. Palampur, India: Department of Entomology, CSKHPKV. *72 p.p.*

Si-ming, C. A. I. 2007. Bionomics and Control of *Selepa celtis* Moore. *Forest Research*: 2007–05.

Singh, A. K., and M. Kumar. 2003. Efficacy and Economics of Neem Based Products against Cotton Jassid, *Amrasca biguttulla biguttulla* Ishida in Okra. *Crop Research-Hisar* 26, no. 2: 271–4.

Singh, A., J. Singh, K. Singh, and P. Rani. 2018. Host Range and Biology of *Amrasca biguttula biguttula* (Hemiptera, Cicadellidae). *International Journal of Environment* 8: 19–24.

Singh, B. U., and H. C. Sharma. 2002. Natural Enemies of Sorghum Shoot Fly, *Atherigona soccata* Rondani (Diptera: Muscidae). *Biocontrol Science & Technology* 12, no. 3: 307–23.

Singh, R., and R. A. Agarwal. 1988. Influence of Leaf-Veins on Ovipositional Behaviour of Jassid, *Amrasca biguttula biguttula* (Ishida). *Journal of Cotton Research Development* 2: 41–8.

Singh, S. R., L. E. N. Jackai, J. H. R. Santos, and C. B. Adalla. 1990. Insect Pests of Cowpea. In *Insect Pests of Tropical Food Legumes. Chichester*, edited by S. R. Singh. Nigeria: John Wiley & Sons Ltd: 43–89.

Singh, S., D. P. Choudhary, H. C. Sharma, R. S. Mahla, Y. S. Mathur, and D. B. Ahuja. 2008. Effect of Insecticidal Modules against Jassid and Shoot and Fruit Borer in Okra. *Indian Journal of Entomology* 70, no. 3: 197–9.

Sivakumar, G., M. Kannan, V. R. Babu, M. Mohan, Surabhi Kumari, R. Rangeshwaran, T. Venkatesan, C. R. Ballal, and R. Chandish. 2020. Characterization and Field Evaluation of Tetrahedral and Triangular Nucleopolyhedrovirus of *Spilosoma obliqua* (SpobNPV) Strain NBAIR1 Against Jute Hairy Caterpillar. *J. Egyptian Journal of Biological Pest Control* 30, no. 82.

Smith, C. M., and E. V. Boyko. 2007. The Molecular Bases of Plant Resistance and Defense Responses to Aphid Feeding: Current Status. *Entomologia Experimentalis & Applicata* 122, no. 1: 1–16.

Soomro, S., and M. S. Wagan. 2005. Notes on Subfamily Calliptaminae (Acrididae: Acridoidea: Orthoptera) of Pakistan, with the Description of One New Species. *Pakistan Journal of Zoology* 37, no. 3: 229–36.

Southgate, B. J. 1979. Biology of the Bruchidae. *Annual Review of Entomology* 24, no. 1: 449–73.

Spinola, M. 1808. *New or rarer species of Ligurian insects (Hymenoptera), discovered in the Ligurian field.* Genuae, sumptibus auctoris, typis Yves Gravier, Italy. 262 pp.

Srinivasan, R. 2008. Susceptibility of Legume Pod Borer (LPB), *Maruca vitrata* to Delta-Endotoxins of *Bacillus thuringiensis* (Bt) in Taiwan. *Journal of Invertebrate Pathology* 97, no. 1: 79–81.

Srinivasan, Ramasamy, Manuele Tamò, and Periasamy Malini. 2021. Emergence of *Maruca vitrata* as a Major Pest of Food Legumes and Evolution of Management Practices in Asia and Africa. *Annual Review of Entomology* 66: 141–61.

Srinivasan, R., M. Tamò, P. A.-C. Ooi, and W. Easdown. 2007. Biocontrol News and Information 28: 34–7.

Srinives, P., P. Somta, and R. Somta. 2007. Genetics and Breeding of Resistance to Bruchids (*Callosobruchus* spp.) in Vigna Crops: A Review. *Nu Science Journal* 4, no. 1: 01–17.

Stebbing, E. 1903. Insect Pests of Sugarcane in India. *Indian Museum Notes* 5: 86–7.

Stein, A. F., R.R. Draxler, G.D. Rolph, B.J. B. Stunder, M. D. Cohen, and F. Ngan. 2015. Noaa's Hysplit Atmospheric Transport and Dispersion Modeling System. *Bulletin of the American Meteorological Society* 96, no. 12: 2059–77.

Street, J. R., R. Randell, and G. Clayton. 1978. Greenbug Damage Found on Kentucky Bluegrass. *Weeds, Trees and Turf* 17: 26.

Sufyan, M., A. Abbasi, W. Wakil, M. D. Gogi, M. Arshad, and A. Nawaz. 2019. Efficacy of *Beauveria bassiana* and *Bacillus thuringiensis* against Maize Stem Borer *Chilo partellus* (Swinhoe) (Lepidoptera: Pyralidae). *Gesu. Pflan* 71, no. 3: 197–204.

Sultana, P., M.J. Alam, K. Das, M.A.K. Azad, and M.T. Islam. 2016. Eco-friendly Control of Okra Jassid Using Botanicals from Jute (*Corchorus capsularies*). *Bangladesh Journal* 30: 65–70.

Sutaria, V. K., M.N. Motka, D.M. Jethva, and D.R. Ramoliya. 2010. Field Efficacy of Insecticides against Jassid, *Empoasca kerri* (Pruthi) in Soybean. *Annals of Plant Protection Sciences* 18, no. 1: 94–7.

Swaminathan, R., T. Hussain, and K.K. Bhati. 2007. Influence of Crop Diversity on Host Preference by Major Insect Pests of Kharif Pulses. *Indian. Journal of Applied Entomology* 21: 59–62.

Takahashi, H., J. Mitsui, N. Takakusa, M. Matsuda, H. Yoneda, J. Suzuki, K. Ishimitsu, and T. Kishimoto. 1992. Ni-25, a New Type of Systemic and Broad Spectrum Insecticide. *Paper Read at Proceedings, Brighton Crop Protection Conference, Pests and Diseases* 1992. Brighton, November 23–26, 1992.

Talamas, E. J., M. Buffington, and K. Hoelmer. 2013. New Synonymy of *Trissolcus halyomorphae* Yang. *Journal of Hymenoptera Research* 33: 113–7.

Tamiru, A., T. J. A. Bruce, C. M. Woodcock, J. C. Caulfield, C. A. O. Midega, C. K. P. O. Ogol, P. Mayon, M. A. Birkett, J. A. Pickett, and Z. R. Khan. 2011. Maize Landraces Recruit Egg and Larval Parasitoids in Response to Egg Deposition by a Herbivore. *Ecology Letters* 14: 1075–83.

Tamò, M., D.Y. Arodokoun, N. Zenz, M. Tindo, C. Agboton, and R. Adeoti. 2002. The Importance of Alternative Host Plants for the Biological Control of Two Key Cowpea Insect Pests, the Pod Borer *Maruca vitrata* (F.) and the Flower Thrips *Megalurothrips sjostedti* (Thrybom). In *Challenges and Opportunities for Enhancing Sustainable Cowpea Production*, edited by C.A. Fatokun, S.A. Tarawali, B.B. Singh, P.M. Kormawa, and M. Tamò. Ibadan: IITA: 81–93.

Tamò, M., I. Godonou, J. Braima, R. Srinivasan, D. Kpindu, C. Kooyman, C. Agboton, B. Datinon, S. Nakamura, T. Adati, N. Maniania, and S. Ekesi. 2011, Aug. 6–11. Honolulu: International Plant Protection Congress.

Taylor, Christopher M., Peter L. Coffey, Bridget D. DeLay, and Galen P. Dively. 2014. The Importance of Gut Symbionts in the Development of the Brown Marmorated Stink Bug, *Halyomorpha halys* (Stål). *PLOS One* 9, no. 3: e90312.

Tazerouni, Z., M. Rezaei, and A. A. Talebi. 2019. Cowpea: Insect Pest Management. In *Agricultural Research Updates*, vol. 26, edited by P. Gorawala and S. Mandhatri. Nova Science Publishers, Inc., New York, USA.

Thakur, P., G. Bhandari, J. Shrestha, and B.B. Achhami. 2018. Effect of Planting Dates and Varieties on Infestation of Maize Stem Borer *Chilo partellus* (Swinhoe). *Int. J. App. Bio* 2, no. 1: 22–8.

Thirumalaraju, G. T. 1984. *Bionomics and Control of Cotton Jassid,* Amrasca biguttula biguttula *(Ishida) (Homoptera: Cicadellidae) and Screening of Cotton Varieties for Their Resistance to the Pest. Moo. & S.* (Agri.) [Thesis]. India: University of Agricultural Sciences, Dharwad.

TNAU, AGRITECH Portal. 2016. *Crop Protection.* https://agritech.tnau.ac.in/crop_protection/potato/potato_5.html.

Tofangsazi, N., K. Kheradmand, S. Shahrokhi, and A. Talebi. 2012. Effect of Different Constant Temperatures on Biology of *Schizaphis graminum* (Rondani)(Hemiptera: Aphididae) on Barley, Hordeum vulgare L. (Poaceae) in Iran. *Journal of Plant Protection Research* 52, no. 3: 319–23.

Tran, B. M. D., and P. F. Credland. 1995. Consequences of Inbreeding for the Cowpea Seed Beetle, *Callosobruchus maculatus* (f.) (Coleoptera: Bruchidae). *Biological Journal of the Linnean Society* 56, no. 3: 483–503.

Traore, F., C. L. Dabire-Binso, N. M. Ba, A. Sanon, and B. R. Pittendrigh. 2013. Feeding Preferences of the Legume Pod Borer *Maruca vitrata* (Lepidoptera: Crambidae) Larvae and Suitability of Different Flower Parts for Larval Development. *International Journal of Tropical Insect Science* 33, no. 2: 107–13.

Umeozor, O.C. 2005. Effect of the Infection of *Callosobruchus maculatus* (Fab.) on the Weight Loss of Stored Cowpea (*Vigna unguiculata* (L.) Walp.). *J. App. Sci. Enviro. Manag* 9, no. 1: 169–72.

USDA. 2015. *'Melanoplus differentialis fact sheet'. Grasshoppers: Their Biology, Identification, and Management*. Retrieved July 7 2015. United States Department of Agriculture.

Utida, S. 1972. Density Dependent Polymorphism in the Adult of *Callosobruchus maculatus* (Coleoptera, Bruchidae). *Journal of Stored Product Research* 8, no. 2: 111–25.

Vaishampayan, S. Jr., and H. N. Singh. 1994. Seasonal Variation in the Activity and Reproductive Condition of Gram Cutworm *Agrotis ipsilon* Huf. Caught on Light Trap at Varanasi. *Legume Research* 17, no. 3/4: 213–6.

Vakhide, N., and S. A. Safavi. 2014. Biology and Fertility Life Table of the Greenbug, *Schizaphis graminum* (Hemiptera: Aphididae) on the Resistant Winter Wheat Cultivar (Pishgam) in Iran. *Archives of Phytopathology & Plant Protection* 47, no. 3: 355–65.

Valenzuela, H. R., and D. R. Bienz. 1989. Asparagus Aphid Feeding and Freezing Damage Asparagus Plants. *Journal of the American Society for Horticultural Science* 114, no. 4: 578–81.

Van Den Berg, J. 2009. Case Study: Vetiver Grass as Component of Integrated Pest Management Systems. http://www.vetiver.org/ETH_WORKSHOP_09/ETH_A3a.pdf.

Van den Berg, J., and A. Erasmus. 2015. Text extracted with permission from the editors from: Prinsloo, G. L. & Uys. M. R. Van. In *Insects of Cultivated Plants and Natural Pastures in Southern Africa*, edited by V. M. Entomological Society of Southern Africa, Africa.

Van Veen, F. J., C. B. Müller, J. K. Pell, and H. C. Godfray. 2008. Food Web Structure of Three guilds of Natural Enemies: Predators, Parasitoids and Pathogens of Aphids. *Journal of Animal Ecology* 77, no. 1: 191–200.

Varma, A. 1986. The Pyrilla, *Pyrilla perpusilla* Walker. In *Sugarcane Entomology in India*, edited by H. David, S. Easwaramoorthy, and R. Jayanthi. India: Sugarcane Breeding Institute. Coimbatore: 277–95.

Vásquez, E., D. Dean, D. Schuster, and P. Van Etten. 2005. A Laboratory Method for Rearing *Catolaccus hunteri* (Hymenoptera: Pteromalidae), a Parasitoid of the Pepper Weevil (Coleoptera: Curculionidae). *Florida Entomologist* 88, no. 2: 191–4.

Vasquez, J. C., and E. A. Buss. 2006. Southern Chinch Bug Feeding Impact on St. Augustinegrass Growth Under Different Irrigation Regimes. *Applied Turfgrass Science*.

Vendramim, J. D., M. C. V. D. Ferraz, and J. R. P. Parra. 1982. Biologia Comparada de *Agrotis subterranea* (F.) em Meios Natural e Artificial. *SOLO (Brazil)* 74, no. 1: 76–80.

Vennila, S., V. K. Biradar, M. Sabesh, and O. M. Bambawale. 2007. *Know Your Cotton Insect Pest Jassids*.

Vereschagina, A., and E. Gandrabur. 2014. Polymorphism and Damage of Aphids (Homoptera: Aphidoidea). *International Journal of Biology* 6, no. 4: 124–37.

Verma, K. S., and A. K. Verma. 2001b. Biology of *Agrotis segetum* Denis and Schiffermüller (Lepidoptera: Noctuidae) on Tomato. *Pest Management & Economic Zoology* 9, no. 1: 97–101.

Verma, K. S., and A. K. Verma. 2002b. Intensity of Plant Infestation by *Agrotis segetum* in Tomato. *Insect & Environment* 8, no. 1: 16–7.

Verma, K. S. 2015. The Bioecology and Management of Cutworms in India group meeting of All India Network Project on Soil Arthropod Pests, Palampur, Himachal Pradesh, India vol. XVIII: 19–20.

Vickerman, G. P., and S. D. Wratten. 1979. The Biology and Pest Status of Cereal Aphids (Hemiptera: Aphididae) in Europe: A Review. *Bulletin of Entomological Research* 69, no. 1: 1–32.

Viraktamath, C. A. 2006. *Final Report of Emeritus Scientist Project on Taxonomic Studies on the Economically Important Leafhoppers (Hemiptera: Cicadellidae) of the Indian Subcontinent. Department of Entomology, University of Agricultural Sciences*: 65–70. Bangalore: GKVK.

von Dohlen, Carol D., Carol A. Rowe, and Ole E. Heie. 2006. A Test of Morphological Hypotheses for Tribal and Subtribal Relationships of Aphidinae (Insecta: Hemiptera: Aphididae) Using DNA Sequences. *Molecular Phylogenetics & Evolution* 38, no. 2: 316–29.

W, D. D., S. N. A, and A. N. A. J. 2010. Effects of Insecticide Spray Application on Insect Pest Infestation and Yield of Cowpea (*Vigna unguiculata* (L.) Walp.) in the Transkei, South Africa. *African Journal of Biotechnology* 9, no. 11: 1673–9.

Walczak, F. 2002. Beware of Noctuids and Other Soil Pests. *Ochrona Roslin* 46, no. 8: 8–10.

Walker, K. 2009. *Brown Marmorated Stink Bug (Halyomorpha halys)*, Accessed November 25 2011. http://www.padil.gov.au.

Wang, B. 2022. Identification of Crop Diseases and Insect Pests Based on Deep Learning. *Scientific Programming* 2022: 1–10.

Wiedemann, C. R. W. 1824. *Munus Rectoris in Academia Christiana Albertina Aditurus Analecta Entomologica Ex Museo Regio Havniensi*. Kiliae: Eregio Typoguapheo Scholarum: 60.

Wiman, N. G., V. M. Walton, P. W. Shearer, S. I. Rondon, and J. C. Lee. 2015. Factors Affecting Flight Capacity of Brown Marmorated Stink Bug, *Halyomorpha halys* (Hemiptera: Pentatomidae). *Journal of Pest Science* 88, no. 1: 37–47.

Xu, J., D. M. Fonseca, G. C. Hamilton, K. A. Hoelmer, and A. L. Nielsen. 2014. Tracing the Origin of US Brown Marmorated Stink Bugs, *Halyomorpha halys*. *Biological Invasions* 16, no. 1: 153–66.

Yadav, L. S., J. P. Chaudhary, and P. R. Yadav. 1984. Relative Abundance of Important Noctuids Moth Light Trap Infesting Chickpea in Haryana. *Bulletin of Entomology* 25, no. 2: 103–10.

Yadav, D. N., and R. Patel. 1990. Effect of Some Botanicals on Oviposition of *Chrysoperla scelestes* and Their Ovicidal Action. In *Proceedings of Symposium on Botanical Pesticide in Integrated Pest Management, Rajamundrai, India*, edited by M. S. Chari and G. Ramaprasad: 166–9.

Yakir, D. B., M. Chen, S. Y. Sinev, and V. Seplyarsky. 2013. *Chilo partellus* (Swinhoe) (Lepidoptera: Pyralidae) a New Invasive Species in Israel. *Journal of Applied Entomology* 137, no. 5: 308–400.

Yang, Z. Q., Y. X. Yao, L. F. Qiu, and Z. X. Li. 2009. A New Species of *Trissolcus* (Hymenoptera: Scelionidae) Parasitizing Eggs of *Halyomorpha halys* (Heteroptera: Pentatomidae) in China with Comments on Its Biology. *Annals of the Entomological Society of America* 102, no. 1: 39–47.

Yonow, T., D. J. Kriticos, N. Ota, J. V. D. Berg, and W. D. Hutchison. 2017. The Potential Global Distribution of *Chilo partellus*, Including Consideration of Irrigation and Cropping Patterns. *Journal of Pest Science* 90: 459–77.

Yousuf, M., and P. Ray. 2009. Braconids as Potential Bio Control Agents on Insect Pests: An Overview. *5th International Conference on Biopesticides: Stakeholder's Perspectives*. 26–30 April 2009, New Delhi, pp. 137–38.

Yuan, L., Y. Huang, R. W. Loraamm, C. Nie, J. Wang, and J. Zhang. 2014. Spectral Analysis of Winter Wheat Leaves for Detection and Differentiation of Diseases and Insects. *F. Crop. Res.* 156: 199–207.

Yule, S., and R. Srinivasan. 2013. Evaluation of Bio-Pesticides against Legume Pod Borer, *Maruca vitrata* Fabricius (Lepidoptera: Pyralidae), in Laboratory and Field Conditions in Thailand. *Journal of Asia-Pacific Entomology* 16, no. 4: 357–60.

Zhu, Gengping, Wenjun Bu, Yubao Gao, and Guoqing Liu. 2012. Potential Geographic Distribution of Brown Marmorated Stink Bug Invasion (*Halyomorpha halys*). *PLOS One* 7, no. 2: e31246.

10 Current Status of Fall Armyworm and Maize Fodder Decline Foreseen

Rashad Rasool Khan, Muhammad Umair Sial, and Muhammad Arshad

10.1 INTRODUCTION

For almost 200 years, *Spodoptera frugiperda* has been reported in the United States (Edosa and Dinka, 2021). The pest was first found in numerous African countries in 2016, and it has since spread throughout almost the entire African continent (Allen *et al.*, 2021), as well as in various Asian countries in 2018 (Hussain *et al.*, 2021), with almost all Asian maize-producing countries now under *S. frugiperda* risk (Paredes-Sanchez *et al.*, 2021). *S. frugiperda* has lately spread to Europe and Australia (Plessis *et al.*, 2020; Parra *et al.*, 2022). The insect has infested crops in over 109 countries worldwide (Zhao *et al.*, 2022). The pest is capable of causing damage to 353 host plants (Badhai *et al.*, 2020). Conversely, maize is *S. frugiperda*'s preferred crop (Chimweta *et al.*, 2020). The pest life cycle consists of four stages (egg, larva, pupa, and adult) and is extremely fertile (Sagar *et al.*, 2020). It feeds on all stages of plant development but is considered most detrimental at the larval stage (Badhai *et al.*, 2020). The pest is economically significant due to its voracity (Chen *et al.*, 2021), high reproduction (Zhang *et al.*, 2021), prolonged adult dispersal (Deshmukh *et al.*, 2021a), multiple generations per year, and a lack of diapause (Edosa and Dinka, 2021). Because of these characteristics, *S. frugiperda* is a potentially hazardous pest to maize and other crops. The economic losses in Africa are US$9.4 billion (Eschen *et al.*, 2021).

It has been reported in most countries like Africa, Iran, India, and China. Additionally, the fall armyworm (FAW) has been reported in several parts of Pakistan. Between August and November 2019, the prevalence of FAW was evaluated in Faisalabad, Chiniot, Gojra, and Ameen Pur Bangla. FAW was found around the end of October, larvae were collected, and their rearing was conducted. This report confirmed the presence of the FAW in Faisalabad (Naeem-Ullah *et al.*, 2019).

FAW is a foliage arthropod, and it is considered detrimental to plants. Young larvae typically damage leaves by poking holes in them and feeding mostly on the epidermis of the leaf. Dead heart results from feeding on young plants through the whorl. The production and quality of older plants are decreased by the larger larvae in the whorls feeding on the cobs or kernels of maize (Abrahams *et al.*, 2017). The caterpillars feed on foliage throughout all their life stages. However, they prefer young plants for feeding (Bissiwu *et al.*, 2016). Due to feeding in the plant's whorls, rows of pores are generated in the leaves. This can occasionally cause considerable defoliation and reduce the plant's ability to expand (Capinera, 2002). In particular, the corn strain (C) and rice strain (R) are considered the two strains of *S. frugiperda* (Unbehend *et al.*, 2013).

Because of its powerful flight, FAW is rapidly spreading. Furthermore, the FAW is active all year and does not hibernate in the winter. Even at temperatures below 18°C, FAW can continue

DOI: 10.1201/b23394-10

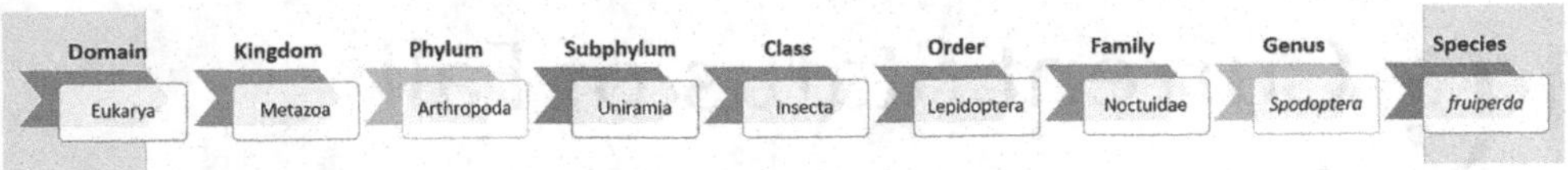

FIGURE 10.1 Taxonomy hierarchy of *Spodoptera frugiperda*.

the developing process. Because of these distinguishing characteristics, FAW has become more lethal (Garcia *et al.*, 2018).

10.2 PEST IDENTIFICATION AND BIOLOGY

10.2.1 Taxonomy

10.2.1.1 Morphological Identification

S. frugiperda can be differentiated from the other species of the same genus and three other prevalent noctuid pests of corn by its soft body appearance and the presence of a noticeable black pinnacle on the dorsal body segments of larvae (Figure 10.2). These other species included *S. exempta* (Walker), *S. exigua* (Hübner), *S. litura* (Fabricius), *S. mauritia* (Boisduval), and *Helicoverpa armiger* (Doubleday). Except *for Chrysodeixis eriosoma*, all species share markings along the inverted Y-shaped fronto-clypeal suture (Figure 10.2) and ecdysial line, although they vary in breadth and color. FAW occurrence was noted in 17 municipalities across 10 provinces. Preliminary findings suggested that this pest damages only non-Bt corn. It can infest all phases of corn growth, but it seems to favor earlier stages of the host's development (Navasero *et al.*, 2019). However, at the first, second, and third instar stages, it is particularly perplexing to identify species of FAW morphologically. Although it can be identified at the fourth and fifth developmental instars based on obvious morphological characteristics, the economic loss has already gone out of control for the target crop. Therefore, accurate early identification of this pest is necessary.

10.2.1.2 Molecular Identification

Although genetic identification of taxa may be difficult and time-consuming, which constitutes a major limitation for biodiversity evaluations, its use as an additional identification method that can give intermediate evidence for morphological identification is urged. Even though regular genetic identification of taxa may be challenging and time-consuming (Hajibabaei *et al.*, 2007). Nevertheless, DNA sequencing, which uses certain genetic markers, is a method that is possible and helpful for identifying species quickly and accurately (Kress and Erickson, 2008; Pereira *et al.*, 2008). It has been suggested that, due to the rise in globalization, invasive species should be recognized by DNA barcoding using clear tags of database files (Figure 10.3) for putative specimen sources as opposed to interception sites (Madden *et al.*, 2019).

Authentic species identification for FAW control is difficult in the early attack of pests. A quick and accurate method for the species-specific identification of this pest is bioinformatics which created and discovered the Sf00067, FAW-specific gene; It was recommended that the loop-mediated isothermal amplification (LAMP) test is the best setting for an Sf00067. FAW captured in Korea and FAWs collected in Benin, Africa, established Sf00067 6 primer LAMP (Sf6p-LAMP) assay was used to identify the distinct strains. An FAW diagnostic test can be completed in 30 minutes from the extraction of DNA from an egg or a first instar larva to the species level (Osabutey *et al.*, 2022).

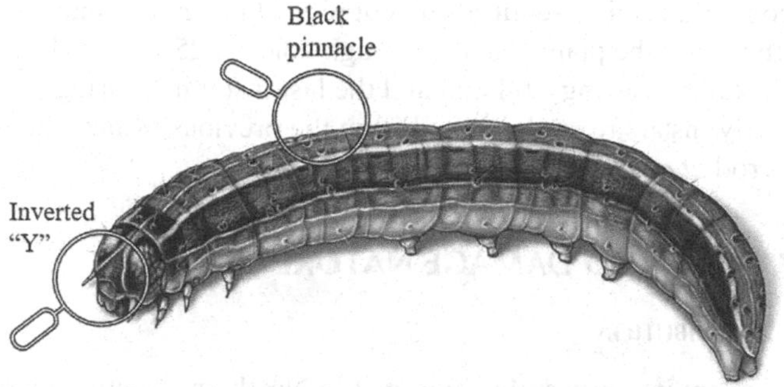

FIGURE 10.2 Morphological identification of *S. frugiperda* with inverted "Y" and black pinnacle on the dorsal surface.

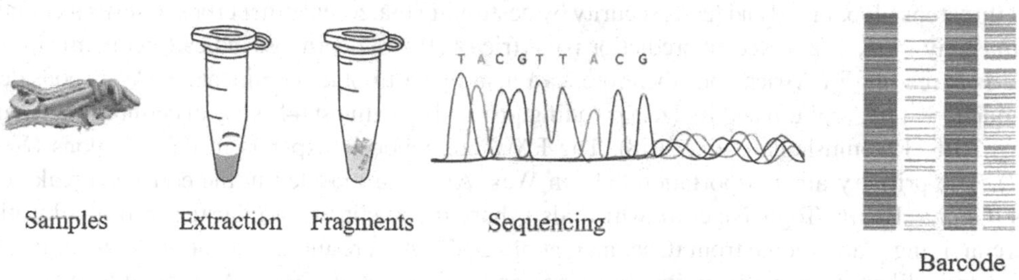

FIGURE 10.3 Schematic representation of various steps for the molecular-based identification of FAW.

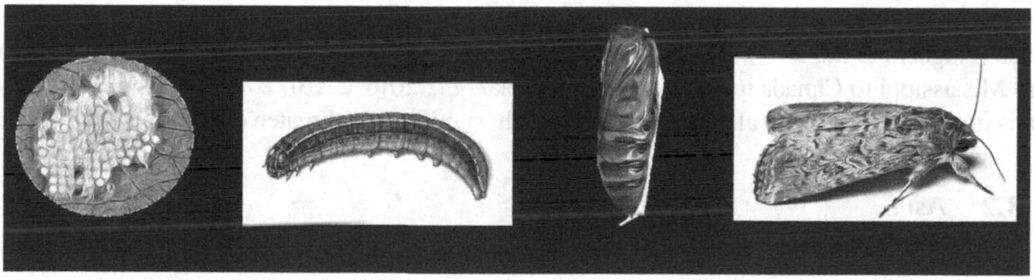

FIGURE 10.4 Different stages of FAW life cycle.

10.2.2 Life Cycle Description

FAW has caused an annual loss of US$2400–4000 million in Asia due to its infection of 40,000 hectares of 20% maize in Sri Lanka (IPPC, 2018). The fall armyworm larva survives by spending most of its life in the whorl of maize, minimizing its exposure to insecticides (FAO, 2018). The FAW life cycle begins with an egg and continues through the stages of caterpillar, pupa, and adult (Figure 10.4). The life cycle changes according to climate and other factors, with the life cycle completing in 30 days during the summer, 60 days in the spring and fall, and up to 80–90 days during the winter season (James and Engelke, 2010).

After 3–5 days, depending on temperature, the young caterpillars in the second and third instars consume the leaf whorl, while the adult caterpillars in the fourth and sixth instars prefer to eat around the cob silk, leaving semitransparent areas known as windows. As the caterpillar grows, it climbs to the top of the plant, leaving a rough hole. At 25°C, larval development takes 22 days, with the first instar measuring 1.08 mm and the last instar measuring 34.15 mm. L2–L3 are similar in color to early instars recently molted from the previous instar. The final three, L4–L6, are usually dark (Jarrod *et al.*, 2015).

10.3 DISTRIBUTION AND DAMAGE NATURE

10.3.1 Global Distribution

S. frugiperda is a virulent insect pest that emerged in North and South America but has rapidly spread across sub-Saharan Africa in the past 2 years. Before 2015, the pest was only spotted in the Americas; however, in 2016, a severe epidemic of FAW in maize was observed in several African nations, including Sao Tome, Nigeria, Benin, and Togo (Goergen *et al.*, 2016). The first verified cases of FAW were reported in West Africa in early 2016 (Goergen *et al.*, 2016; Cock *et al.*, 2017).

It threatens global food and feeds security by destroying maize and other crops in several countries (Ramasamy *et al.*, 2022). Recent predictions of Africa's 20–50% corn loss indicate economic impact. FAW is spreading in Africa and, therefore, can also move to other continents. FAW is considered the major pest of sorghum and maize, spreading promptly to almost 44 African countries (Prasanna *et al.*, 2018; Rwomushana *et al.*, 2018). The FAW continuously expands in these regions (Njeru, 2017). The primary air transportation hubs in West Africa are located in the early outbreak countries (Ghana, Benin, Togo, Nigeria), which also share many climate similarities with the locations where arriving planes come from (Chapman *et al.*, 2017). As a result, this location is one of those in Africa most likely to serve as an invasion's epicenter (Early *et al.*, 2016). According to speculation (Cock *et al.*, 2017), the FAW may have arrived in Africa as a stowaway aboard a passenger plane. Because prevailing winds are typically from east to west, it is thought implausible that the armyworm could have dispersed on its own. According to molecular evidence from samples from Togo, the eastern United States, the Caribbean, and the Lesser Antilles are where fall armyworm originates in Africa (Nagoshi *et al.*, 2017). With the aid of a low-level jet stream, adult moths of FAW may travel from Mississippi to Canada in 30 hours (Westbrook *et al.*, 2016; CABI, 2018). The highly migratory pest is quickly spreading to all continents and might significantly threaten crop output worldwide.

10.3.2 Asia

The Lepidopteran (Noctuidae) pest known as FAW was originally discovered in Georgia in 1797. Several outbreaks then took place throughout the Americas in the 19th century. It originated in West Africa and was first identified outside of the Americas in late 2016, and within 3 years, it spread to more than 40 African countries (Sisay *et al.*, 2019). After conquering Africa, it first appeared in Asia in May 2018 from Karnataka, India. Beyond South Asia, it has expanded to Southeast Asia and even China (as of March 2020) and, most recently, Australia (2020) (Lamsal *et al.*, 2020; Deshmukh *et al.*, 2021b).

In Asia, FAW was initially noted in India and then in Bangladesh, China, Myanmar, Thailand, and South Asian nations (FAO, 2019).

10.3.3 Host Range

The highly entomopathogenic, migratory fall armyworm is native to the subtropical and tropical areas of the United States. It is regarded as the most significant pest of maize in Canada, Brazil,

Argentina, Chile, and other countries. *S. frugiperda* reduces grain yield in Brazil by up to 34%, resulting in an annual loss of US$400 million (Lima *et al.*, 2010). The host range of the FAW is highly broad, with records on more than 353 host plants (Montezano *et al.*, 2018). The most frequently harmed plants in the Americas are Sudan grass, other fodder grasses, grain, millet, soybean, nut, cotton, field maize, and sorghum (CABI, 2018; Montezano *et al.*, 2018).

According to Early *et al.* (2018), FAW caterpillars are significant pests of cereals and fodder grasses and have been seen to consume 186 plant species from 42 families. Among the other commercially important plants in Africa severely harmed by FAW are rice, sugarcane, sorghum, beet, tomato, potato, cotton, and pasture grasses (Figure 10.5) (Day *et al.*, 2017).

10.3.4 Pest Strains

In Africa, the species maize and rice strains have been found (Cock *et al.*, 2017; Nagoshi *et al.*, 2017). Considering their host and mating differences, both strains are unlikely to be introduced simultaneously (Hanniger *et al.*, 2017). It is unknown if single or multiple introductions triggered the invasion. Morphologically identical FAW subpopulations have distinct host plant preferences. The C-strain prefers maize, sorghum, and other large grasses, while the R-strain prefers Bermuda grass, rice, and other small grasses. In Karnataka, India, FAW has been found on maize. Telangana, Tamil Nadu, Madhya Pradesh, and Gujarat are now infected. India's FAW strain matched the rice's (Mahadevaswamy *et al.*, 2018). In Asia, the pest eats sorghum, maize, and sweet corn (Sharanabasappa *et al.*, 2018).

10.3.5 Environmental Impact

It is speculated that FAW traveled across continents to reach Asia. According to Lamsal *et al.* (2020), a tropical and warm environment with mean annual temperatures between 17 and 35°C and mean annual rainfall between 0 and 400 mm is suitable for FAW. Due to its inability to endure prolonged exposure to low temperatures below 10°C, the pest is known to go from warmer climates to overwinter (Burtet *et al.*, 2017). This pest is a more effective colonizing agent than its rivals. A sizable portion of South Asia has a tropical wet and dry, humid subtropical, and semiarid climate, offering a pleasant environment all year. It economically threatens cultivated species of grasses like rice, maize, sorghum, sugarcane, and other crops (Bateman *et al.*, 2018). Both the geographic distribution and quantity of insect pests are significantly impacted by climate change. Climate variables, including precipitation, temperature, and humidification, directly impact an insect's ability to reproduce, spread, invade, migrate, and adapt (Bale *et al.*, 2002).

10.3.6 Damage Pattern

The vicious pest FAW inhabited America's tropical and subtropical regions. It was initially discovered in Kenya in 2016 and has been seriously harming crops since then (De Groote *et al.*, 2020). In its natural area, 353 plant and crop species from 76 land plant families are known to be the targets of this polyphagous pest, some of which are the most frequently affected (Montezano *et al.*, 2018). The FAW's larval stage is the most harmful because its caterpillars consume tender leaf whorls and tips, reducing maize crop yield and high-grain output (Mutyambai *et al.*, 2022). The early maize seedling's base can be severed by late larval instars, killing the entire plant (Harrison *et al.*, 2019).

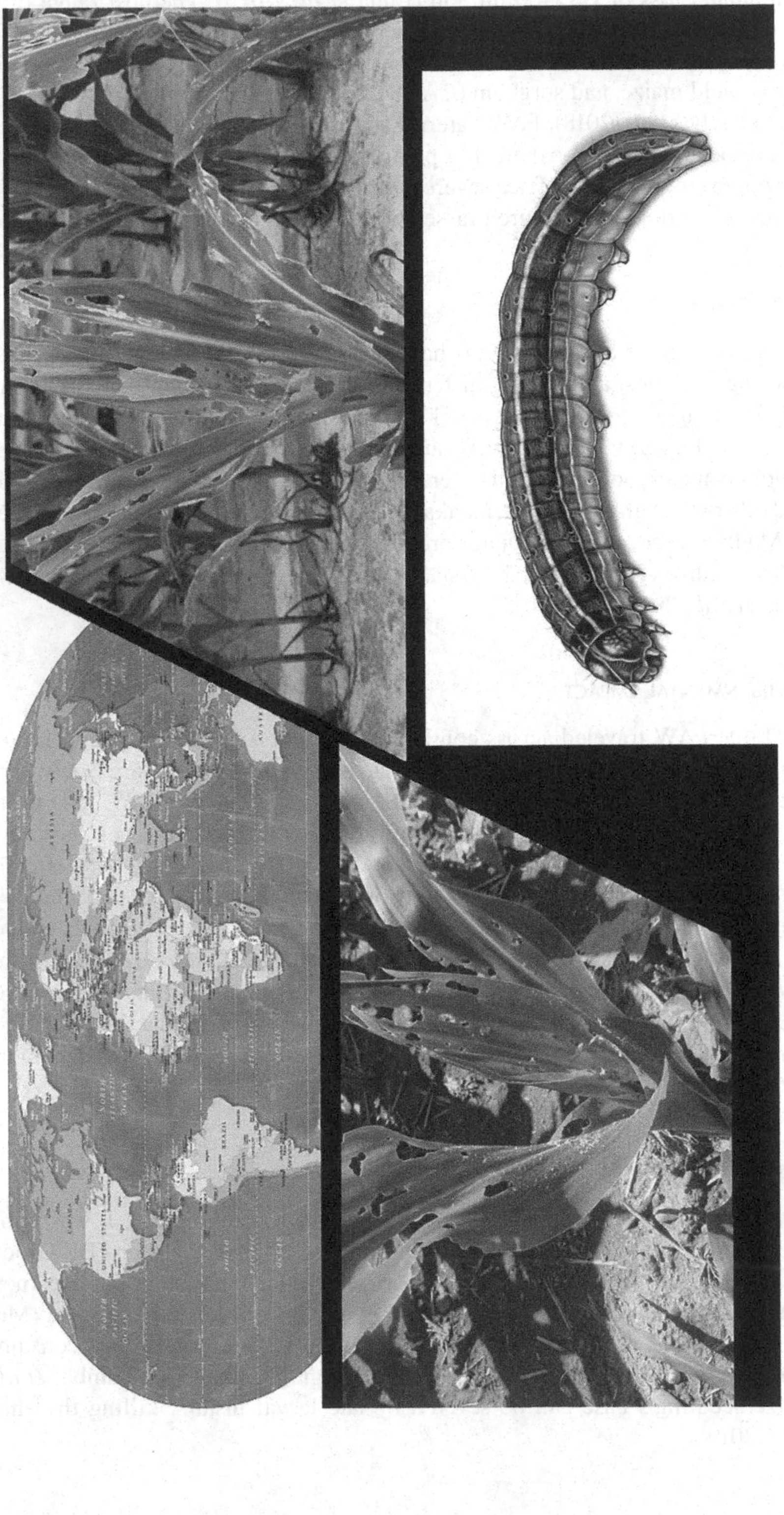

FIGURE 10.5 Distribution and damage pattern of FAW on maize.

10.4 FAW AND MAIZE FODDER DECLINE FORESEEN

10.4.1 Global Scenario of Maize Fodder Losses

Maize has the highest potential for output of any cereal; maize is known as the "Queen of Cereals" worldwide. With more than 36% of worldwide production, the world's largest suppliers of corn are the United States and the engine of the US economy. After rice and wheat, corn is India's third-most significant cereal crop grown year-round. Ten percent of the nation's entire production of food grains comes from it. Numerous industrial products, including those in the carbohydrates, amino group, fatty acids, alcoholic drinks, film, textile, gum, packaging, food additives, cosmetics, medicinal, and paper sectors, utilize it as a basic raw ingredient (https://apeda.gov.in). In 2020, there were 194.77 million ha (mha), 1148 (mmt), and 5.74 tonnes produced per ha worldwide of maize crop. The total productivity in India was 29 million tones and 2945 kg/ha. About 1.34 mha of maize is farmed in Madhya Pradesh, with an average yield of 3.91 million tones and productivity of 2922 kg/ha (www.mospi.gov.in).

10.4.2 Maize Fodder Losses in Asia

There were numerous reports of it in the Indian subcontinent in 2018, and it was once more discovered to be primarily linked to maize crops throughout the nation. FAW has spread to most of India's regions in less than 2 years and has been observed to target millets, maize, sorghum, and sugarcane (Swamy *et al.*, 2018; Babu *et al.*, 2019). The alarming rate at which FAW spreads throughout these extensive areas and diverse landscapes is dangerous to the various nations' efforts to ensure food security. With the loss in yield, it can generate major harm. For instance, in Africa, the total yearly economic losses from the 12 maize countries were predicted to vary from US$2531 to 6312 million without adequate regulation (Day *et al.*, 2017). Due to this serious harm to the world's crop productivity, this species has drawn much interest due to its propensity for rapid reproduction. Until 2015, only North and South American reports of this pest were made. However, in 2016, the first record of its presence in Africa was made, where it seriously harmed the production of maize (Goergen *et al.*, 2016). Additionally, with initially discovered in South Asia, specifically India, in 2018 (Shylesha *et al.*, 2018). As a result, this pest has a high risk for further spread and poses a global problem to worldwide food security by destabilizing crop production (Westbrook *et al.*, 2016).

10.4.3 Maize Fodder Losses Foreseen

Despite its great productivity, various abiotic and biotic stressors lead maize yields to fall short of their potential (Assefa and Ayalew, 2019). According to global climate change projections, droughts are expected to worsen and last longer in future decades in places where maize, wheat, and other significant crops are grown (Vincent *et al.*, 2013). According to Tesfaye *et al.* (2017), the end quantity of maize lessens by 1–1.8% under favorable and desiccation conditions, respectively, when the temperature increases one degree on days when the temperature exceeds 30°C, whereas the yield may be reduced by up to 40% under combined drought and heat (Lobell *et al.*, 2011). Africa's maize output is anticipated to be impacted, and over 40% of the continent's maize-producing area currently experiences intermittent drought stress (Fisher *et al.*, 2015). Additionally, climate change may increase the frequency of pests and diseases in livestock and crops, as well as often occurring droughts and hot waves (IPCC, 2013). Insect metabolism and consumption rates increase as the temperature rises to levels within the ideal range of development. As they grow bigger and faster, they breed more quickly and experience lower mortality rates (Bentz *et al.*, 2019; Ngumbi, 2020). For instance, the fall armyworms' ideal temperature for the egg to adult development was 26–30°C, and the larval development was quickest at 30°C.

According to a model, a 2°C increase in the average world surface temperature would result in an average 31% increase in maize production losses due to pest pressure or a loss of 62 Tg each year (Du Plessis *et al.*, 2020). Like this, FAW's egg, larval, and pupal development times are significantly reduced when the temperature is raised from 19 to 31°C (Huang *et al.*, 2021). As a result of the fall armyworm's damaging effects on crop productivity and its powerful capacity for reproduction, it is crucial to determine its future distribution patterns and controlling variables. Such understanding could aid in developing techniques to stop it from spreading and lessen the risk it poses to the world's food security.

10.5 FIELD MEASUREMENT AND MONITORING

10.5.1 Conventional Scouting and Surveillance

If suitable monitoring techniques, indicators of the presence and absence of a certain pest, are in place, FAW is better handled. Monitoring-based Integrated Pest Management (IPM) is fundamental to a good FAW management strategy. Different methods, including routine field inspection, pheromone traps, and light traps, can be used to check the FAW (FAO, 2017a, Haftay and Fissiha, 2020). The key to managing fall armyworms is spotting an infestation before it harms the economy. Agricultural fields must be carefully inspected every 3–4 days throughout the first 40 days following planting. Applying control measures is important if FAW is found during scouting. For example, it is suggested to use an efficient management technique for maize if 5% of seedlings are removed or 20% of tiny plant whorls have FAW (Fernandez, 2002; Assefa and Ayalew, 2019).

Light sources draw the attention of FAW moths. If moths are captured in light traps, a pest may present that could harm the crops. Therefore, one of the monitoring methods for FAW could be using night-time light traps. Thus, the management techniques stated that 3.4–3.8 may be used individually or as part of a comprehensive approach (Fernandez, 2002).

Farmers in Ethiopia and Kenya utilize mud in corn whorls, plant extracts, handpicking larvae, synthetic pesticides, and plant extracts as FAW management measures (Kumela *et al.*, 2019). However, the most typical means of control used by farmers in Zambia and Ghana were pesticide spraying and handpicking larvae (Tambo *et al.*, 2020). These results demonstrated a rising trend in pesticide use against FAW; however, before the spread of FAW, most smallholder maize farmers in Africa employed cultural control techniques rather than chemical pesticides to control insects (Hruska, 2019). Additionally, most of Africa's least expensive and widely used pesticides fall under the category of in the Americas, FAW has developed resistance to broad-spectrum synthetic insecticides (Day *et al.*, 2017). Their use would stifle the natural enemies of FAW and their impact on FAW populations (Harrison *et al.*, 2019).

The push–pull technology also called PPT, requires intercropping crops with a plant that repels or deters insect pests, such as desmodium, *Desmodium uncinatum* J., and placing an attractive plant that is highly evident and attractive to a pest, such as Napier grass, *Pennisetum purpureum* Schumach (Poaceae). In the PPT, volatile chemicals released by the push plant can repel the female FAW moths, whereas volatile chemicals produced by the pulled plant are more alluring to adult FAW and stem-borer moths than maize, facilitating their concentration in the pulled plant (Midega *et al.*, 2018; Khan *et al.*, 2011).

10.5.2 Forecasting and GIS Approaches

Due to rising globalization, it is advised to identify invasive species utilizing DNA barcoding and to use database records with clear annotations for suspicious specimen origin versus interception location (Madden *et al.*, 2019). Nonetheless, its use as an additional identification method that can offer supplemental proof to back up morphological identification is advised because routine molecular identification of many species could be challenging and momentous, which poses a constraint on biodiversity surveys (Hajibabaei *et al.*, 2007). However, utilizing precise genetic markers and DNA microarrays is an exciting and efficient approach for quick and accurate specimen identification (Kress and Erickson, 2008; Pereira *et al.*, 2008). A key element of security and bio-surveillance programs is preventing invasive pests at ports of entry, which necessitates the identification of the pest species to make quarantine choices. However, it is particularly challenging to morphologically identify the Noctuid pest species of FAW at the first, second, and third instar stages. Although it can be recognized at the fourth and fifth developmental instars based on obvious morphological characteristics, the economic harm has already outgrown control for the target crop. Therefore, accurate early identification of this pest is necessary.

10.5.3 Molecular Surveillance

The *S. frugiperda* comprises two strains modified to feed on various host plants. The corn strain (C-strain) prefers to eat maize, cotton, and sorghum, whereas the rice strain (R-strain) prefers rice and several pasture grasses. Although there is no morphological difference between these strains, there are variances in their mating habits, allelochemical compositions, adaptations, and DNA backgrounds (Dumas *et al.*, 2015). Only two publications from India regarding the molecular characterization of FAW from Asia showed that rice has an R-strain present (Sharanabasappa *et al.*, 2019; Maruthadurai and Ramesh, 2020). However, given how little is known, the molecular analysis of *S. frugiperda* requires more information. Here, we provide the FAW's initial molecular analysis results from a corn-growing region in Sindh, Pakistan. The present analysis was conducted to increase understanding of the geographical distribution of FAW in Pakistan and to analyze the various locations of Pakistan for the presence of FAW. Both morphological and molecular analyses supported the presence of *S. frugiperda*. Therefore, activities like surveillance, awareness campaigns, and capacity-building initiatives are quite important during this early invasion stage.

10.6 MANAGEMENT APPROACHES

10.6.1 Cultural and Mechanical

Yigezu and Wakgari (2020) summarized the cultural methods to manage and control maize yield and FAW infection losses. This would include handpicking and killing larvae, covering maize plants in sand or wood ash, soaking plants in tobacco tinctures, deep plowing to kill pupae that have overwintered, early planting, eliminating ratoon vegetation, scorching infected vegetation after cultivation, intercropping with non-host plants and using new varieties (Kebede and Shimalis, 2018).

According to studies by Baudron *et al.* (2019), smallholder farmers can combat FAW by using additional cultural practices, including weeding and fallow periods. The majority of the weeds in the research region were FAW hosts belonging to the Graminaceous family, which is probably why the results showed that frequent weeding reduced FAW damage. It has been observed that producing maize with zero or minimal tillage lessens FAW damage since it encourages predatory organisms to multiply in numbers (Rivers *et al.*, 2016). According to Kebede and Shimalis (2018), agroforestry could lessen the effects of FAW by increasing the population of natural enemies and enhancing the resistance of maize crops to infestation. To control worms and pupae, it is necessary to plow deep and burn agricultural leftovers in affected fields. It is crucial to conduct an additional study on reduced and zero tillage to mitigate FAW in smallholder farmer systems because these techniques are widely used in Asia.

Smallholder farmers typically inject mud or sand into the leaf whorls to combat the insect. Some farmers in southern Africa applied grains of sand as individual treatments or in combination with fertilizer containing ammonium nitrate to effectively manage FAW (Hruska, 2019). It is believed that while the ammonium nitrate fertilizer dries out the worm, killing it, the sand scarifies the insect's body, making it more susceptible to infection by natural infections. Farmers also employed spraying sugar solution and fish soup to manage FAW (Harrison *et al.*, 2019). In the field, the techniques aim to draw in and increase populations of predatory ants, parasitoids, solitary wasps, and other adversaries known as natural enemies.

FAW can be managed with generally positive crop management. The maize crop's nutritional and water state determine its capacity to tolerate FAW attack. Farmers indicated that FAW attacks were more severe in fields without fertilizer (Kansiime *et al.*, 2019). The farmers also noted that timely fertilizer applications and early planting lessened the severity of FAW. Compared to a water-stressed and nutritionally inadequate maize crop, a well-fertilized and well-watered crop may resist or recover quickly and considerably better from FAW attack. Low levels of FAW damage can be repaired in the maize crop, although this greatly relies on the crop's nutritional quality and growth stage (Hruska, 2019).

10.6.2 Biological Control Practices

Predators, parasitoids, and diseases are a few of the FAW's natural enemies that control the population numbers of this species. In some instances, intercropping fosters the emergence and expansion of biological that is sufficient to manage FAW (FAO, 2017b). This observation led to the spread of the PPT that is now advised for FAW management (ICIPE, 2018). The International Centre of Insect Physiology and Ecology (ICIPE) developed the technology, which is now being promoted and used in various SSA nations to manage FAW. The method was created to control stem borers in maize.

The PPT calls for edging the intercrop with *Brachiaria* "Mulato II" and intercropping maize with greenleaf desmodium (Midega *et al.*, 2018). Desmodium repels the moths that are simultaneously drawn (tugged) by pheromones produced by the main crop, protecting the maize in the process. FAW infestation can be decreased by at least 80% in a field where the technique is being used, according to ICIPE (2018) and Midega *et al.* (2018).

Unsprayed fields have been shown to have natural parasitism levels of over 44% (FAO, 2017a), which has consequences for the formulation and implementation of control methods. Five native species of parasitoids were found in an analysis by Sisay *et al.* (2018), some of 45.3% of parasitism levels. This may include *Cotesia icipe*, *Charops ater* (Szépligeti), *Chelonus curvimaculatus*, and *Palexorista zonata* (Curran). The most parasitism has been seen in several other investigations that have discovered natural enemies of FAW. According to Kenis *et al.* (2019), *Telenomus remus* (Dixon) is found in at least five nations. Agboyi *et al.* (2020) reported ten species of parasitoids. In addition, three FAW predator species and seven parasitoid species were found by Koffi *et al.* (2020).

Geographical fluctuations in organisms' occurrence and parasitism levels have been observed (Koffi *et al.*, 2020), which can be attributed to variations in different regions, techniques, crop types, and developmental stages (Hay-Roe *et al.*, 2016). Most nations implement mass parasitoid and predator-rearing and release programs, successfully controlling FAW and other pests (Soares *et al.*, 2012).

According to Molina-Ochoa *et al.* (2003), entomopathogens damage pests and naturally control FAW populations. According to Assefa and Ayalew (2019), the FAW was vulnerable to 16 different entomopathogen species, most viruses, and bacteria. They emphasized how geographic location, agricultural techniques, and insecticide use influenced biocontrol agents' presence and dispersion, particularly entomopathogens. For effective use in FAW control techniques, exploring the existence and distribution of these entomopathogens is still necessary. Identifying the FAW larvae that are destroyed by viruses and fungi is simple. Virus-killed larvae become soft and typically hang their heads from the leaves, but fungus-killed larvae become rigid and appear with a green tint (FAO, 2018).

10.6.3 Host Plant Resistance

Host plant resistance refers to the heritable traits that make a plant less likely to be a host for an insect. Identifying and developing FAW-resistant germplasm are crucial to host plant resistance, according to (Prasanna *et al.*, 2018). Transgenic resistance uses an exogenous gene or gene combination to produce the host plant FAW-resistant. Fall armyworm tolerant cultivars, whether genetically resistant or transgenic, offer a beneficial and economical way to reduce crop losses. Additionally, FAW-resistant maize cultivars promote IPM strategies (Tefera *et al.*, 2016).

When establishing a breeding plan to integrate FAW resistance into superior maize germplasm, breeders consider the origin and degree of FAW resistance and the probability of resistance through time. Transgenic insect pests like FAW can overcome monogenic or oligogenic resistance. Breeding for FAW resistance is a "victory less" battle between the host and the ever-evolving pest. To increase host plant resistance, breeding programmers should identify, use, and integrate several conventional or transgenic resistance traits (Prasanna *et al.*, 2018).

Prasanna *et al.* (2021) reported an extensive assessment of FAW resistance in the host species, particularly in corn. It includes detailed instructions for mass raising FAW, testing genetic material with a simulated FAW infection, and information on the likelihood of FAW resistance in tropical and temperate corn germplasm discovered or developed previously in the United States. We present a report on the emergence of native FAW resistance in cereal grains, illustrate the use of genetically modified (GM) maize—specifically, Bt maize—for FAW control in Asia and Africa, and the importance of implementing effective insect tolerance management strategies for Bt maize, and suggest reasonable steps for integrating host plant resistance as a key component of an IPM-based strategy for viable microbial control.

10.6.4 Bio and Synthetic Pesticides

By defending crops from harmful insect pests, insecticides are critical in boosting yield. However, the uncontrolled use of conventional and broad-spectrum insecticides causes environmental degradation, resistance to development, and recurrence. Therefore, choosing new, secure, less enduring products is essential for sustainable and eco-friendly pest management. Insecticides based on benzophenyl urea are a good fit in this regard. These prevent insects from synthesizing chitin, making them strong candidates for integrated pest management.

The crucial compounds among the chitin synthesis inhibitors are benzophenylureas (BPUs). These are strong insecticides that unfavorably control insect growth so that they can be contained before threatening humans. These BPUs are reasonably harmless for higher animals and plants because they do not have chitin in their body structures. An insect's endocuticle is damaged and loses its flexibility when exposed to these insecticides. Due to their high levels of action selection, BPUs can play a significant role in any IPM program. BPU insecticides were reported to hold a 3.6% market share globally in 2011, and the trend is rising (Sun *et al.*, 2015). Several compounds are approved for usage in India's crops, homes, and public health by the Central Insecticide Board & Registration Committee. These have consistent effects on insects and are extremely stable in crop environments (Douris *et al.*, 2016).

Despite evidence that chemical insecticides can effectively control pests, transgenic crops like Bt maize and examples of pesticide resistance have been documented (Al-Sarar *et al.*, 2006;

Frizzas *et al.*, 2014). Overall, synthetic chemical pesticides in Africa have had poor results since farmers' access, understanding of proper use, consistency, purchasing power, and choice of pesticide products determine how well they are used (Midega *et al.*, 2012). Additionally, FAW larvae are spread out lower in the maize plant canopy, protecting them from common insecticide applications. Furthermore, most smallholder farmers in Africa cannot afford pesticides. IPM programs that use natural enemies instead of pesticides are more suitable and economical for African farmers (Day *et al.*, 2017; Midega *et al.*, 2018).

10.6.5 Biotechnology

Bt maize is one of the main techniques used to control FAW prevalence. FAW has been successfully managed by Bt maize events expressing the Cry1Ab, Cry1F, Cry2Ab2, Cry1A.105, and Vip3Aa20 proteins (Botha *et al.*, 2019; Moscardini *et al.*, 2020). In the GM Approval Database maintained by the International Service for the Acquisition of Agri-Biotech Applications (ISAAA). It only took 3–4 years to find resistance in FAW when the Bt maize TC1507 (producing Cry1F protein) was introduced into Brazil in 2008 (Farias *et al.*, 2014) and Puerto Rico in 2003 (Storer *et al.*, 2010). Commercial Bt maize TC1507 cultivation started in Argentina in 2005, and potential FAW resistance was first noted in 2013 (Chandrasena *et al.*, 2018). Despite reports of the evolution of field tolerance in some FAW populations, Bt maize may hinder the survival and expansion of FAW (Silva *et al.*, 2018). Because there are few choices for controlling this insect, farmers in areas where GM foods have not been widely embraced, Bt maize's ability to combat FAW may be crucial (Carzoli *et al.*, 2018).

Bt proteins are synthesized in transgenic crops in a tissue- and time-specific way (Bakhsh *et al.*, 2011; Székács *et al.*, 2012). The unpredictable or sporadic expression of Bt proteins in plants, associated with their effectiveness in warding off insect pests, has given rise to several issues (Erasmus *et al.*, 2019). The Bt protein must be expressed in sufficient amounts in the required plant parts at the required time of the growing season for Bt maize to be sustainable. It has been suggested that the primary factors influencing the Cry1Ab Contents of GM maize MON810 are plant tissue and plant growth. It is uncertain whether Bt protein concentrations directly correlate with how well Bt crops repel insect pests (Girón-Calva *et al.*, 2020). The biological activity of the detected protein cannot be determined by merely assessing Cry protein concentrations (Svobodová *et al.*, 2017; Lohn *et al.*, 2020).

Planting a trap crop attracts the pests of the main crop that protect it from damage. It is helpful to reduce the occurrence of pests on the main crops like wheat and maize (Wu *et al.*, 2008). In the United States, it was demonstrated that in interplanting *Lygus hesperus* populations, insecticide application was decreased by planting alfalfa strips in cotton fields (Godfrey and Leigh, 1994). By cultivating transgenic cotton as a trap crop, *Helicoverpa armigera* (Hübner) was controlled in China. Moreover, for the *Chilo sacchariphagus* sugarcane stem borer, *Erianthus arundinaceus* is a net product Nibouche *et al.* (2019). *Apolygus lucorum* (Meyer-Dur) has a trap crop velvetleaf (Abutilon theophrasti Medicus) for Bt-cotton (Lu *et al.*, 2009; Lin *et al.*, 2015). These studies showed trap cropping merits investigation as a FAW management method in China.

Planting transgenic/Bt insect-resistant maize types is a very successful measure and an alternative to pesticides for reducing FAW damage. By exposing pupae to sunshine and predatory birds before sowing, traditional preplanting techniques like deep plowing can reduce the FAW population (Prasanna *et al.*, 2018). The bioindicators of FAW, such as oviposition selectivity (Tellez-Rodriguez *et al.*, 2014), larval distribution (Malaquias *et al.*, 2017), control efficiency (Botha *et al.*, 2019), and fitness implications, are frequently influenced by Bt maize (Jakka *et al.*, 2014). Since 2003, this pest has been controlled commercially using transgenic maize expressing bacterial Bt proteins (such as Cry1F; Siebert *et al.*, 2008). As mentioned, Puerto Rico first detected the FAW population's resistance to transgenic maize containing the Cry1F toxin in 2010 (Storer *et al.*, 2010). The advent of Bt-resistant FAW populations highlights the urgent need to develop new insecticidal targets.

10.7 FUTURE PERSPECTIVES

Upcoming acts to automatically monitor population density and know when to take action to maintain it below the threshold level, a system with the Internet of Things is vitally necessary. Süto (2021) demonstrated automatic insect counting using a system-based sticky paper and a deep-learning algorithm. Legumes and maize must be intercropped to reduce yield loss experienced by small-scale farmers and to improve soil nutrients through nitrogen fixation. Push–pull farming is essential for farmers with limited resources to increase yield using little input and maintenance. Research on local plants that produce chemicals and act as push-and-pull crops should be conducted to promote the push–pull agricultural system in all African nations. The local populations' identification of these crops will encourage the adoption and spread of the farming method.

The use of plant extract, however, can be substituted for synthetic pesticides and cultural practices or employed in monocropping to combat FAW. The governments of developing countries should take a commendable first step: creating biopesticides from fungal or bacteria strains to suppress FAW. Furthermore, the FAW population could be managed effectively if farmers were informed of weather and climatic forecasts. They will concentrate on early maize planting and when to apply synthetic or biological pesticides. Particularly beneficial in this climate change period. Farmers in the tropics and subtropics are urged to pursue organic farming and agroforestry to temper the environment's harsh and unstable nature. Agricultural extension and advising services are badly needed to control the pest. These services include educating farmers about new and improved technology and giving researchers feedback (Amungwa, 2018). Extension agents share study findings with farmers and refer farmers' problems to researchers for additional investigation. Developing nations' governments are urged to place these agents in rural areas to give farmers the knowledge they need to boost productivity, especially when effectively communicating research findings. Studies in Kenya's Wareng District showed that when extension officers often visited farmers' farms, there was an increase in productivity (Neven and Reardon, 2006).

REFERENCES

Abrahams, P., T. Beale, M. Cock, N. Corniani, R. Day, J. Godwin, S. Murphy, G. Richards and J. Vos. 2017. Fall armyworm status. Impacts and control options in Africa. Preliminary Evidence Note. 14 p. CABI, UK. Fall armyworm: Impacts and implications for Africa Evidence Note, September 2017.

Agboyi, L. K., G. Goergen, P. Beseh, M. Samuel, V. Clottey, R. Glikpo, A. Buddie, G. Cafa, L. Offord, R. Day, I. Rwomushana and M. Kenis. 2020. Parasitoid complex of fall armyworm *Spodoptera frugiperda* in Ghana and Benin. *Insects*. 11, 2–15.

Allen, T., M. Kenis and L. Norgrove. 2021. *Eiphosoma laphygmae*, a classical solution for the biocontrol of the fall armyworm *Spodoptera frugiperda*? *J. Plant Dis. Protect.* 128, 1141–1156.

Al-Sarar, A., F. R. Hall and R. A. Downer 2006. Impact of spray application methodology on the development of resistance to cypermethrin and spinosad by fall armyworm *Spodoptera frugiperda* (JE Smith). *Pest Manag. Sci.: Formerly Pestic. Sci.* 62(11), 1023–1031.

Amungwa, F. A. 2018. Appraisal of innovations in agricultural extension and advisory services in Cameroon. *J. Adv. Plant Sci.* 1(2), 1–9.

Assefa, F. and D. Ayalew. 2019. Status and control measures of fall armyworm (*Spodoptera frugiperda*) infestations in maize fields in Ethiopia: A review. *Cogent Food & Agric.* 5(1), 1641902.

Babu, S. R., R. K. Kalyan, S. Joshi, C. M. Balai, M. K. Mahla and P. Rokadia. 2019. Report of an exotic invasive pest the fall armyworm, *Spodoptera frugiperda* (JE Smith) on maize in Southern Rajasthan. *J. Entomol. Zool. Stud.* 7(3), 1296–1300.

Badhai, S., A. K. Gupta and B. Koiri. 2020. Integrated management of fall armyworm (*Spodoptera frugiperda*) in maize crop. *Rev. Food Agric.* 1, 27–29.

Bakhsh, A., K. Shahzad and T. Husnain 2011. Variation in the spatio-temporal expression of insecticidal genes in cotton. *Czech J. Genet. Plant Breed.* 47(1), 1–9.

Bale, J. S., G. J. Masters, I. D. Hodkinson, C. Awmack, T. M. Bezemer, V. K. Brown, J. Butterfeld, A. Buse, J. C. Coulson, J. Farrar, J. E. Good, R. Harrington, S. Hartley, T. H. Jones, R. L. Lindroth, M. C. Press, I. Symrnioudis, A. D. Watt and J. B. Whittaker. 2002. Herbivory in global climate change research: Direct effects of rising temperature on insect herbivores. *Glob. Chang. Biol.* 8, 1–16.

Bateman, M. L., R. K. Day, B. Luke, S. Edgington, U. Kuhlmann, M. J. W. Cock. 2018. Assessment of potential biopesticide options for managing fall armyworm (*Spodoptera frugiperda*) in Africa. *J. Appl. Entomol.* 142(9), 805–819. doi: 10.1111/jen.12565

Baudron, F., M. A. Zaman-Allah, I. Chaipa, N. Chari and P. Chinwada. 2019. Understanding the factors influencing fall armyworm (*Spodoptera frugiperda* JE Smith) damage in African smallholder maize fields and quantifying its impact on yield. A case study in Eastern Zimbabwe. *Crop Protection.* 120, 141–150.

Bentz, B. J., A. M. Jonsson, M. Schroeder, A. Weed, R. A. I. Wilcke and K. Larsson. 2019. *Ips typographus* and *Dendroctonus ponderosae* models project thermal suitability for intra- and inter-continental establishment in a changing climate. *Front. For. Glob. Change.* 2, 1. https://doi.org/10.3389/ffgc.2019.00001.

Bissiwu, P., M. J. Pérez and N. T. Walter. 2016. Control Efficacy of Spodoptera frugiperda using the Entomopathogens *Heterorhabditis bacteriophora* and *Metarhizium anisopliae* with Insecticide Mixtures in Corn. (Unpublished master's thesis). University of Earth, Guácimo, Limón, Costa Rica.

Botha, A., A. Erasmus, H. du Plessis and J. Van den Berg. 2019. Efficacy of Bt maize for control of *Spodoptera frugiperda* (Lepidoptera: Noctuidae) in South Africa. *J. Econ. Entomol.* 112(3), 1260–1266.

Burtet, L. M., O. Bernardi, A. A. Melo, M. P. Pes, T. T. Strahl and J. V. C. Guedes. 2017. Managing fall armyworm, *Spodoptera frugiperda* (Lepidoptera: Noctuidae), with Bt maize and insecticides in Southern Brazil. *Pest. Manag. Sci.* 73(12), 2569–2577. doi: 10.1002/ps.4660

CABI. 2018. Invasive Species Compendium 2018. www.cab.org/isc/datasheet

Capinera, J. L. 2002. Fall armyworm, *Spodoptera frugiperda* (JE Smith) (Insecta: Lepidoptera: Noctuidae). EENY098/IN255, rev. 7/2000. *EDIS*, 2002(7).

Carzoli, A. K., S. I. Aboobucker, L. L. Sandall, T. T. Lübberstedt and W. P. Suza. 2018. Risks and opportunities of GM crops: Bt maize example. *Glob. Food Sec.* 19, 84–91.

Chandrasena, D. I., A. M. Signorini, G. Abratti, N. P. Storer, M. L. Olaciregui, A. P. Alves and C. D. Pilcher. 2018. Characterization of field-evolved resistance to *Bacillus thuringiensis* derived Cry1F δ-endotoxin in Spodoptera frugiperda populations from Argentina. *Pest Manag. Sci.* 74(3), 746–754.

Chapman, D., B. V. Purse, H. E. Roy and J. M. Bullock. 2017. Global trade networks determine the distribution of invasive non-native species. *Global. Ecol. Biogeogr.* 26, 907–917. doi: 10.1111/geb.12599.

Chen, Q., X. Liu, S. Cao, B. Ma, M. Guo, J. Shen and G. Wang. 2021. Fine structure and olfactory reception of the labial palps of *Spodoptera frugiperda. Front. Physiol.* 12, 680697.

Chimweta, M., I. W. Nyakudya, L. Jimu and A. B. Mashingaidze. 2020. Fall Armyworm *Spodoptera frugiperda* (J.E. Smith) Damage in maize: Management options for flood-recession cropping smallholder farmers. *Intern. J. Pest Manag.* 66, 142–154.

Cock, M. J. W., P. K. Beseh, A. G. Buddie, G. Cafá and J. Crozier. 2017. Molecular methods to detect *Spodoptera frugiperda* in Ghana, and implications for monitoring the spread of invasive species in developing countries. *Sci. Rep.* 7(4103), 10 pp. doi: 10.1038/s41598-017-04238-y

Day, R., Abrahams, P., Bateman, M., Beale, T., Clottey, V., Cock, M., ... & Witt, A. (2017). Fall armyworm: impacts and implications for Africa. *Outlooks on Pest Management, 28*(5), 196–201.

Day, R., P. Abrahams, M. Bateman, T. Beale, V. Clottey, M. Cock, Y. Colmenarez, N. Corniani, R. Early, J. Godwin and J. Gomez. 2017. Fall armyworm: Impacts and implications for Africa. *Outlooks Pest Manag.* 28(5), 196–201. doi: 10.1564/v28_oct_02

De Groote, H., S. C. Kimenju, B. Munyua, S. Palmas, M. Kassie and A. Bruce. 2020. Spread and impact of fall armyworm (*Spodoptera frugiperda* J.E. Smith) in maize production areas of Kenya. *Agric. Ecosyst. Environ.* 292, 106804.

Deshmukh, S. S., B. M. Prasanna, C. M. Kalleshwaraswamy, J. Jaba and B. Choudhary. 2021a. Fall armyworm *Spodoptera frugiperda* (J.E. Smith). *Ind. J. Entomol.* 8: 349–372.

Deshmukh, S. S., B. M. Prasanna, C. M. Kalleshwaraswamy, J. Jaba, B. Choudhary. 2021 b. Fall Armyworm (*Spodoptera frugiperda*). Omkar (eds) *In Polyphagous Pests of Crops*. Springer, Singapore. https://doi.org/10.1007/978-981-15-8075-8_8

Douris, V., D. Steinbach, R. Panteleri, I. Livadaras, J. A. Pickett, T. Van Leeuwen, . . . J. Vontas. 2016. Resistance mutation conserved between insects and mites unravels the benzoylurea insecticide mode of action on chitin biosynthesis. *PNAS*. 113(51), 14692–14697.

Du Plessis, H., M. L. Schlemmer and J. Van den Berg. 2020. The effect of temperature on the development of *Spodoptera frugiperda* (Lepidoptera: Noctuidae). *Insects*. 11, 228. https://doi.org/10.3390/insects11040228

Dumas, P., F. Legeai, C. Lemaitre, E. Scaon, M. Orsucci, K. Labadie, S. Gimenez, A. L. Clamens, H. Henri, F. Vavre and J. M. Aury. 2015. *Spodoptera frugiperda* (Lepidoptera: Noctuidae) host-plant variants: Two host strains or two distinct species?. *Genetica*. 143, 305–316.

Early, R., Bradley, B. A., Dukes, J. S., Lawler, J. J., Olden, J. D., Blumenthal, D. M., ... & Tatem, A. J. (2016). Global threats from invasive alien species in the twenty-first century and national response capacities. *Nature communications, 7*(1), 12485. doi: 10.1038/ncomms12485 http://www.nature.com/articles/ncomms12485#supplementary-information

Early, R., González-Moreno, P., Murphy, S. T., and Day, R. (2018). Forecasting the global extent of invasion of the cereal pest Spodoptera frugiperda, the fall armyworm. *BioRxiv*, 391847.

Erasmus, R., R. Pieters, H. Du Plessis, A. Hilbeck, M. Trtikova, A. Erasmus and J. Van den Berg. 2019. Introgression of a cry1Ab transgene into open pollinated maize and its effect on Cry protein concentration and target pest survival. *PLOS One*. 14(12), e0226476.

Edosa, T. T. and T. D. Dinka. 2021. Current and future potential distribution, risk and management of *Spodoptera frugiperda*. *J. Innov. Agric*. 8: 14–23.

Eschen, R., T. Beale, J. M. Bonnin, K. L. Constantine, S. Duah, E. A. Finch, F. Makale, W. Nunda, A. Ogunmodede, C. F. Pratt, E. Thompson, F. Williams, A. Witt and B. Taylor. 2021. Towards estimating the economic cost of invasive alien species to African crop and livestock production. *CABI Agric. Biosci.* 2. doi: 10.1186/s43170-021-00038-7

FAO. 2017a. *Sustainable Management of the Fall Armyworm in Africa*. Rome: FAO (Food and Agriculture Organization of the United Nations).

FAO. 2017b. *The Future of Food and Agriculture; Trends and Challenges*. Rome: Food and Agricultural Organizations of the United Nations.

FAO. 2018. *Integrated Management of the Fall Armyworm on Maize: A Guide for Farmer Field Schools in Africa*. Rome: Food and Agriculture Organization.

FAO. 2019. *Briefing Note on FAO Actions on Fall Armyworm*. Rome: FAO.

Farias, J. R., D. A. Andow, R. J. Horikoshi, R. J. Sorgatto, P. Fresia, A. C. dos Santos and C. Omoto. 2014. Field-evolved resistance to Cry1F maize by *Spodoptera frugiperda* (Lepidoptera: Noctuidae) in Brazil. *Crop Prot*. 64, 150–158.

Fernandez, J. 2002. Nota corta: Estimación de umbrales económicos para *Spodoptera frugiperda* (JE Smith) (Lepidoptera: Noctuidae) en el cultivo del maíz. *Invest. Agric. Prod. Prot. Veg*. 17, 467–474.

Fisher, M., T. Abate, R. W. Lunduka, W. Asnake, Y. Alemayehu and R. B. Madulu. 2015. Drought tolerant maize for farmer adaptation to drought in sub-Saharan Africa: Determinants of adoption in eastern and southern Africa. *Clim. Change*. 133, 283–299.

Frizzas, M. R., S. Silveira Neto, C. M. D. Oliveira and C. Omoto. 2014. Genetically modified corn on fall armyworm and earwig populations under field conditions. Ciência Rural, Santa Maria, 44(2), pp. 203–209, ISSN 0103-8478.

Garcia, A. G., W. A. C. Godoy, J. M. G. Thomas, R. N. Nagoshi and R. L. Meagher. 2018. Delimiting strategic zones for the development of fall armyworm (Lepidoptera: Noctuidae) on corn in the state of Florida. *J. Econ. Entomol*. 111(1), 120–126.

Girón-Calva, P. S., R. M. Twyman, R. Albajes, A. M. Gatehouse and P. Christou. 2020. The impact of environmental stress on Bt crop performance. *Trends Plant Sci*. 25(3), 264–278.

Godfrey, L. D. and T. F. Leigh 1994. Alfalfa harvest strategy effect on Lygus bug (Hemiptera: Miridae) and insect predator population density: Implications for use as trap crop in cotton. *Environ. Entomol*. 23(5), 1106–1118.

Goergen, G., P. L. Kumar, S. B. Sankung, A. Togola and M. Tamò. 2016. First report of outbreaks of the fall armyworm *Spodoptera frugiperda* (JE Smith) (Lepidoptera, Noctuidae), a new alien invasive pest in West and Central Africa. *PLOS One*. 11(10), e0165632. doi: 10.1371/journal.pone.0165632

Haftay, G. G. and G. G. Fissiha. 2020. Effect of integrating night-time light traps and push-pull method on monitoring and deterring adult fall armyworm (*Spodoptera frugiperda*). *Int. J. Entomol. Res.* 5(1), 28–32.

Hajibabaei, M., G. A. C. Singer, E. L. Clare and P. D. N. Hebert. 2007. Design and applicability of DNA arrays and DNA barcodes in biodiversity monitoring. *BMC Biol.* 5, 24.

Hänniger, S., Dumas, P., Schöfl, G., Gebauer-Jung, S., Vogel, H., Unbehend, M., ... and Groot, A. T. 2017. Genetic basis of allochronic differentiation in the fall armyworm. *BMC Evol. Biol.* 17, 1–14. doi: 10.1186/s12862-017-0911-5

Harrison, R. D., C. Thierfelder, F. Baudron, P. Chinwada, C. Midega and U. Schaffner. 2019. Agro-ecological options for all fall armyworm (*Spodoptera frugiperda* JE Smith) management: Providing low-cost, smallholder friendly solutions to an invasive pest. *J. Environ. Manag.* 243, 318–330.

Hay-Roe, M. M., R. L. Meagher, R. N. Nagoshi and Y. Newman. 2016. Distributional patterns of fall armyworm parasitoids in a corn field and a pasture field in Florida. *Biol. Control.* 96, 48–56.

Hruska, A. J. 2019. Fall armyworm (*Spodoptera frugiperda*) management by smallholders. *CABI Reviews*, 1–11.

Huang, L. L., F. S. Xue, C. Chen, X. Guo, J. J. Tang, L. Zhong and H. M. He. 2021. Effects of temperature on life history traits of the newly invasive fall armyworm, *Spodoptera frugiperda* in Southeast China. *Ecol. Evol.* 11(10), 5255–5264. https://doi.org/10.1002/ece3. 7413.

Hussain, A. G., J. T. Wennmann, G. Goergen, A. Bryon and V. I. D. Ros. 2021. Viruses of the fall armyworm *Spodoptera frugiperda*: A review with prospects for biological control. *Viruses*. 13: doi: 10.3390/v13112220

ICIPE (International Centre of Insect Physiology and Ecology). 2018. *Combating the Fall Armyworm in Africa*—Food and Agriculture Organization. 2018. *Integrated Management of the Fall Armyworm on Maize: A Guide for Farmer Field Schools in Africa*. Rome, Italy. *The European Union (EU) Provides New Funding to ICIPE*. Nairobi, Kenya: ICIPE.

International Plant Protection Convention (IPPC). 2018. Fall armyworm–an emerging food security global threat. www.ippc.int/en/news/fallarmyworm-an-emerging-food-security-globalthreat.

IPCC. 2013. Summary for policymakers. In T. F. Stocker, D. Qin, G. K. Plattner, M. Tignor, S. K. Allen, J. Doschung, A. Nauels, Y. Xia, V. Bex and P. M. Midgley (Eds.), *Climate Change 2013: The Physical Science Basis. Contribution of Working Group I to the Fifth Assessment Report of the Intergovernmental Panel on Climate Change* (pp. 3–29). Cambridge University Press, Cambridge, United Kingdom and New York.

Jakka, S., V. Knight and J. Jurat-Fuentes. 2014. Fitness costs associated with field-evolved resistance to Bt maize in *Spodoptera frugiperda* (Lepidoptera: Noctuidae). *J. Econ. Entomol.* 107(1), 342–351.

James, R. and M. Engelke. 2010. Resistance in Zoysia grass (*Zoysia* spp.) to the fall armyworm (*Spodoptera frugiperda*) (Lepidoptera: Noctuidae). *Fla. Entomol.* 93(2), 254–259.

Jarrod, H. T., G. M. Lorenz III and B. R. Leonard. 2015. Fall armyworm (Lepidoptera: Noctuidae) ecology in southeastern cotton. *J. Integr. Pest Manag.* 6(1), 10.

Kansiime, M. K., I. Mugambi, I. Rwomushana, W. Nunda, J. Lamontagne-Godwin, H. Rware and R. Day. 2019. Farmer perception of fall armyworm (*Spodoptera frugiderda* JE Smith) and farm-level management practices in Zambia. *Pest Manag. Sci.* 75(10), 2840–2850.

Kebede, M. and T. Shimalis. 2018. Outbreak, distribution and management of fall armyworm, *Spodoptera frugiperda* JE Smith in Africa: The status and prospects. *Acad. Agric. J.* 3, 551–568.

Kenis, M., H. Du Plessis, J. Van den Berg, M. N. Ba, G. Goergen, K. E. Kwadjo and A. Polaszek. 2019. *Telenomus remus*, a candidate parasitoid for the biological control of *Spodoptera frugiperda* in Africa, is already present on the continent. *Insects*. 10(4), 92.

Khan, Z., C. Midega, J. Pittchar, J. Pickett and T. Bruce. 2011. Push-pull technology: A conservation agriculture approach for integrated management of insect pests, weeds and soil health in Africa. *Int. J. Agri. Sust.* 9(1), 162–170.

Koffi, D., R. Kyerematen, V. Y. Eziah, K. Agboka, M. Adom, G. Goergen and R. L. Meagher. 2020. Natural enemies of the fall armyworm, *Spodoptera frugiperda* (JE Smith) (Lepidoptera: Noctuidae) in Ghana. *Florida Entomologist*. 103(1), 85–90.

Kress, W. J. and D. L. Erickson. 2008. DNA barcodes: Genes, genomics, and bioinformatics. *PNAS*. 105(8), 2761–2762.

Kumela, T., J. Simiyu, B. Sisay, P. Likhayo, E. Mendesil, L. Gohole and T. Tefera. 2019. Farmers' knowledge, perceptions, and management practices of the new invasive pest, fall armyworm (*Spodoptera frugiperda*) in Ethiopia and Kenya. *Int. J. Pest Manag.* 65, 1–9. https://doi.org/10.1080/09670874.2017.1423129.

Lamsal, S., S. Sibi and S. Yadav. 2020. Fall armyworm in South Asia: Threats and management. *Asian J. Adv. Agric. Res.* 13, 21–34.

Lima, M. S., P. S. L. Silva, O. F. Oliveira, K. M. B. Silva and F. C. L. Freitas. 2010. Corn yield response to weed and fall armyworm controls. *Planta Daninha.* 28, 103–111.

Lin, K., Y. Lu, P. Wan, Y. Yang, K. A. Wyckhuys and K. Wu. 2015. Simultaneous reduction in incidence of *Bemisia tabaci* (Hemiptera: Aleyrodidae) and *Sylepta derogata* (Lepidoptera: Pyralidae) using velvetleaf, *Abutilon theophrasti* as a trap crop. *J. Pest Sci.* 88(1), 49–56.

Lobell, D. B., M. Bänziger, C. Magorokosho and B. Vivek 2011. Nonlinear heat effects on African maize as evidenced by historical yield trials. *Nat. Clim. Change.* 1(1), 42–45.

Lohn, A. F., M. Trtikova, I. Chapela, J. Van den Berg, H. Du Plessis and A. Hilbeck. 2020. Transgene behavior in *Zea mays* L. crosses across different genetic backgrounds: Segregation patterns, cry1Ab transgene expression, insecticidal protein concentration and bioactivity against insect pests. *PLOS One.* 15(9), e0238523.

Lu, Y., K. Wu, K. Wyckhuys and Y. Guo. 2009. Potential of mungbean, *Vigna radiatus* as a trap crop for managing *Apolygus lucorum* (Hemiptera: Miridae) on Bt cotton. *Crop Prot.* 28(1), 77–81.

Madden, M. J., R. G. Young, J. W. Brown, S. E. Miller, A. J. Frewin and R. H. Hanner. 2019. Using DNA barcoding to improve invasive pest identification at US ports-of-entry. *PLOS One.* 14(9), e0222291.

Mahadevaswamy, H. M., R. Asokan, C. M. Kalleshwaraswamy, P. Sharanabasappa, Y. G. Maruthi, M. S. Shashank, P. R. N. I. Devi, A. Surakasula, S. Adarsha, A. Srinivas, S. Rao, M. Shali Raju, G. Shyam Sunder Reddy and S. N. Nagesh. 2018. Prevalence of "r" strain and molecular diversity of fall armyworm *Spodoptera frugiperda* (JE Smith) (Lepidoptera: Noctuidae) in India. *Indian J. Entomol.* 80, 544–553.

Malaquias, J. B., W. A. Godoy, A. G. Garcia, F. d. S. Ramalho and C. Omoto. 2017. Larval dispersal of *Spodoptera frugiperda* strains on Bt cotton: A model for understanding resistance evolution and consequences for its management. *Sci. Rep.* 7(1), 1–10.

Maruthadurai, R., and Ramesh, R. (2020). Occurrence, damage pattern and biology of fall armyworm, *Spodoptera frugiperda* (JE Smith) (Lepidoptera: Noctuidae) on fodder crops and green amaranth in Goa, India. *Phytoparasitica*, 48(1), 15–23.

Midega, C. A., I. M. Nyang'au, J. Pittchar, M. A. Birkett, J. A. Pickett, M. Borges and Z. R. Khan, 2012. Farmers' perceptions of cotton pests and their management in western Kenya. *Crop Prot.* 42, 193–201.

Midega, C. A., J. O. Pittchar, J. A. Pickett, G. W. Hailu and Z. R. Khan. 2018. A climate-adapted push-pull system effectively controls fall armyworm, *Spodoptera frugiperda* (JE Smith), in maize in East Africa. *Crop Protection.* 105, 10–15.

Molina-Ochoa, J., R. Lezama-Gutierrez, M. Gonzalez-Ramirez, M. Lopez-Edwards, M. A. Rodriguez-Vega and F. Arceo-Palacios. 2003. Pathogens and parasitic nematodes associated with populations of fall armyworm (Lepidoptera: Noctuidae) larvae in Mexico. *Florida Entomologist.* 86(3), 244–253.

Montezano, D. G., A. Specht, A. Sosa-Gómez, D. R. Roque-Specht, V. F. Sousa-Silva, J. C. Paula-Moraes, S. V. Peterson and T. E. Hunt. 2018. Host plants of *Spodoptera frugiperda* (Lepidoptera: Noctuidae) in the Americas. *Afr. Entomol.* 26, 286–300.

Moscardini, V. F., L. H. Marques, A. C. Santos, J. Rossetto, O. A. Silva, P. E. Rampazzo and B. A. Castro. 2020. Efficacy of *Bacillus thuringiensis* (Bt) maize expressing Cry1F, Cry1A. 105, Cry2Ab2 and Vip3Aa20 proteins to manage the fall armyworm (Lepidoptera: Noctuidae) in Brazil. *Crop Prot.* 137, 105269.

Mutyambai, D. M., S. Niassy, P. A. Calatayud and S. Subramanian. 2022. Agronomic factors influencing fall armyworm (*Spodoptera frugiperda*) infestation and damage and its co-occurrence with stemborers in maize cropping systems in Kenya. *Insects.* 13(3), 266.

Naeem-Ullah, U., M. A. Ansari, N. Iqbal and S. Saeed. 2019. First authentic report of *Spodoptera frugiperda* (JE Smith) (Noctuidae: Lepidoptera) an alien invasive species from Pakistan. *Appl. Sci. Bus. Econ.* 6(1), 1–3.

Nagoshi, R. N., Koffi, D., Agboka, K., Tounou, K. A., Banerjee, R., Jurat-Fuentes, J. L., & Meagher, R. L. 2017. Comparative molecular analyses of invasive fall armyworm in Togo reveal strong similarities to populations from the eastern United States and the Greater Antilles. *PLOS One.* 12(7), e0181982, doi: 10.1371/journal.pone.0181982.

Navasero, M. V., M. M. Navasero, G. A. S. Burgonio, K. P. Ardez, M. D. Ebuenga, M. J. B. Beltran, . . . B. L. Caoili. 2019. Detection of the fall armyworm, *Spodoptera frugiperda* (JE Smith) (Lepidoptera: Noctuidae) using larval morphological characters, and observations on its current local distribution in the Philippines. *Philipp. Entomol.* 33(2), 171–184.

Neven, D. and T. Reardon. 2006. Kenyan supermarkets and horticultural farm sector development. Internal Association of Agricultural Economists Conference, February.

Ngumbi, E. N. 2020. How changes in weather patterns could lead to more insect invasions. The Conversation US, Inc. https://theconversation.com/how-changes-in-weather-patterns-couldlead-to-more-insect-invasions-131917.

Nibouche, S., R. Tibere and L. Costet. 2019. *Erianthus arundinaceus* as a trap crop for the sugarcane stem borer *Chilo sacchariphagus*: Field validation and disease risk assessment. *Crop Prot.* 124, 104877.

Njeru, R. 2017. Report on Stakeholders Consultation Meeting on: Fall Armyworm in Africa: Status and Strategy for Effective Management.

Osabutey, A. F., B. Y. Seo, A. Kim, T. A. T. Ha, J. Jung, G. Goergen, . . . Y. H. Koh. 2022. Identification of a fall armyworm (*Spodoptera frugiperda*)-specific gene and development of a rapid and sensitive loop-mediated isothermal amplification assay. *Sci. Rep.* 12(1), 1–10.

Paredes-Sanchez, F. A., G. Rivera, V. Bocanegra-Garcia, H. Y. Martinez-Padron, M. Berrones-Morales, N. Nino-Garcia and V. Herrera-Mayorga. 2021. Advances in control strategies against *Spodoptera frugiperda*—A review. *Mol.* 26: 5587. doi: 10.3390/molecule s26185587

Parra, J. R. P., A. Coelho, J. B. Cuervo-Rugno, A. G. Garcia, R. A. Moral, A. Specht and D. D. Neto. 2022. Important pest species of the *Spodoptera* complex: Biology, thermal requirements and ecological zoning. *J. Pest Sci.* 95: 169–186.

Pereira, F., J. Carneiro and A. Amorim. 2008. Identification of species with DNA-based technology: Current progress and challenges. *Recent Pat. on DNA & Gene Sequences (Discontinued),* 2(3), 187–200.

Plessis, H., M. L. Schlemmer and J. Van den Berg. 2020. The effect of temperature on the development of *Spodoptera frugiperda* (Lep., Noctuidae). *Insects*. 11. https://doi.org/10.3390/insects11040228

Prasanna, B. M., A. Bruce, S. Winter, M. Otim, G. Asea, S. Sevgan and L. Gichuru. 2018. Host plant resistance to fall armyworm. *Fall Armyworm in Africa: A Guide for Integrated Pest Management*, 45–62.

Prasanna, B. M., J. E. Huesing, V. M. Peschke, R. N. Nagoshi, X. Jia, K. Wu and R. Eddy. 2021. *Fall Armyworm in Asia: Invasion, Impacts, and Strategies for Sustainable Management*. Mexico: CIMMYT.

Ramasamy, M., B. Das and R. Ramesh 2022. Predicting climate change impacts on potential worldwide distribution of fall armyworm based on cmip6 projections. *J. Pest Sci.* 95(2), 841–854.

Rivers, A., M. Barbercheck, B. Govaerts and N. Verhulst. 2016. Conservation agriculture affects arthropod community composition in a rainfed maize–wheat system in central Mexico. *Appl. Soil Ecol.* 100, 81–90.

Rwomushana, I., M. Bateman, T. Beale, P. Beseh, K. Cameron, M. Chiluba, V. Clottey, D. R. Davis, R. Early, J. Godwin, P. Gonzalez-Moreno, M. Kansiime, M. Kenis, F. Makale, I. Mugambi, S. Murphy, W. Nunda, N. Phiri, C. Pratt and J. Tambo. 2018. *Fall Armyworm: Impacts and Implications for Africa: Evidence Note Update*. Wallingford: CABI.

Sagar, G. C., B. Aastha and K. Laxman. 2020. An introduction of fall armyworm (*Spodoptera frugiperda*) with management strategies: A review paper. *Nippon J. Environ. Sci.* 1. doi: 10.46266/njes.101

Sharanabasappa, D., C. M. Kalleshwaraswamy, R. Asokan, H. M. Mahadevaswamy, M. S. Maruthi, H. B. Pavithro, K. Hegde, S. Navi, S. T. Prabhu and G. Goergin. 2018. First report of fall armyworm, *Spodoptera frugiperda* (JE Smith) (Lepidoptera: Noctuidae) an alien invasive pest on maize in India. *Pest Management in Horticultural Ecosystems*. 24, 23–29.

Sharanabasappa, S., Kalleshwaraswamy, C. M., Poorani, J., Maruthi, M. S., Pavithra, H. B., & Diraviam, J. (2019). Natural enemies of *Spodoptera frugiperda* (JE Smith) (Lepidoptera: Noctuidae), a recent invasive pest on maize in South India. *Florida Entomologist*, 102(3), 619–623.

Shylesha, A. N., S. K. Jalali, A. Gupta, R. Varshney, T. Venkatesan, P. Shetty, . . . A. Raghavendra. 2018. Studies on new invasive pest *Spodoptera frugiperda* (JE Smith) (Lepidoptera: Noctuidae) and its natural enemies. *J Biol. Control.* 32(3), 1–7.

Siebert, M. W., J. M. Babock, S. Nolting, A. C. Santos, J. J. Adamczyk Jr, P. A. Neese, . . . R. B. Lassiter. 2008. Efficacy of Cry1F insecticidal protein in maize and cotton for control of fall armyworm (Lepidoptera: Noctuidae). *Fla. Entomol.* 91(4), 555–565.

Silva, G. A., M. C. Picanço, L. R. Ferreira, D. O. Ferreira, E. S. Farias, T. C. Souza, . . . E. J. G. Pereira. 2018. Yield losses in transgenic Cry1Ab and non-Bt corn as assessed using a crop-life table approach. *J. Econ. Entomol.* 111(1), 218–226.

Sisay, B., J. Simiyu, P. Malusi, P. Likhayo, E. Mendesil, N. Elibariki and T. Tefera. 2018. First report of the fall armyworm, *Spodoptera frugiperda* (Lepidoptera: Noctuidae), natural enemies from Africa. *J. Appl. Entomol.* 142(8), 800–804.

Sisay, B., Simiyu, J., Mendesil, E., Likhayo, P., Ayalew, G., Mohamed, S., ... & Tefera, T. 2019. Fall armyworm, *Spodoptera frugiperda* infestations in East Africa: Assessment of damage and parasitism. *Insects*. 10(7), 195. doi: 10.3390/insects10070195

Soares, M. A., G. L. D. Leite, J. C. Zanuncio, V. G. M. D. Sá, C. S. Ferreira, S. L Rocha, . . . J. E. Serrão. 2012. Quality control of *Trichogramma atopovirilia* and *Trichogramma pretiosum* (Hym.: Trichogrammatidae) adults reared under laboratory conditions. *Braz. Arch. Biol. Technol.* 55, 305–311.

Storer, N. P., J. M. Babcock, M. Schlenz, T. Meade, G. D. Thompson, J. W. Bing and R. M. Huckaba. 2010. Discovery and characterization of field resistance to Bt maize: *Spodoptera frugiperda* (Lepidoptera: Noctuidae) in Puerto Rico. *J. Econ. Entomol.* 103(4), 1031–1038.

Sun, R., C. Liu, H. Zhang and Q. Wang. 2015. Benzoylurea chitin synthesis inhibitors. *J Agric Food Chem.* 63(31), 6847–6865.

Sütő, J. 2021. Embedded system-based sticky paper trap with deep learning-based insect-counting algorithm. *Electronics* (Switzerland). 10(15).

Svobodová, Z., E. C. Burkness, O. Skoková Habuštová and W. D. Hutchison. 2017. Predator preference for Bt-Fed *Spodoptera frugiperda* (Lepidoptera: Noctuidae) prey: Implications for insect resistance management in Bt maize seed blends. *J. Econ. Entomol.* 110(3), 1317–1325.

Swamy, H. M., R. Asokan, C. M. Kalleshwaraswamy, Y. G. Prasad, M. S. Maruthi, P. R. Shashank, . . . S. N. Nagesh. 2018. Prevalence of "R" strain and molecular diversity of fall army worm *Spodoptera frugiperda* (JE Smith) (Lepidoptera: Noctuidae) in India. *Indian J. Entomol.* 80(3), 544–553.

Székács, A., G. Weiss, D. Quist, E. Takács, B. Darvas, M. Meier, . . . A. Hilbeck. 2012. Inter-laboratory comparison of Cry1Ab toxin quantification in MON 810 maize by enzyme-immunoassay. *Food Agric. Immunol.* 23(2), 99–121.

Tambo, J. A., M. K. Kansiime, I. Mugambi, I. Rwomushana, M. Kenis, R. K. Day and J. Lamontagne-Godwin. 2020. Understanding smallholders's responses to fall armyworm (*Spodoptera frugiperda*) invasion: Evidence from five African countries. *Sci. Total Environ.* 740, 140015.

Tefera, T., S. Mugo and Y. Beyene. 2016. Developing and deploying insect resistant maize varieties to reduce pre-and post-harvest food losses in Africa. *Food Security*. 8(1), 211–220.

Tellez-Rodriguez, P., B. Raymond, I. Moran-Bertot, L. RodriguezCabrera, D. J. Wright, C. G. Borroto and C. Ayra-Pardo. 2014. Strong oviposition preference for Bt over non-Bt maize in *Spodoptera frugiperda* and its implications for the evolution of resistance. *BMC Biol.* 12, 48.

Tesfaye, K., P. H. Zaidi, S. Gbegbelegbe, C. Boeber, D. B. Rahut, F. Getaneh, K. Seetharam, O. Erenstein and C. Stirling. 2017. Climate change impacts and potential benefits of heat-tolerant maize in South Asia. *Theor. Appl. Climatol.* 130, 959–970.

Unbehend, M., S. Hänniger, R. L. Meagher, D. G. Heckel and A. T. Groot. 2013. Pheromonal divergence between two strains of *Spodoptera frugiperda. J. Chem. Ecol.* 39, 364–376.

Vincent, J., Z. Dai, C. Ravel, F. Choulet, S. Mouzeyar, M. F. Bouzidi, . . . P. Martre, 2013. dbWFA: A web-based database for functional annotation of *Triticum aestivum* transcripts. *Database*, 2013, bat014. https://doi.org/10.1093/database/bat014

Westbrook, J. K., R. N. Nagoshi, R. L. Meagher, S. J. Fleischer and S. Jairam, 2016. Modeling seasonal migration of fall armyworm moths. *Int. J. Biometeor.* 60, 255–267. https://doi.org/10.1007/s00484-015-1022-x

Wu, K. M., Y. H. Lu, H. Q. Feng, Y. Y. Jiang and J. Z. Zhao. 2008. Suppression of cotton bollworm in multiple crops in China in areas with Bt toxin–containing cotton. *Science*. 321(5896), 1676–1678.

Yigezu, G. and M. Wakgari. 2020. Local and indigenous knowledge of farmers management practice against fall armyworm (*Spodoptera frugiperda*) (JE Smith) (Lepidoptera: Noctuidae): A review. *J. Entomol. Zool. Stud.* 8(1), 765–770.

Zhang, Z., Y. Batuxi, Y. Jiang, X. Li, A. Zhang, X. Zhu and Y. Zhang. 2021. Effects of different wheat tissues on the population parameters of the fall armyworm (*Spodoptera frugiperda*). *Agron.* 11. doi: 10.3390/agronomy11102044

Zhao, J., A. Hoffmann, Y. Jiang, L. Xiao, Y. Tan, C. Zhou and L. Bai. 2022. Competitive interactions of a new invader (*Spodoptera frugiperda*) and indigenous species (*Ostrinia furnacalis*) on maize in China. *J. Pest Sci.* 95: 159–168.

11 Etiology and Integrated Disease Management of *Zea mays*

Ahmad Nisar, Nabeeha Aslam Khan, Anjum Faraz, Muhammad Kaleem Sarwar, Shahid Majeed, Hayssam Mohamad Ali, Tahir Hussain Awan, and Amer Habib

11.1 INTRODUCTION

Maize belongs to the family Poaceae, a staple crop grown worldwide (Naz *et al.*, 2019), which originated from Central America and Mexico, also known as Indian corn or corn (Ranum *et al.*, 2014). It is an annual crop with 2–3 m in height; the stem consists of nodes and internodes, with 8–21 leaves per plant extending from each internode. The leaves may be lance-like (lanceolate) or linear, having visible primary veins (midrib), and their length ranges from 30 to 100 cm (Gunjan *et al.*, 2021). The female and male inflorescence appear singly on the plant; the female inflorescence is known as the 'ear,' while the male inflorescence is 'tassel' (Bortiri and Hake, 2007). There may be 1–3 spikes (modified ear) per plant. The kernels (maize grains) are 30–1000 per spike and are enclosed in husks and may be yellow, black, white, purple, or red (Gunjan *et al.*, 2021). Maize and cornmeal are staples in many countries worldwide (Gwirtz and Garcia-Casal, 2014). Its ears may be cooked and consumed as a vegetable, or various products like flour and cereals prepared from the kernels. It is also a chief source of starch, which is further processed into high fructose corn syrup and oils. It is also used as livestock feed (Orhun, 2013).

11.2 AGRONOMY

Zea mays can be grown best in tropical and sub-tropical regions with warm, fertile, well-drained, and deep soil having pH ranges from 6.0 to 6.8 (Trowbridge and Bassuk, 2004). It is a highly nutrient-consuming crop (mainly nitrogen), even the well-fertile soil requires to be supplemented with nutrients as it grows. Pollination occurs through wind, plants requiring ample space, moisture, and sunlight to grow (Boutard, 2012). Planting dates may vary for different verities. Supersweet varieties should be sown when the soil temperature reaches almost 65°F (18.3°C), and standard varieties should be sown when the soil temperature is 55°F (12.7°C) (Our Badu-Apraku and Fakorede, 2017). Seeds must be planted almost 1 in. (2.5 cm) deep with 30–36-in. (76–91 cm) row-to-row distance should be 3–4 in. (10–15 cm) apart. As maize is a wind-pollinated crop, it must be implanted in blocks more willingly than in a single long row as it helps transfer the pollen more effectively within the plants. Seedlings should be thinned to a final spacing of 8–12 in. (20–30 cm) when the crop reaches a height of almost 3–4 in. (7.5–10.0 cm) (Khairwal *et al.*, 2007).

 DOI: 10.1201/b23394-11

In terms of nutrient requirements, it is a considerably heavy feeder crop, especially of nitrogen, so enough nutrients should be applied for optimum growth and productivity. Its growth is rapid, and plants attain a high growth rate between 30 and 40 days after planting, and nutrients must be supplemented during this period. To ensure maxim nitrogen utilization, all the fertilizer should be applied before the tasseling period. Nutrient deficiency is quite evident from the visible symptoms; plants' color should be deep green when supplied with enough nitrogen, whereas nitrogen-deficient plants show light green leaves, and phosphorous-deficient plants show purple-tinged leaves. Besides nutrient requirements, plants also need appropriate soil moisture during the growing season to tassel and form silks, and almost every silk produces kernels; however, poor pollination results in incomplete filling of ears (Aylor *et al.*, 2003). Regarding harvesting, each maize stalk normally produces one large ear; however, sometimes, the stalk may produce a second, smaller ear that matures slightly later. Harvesting is recommended at the "milk stage" (production of a milky substance by puncturing well-packed kernels inside the husk), the ears' ripeness can be determined by peeling a little part of the husk. Ears can be overripe soon and may lose their sweetness, so the ears' ripeness should be observed regularly (Veeral and Abirami, 2021).

11.2.1 Growth Constraints

Various biotic growth constraints have been reported to affect maize crop growth and productivity drastically. The infectious diseases caused by fungi, bacteria, viruses, and nematodes are among these factors causing significant yield reduction in productivity.

Fungal Diseases: Fungal plant pathogens cause several severe diseases such as blights, root rot, wilt, die-back, damping-off, mildews, rust, anthracnose, and leaf spots. The primary fungal infection sources are water, soil, farm machinery, animals, contaminated or infected seeds, wind, insects, workers, and invertebrates (Lazarovits *et al.*, 2014). Below we discuss important fungal diseases of maize causing considerable economic losses.

11.2.2 Brown Spot of Maize

Pathogen: *P. maydis* (*P. zeaemaydis*) causes brown spot disease in maize and closely related teosinte (*Zea mays* subsp. *mexicana*).

Symptoms of brown spot of maize: First, on the leaf blade, stalk, leaf sheath, and sometimes on the tassel of the outer ear and husks, small, yellowish, and round to oblong lesions or spots are produced. These lesions might appear in the form of bands on the leaf blade. These spots change from chocolate brown to reddish brown and may enlarge with angular or irregular shapes; these pustules rupture to disseminate the sporangia (18–24 by 30–34 microns). Sporangia may be dark or golden brown (Sinha *et al.*, 2018).

Disease cycle: *P. maydis* can overwinter in maize residue or soil as resting spores (the thick-walled brown sporangia) formed in infected cells. These sporangia are disseminated to susceptible plants by insects, humans, air currents, flowing water, or splashing rain. Until the plants are about 45–50 days old, corn is more sensitive to *Physoderma* infection; afterward, sensitivity decreases gradually. Infection requires free water; sporangium develops to discharge 20–50 swimming zoospores when the temperature is high (23–30°C, 73–90°F), and there is moisture under the leaf sheaths or in the whorl. Before landing, the zoospores move around in the water for one to two hours, assuming an amoeba-like appearance and using tiny infection hyphae to penetrate the young meristematic tissue. After this, the mycelium enters the parenchyma or mesophyll cells, forming large vegetative structures. As the leaf tissue emerges from the whorl, infection frequently occurs in a diurnal cycle, producing lateral bands of alternately healthy and infected leaf tissue. *P. maydis* zoospores can only infect corn tissue during specific times of the day and shortly after being released. Six to 20 days after infection, the onset of symptoms and the germination of fresh sporangia occur, completing the disease cycle (Zibani *et al.*, 2022).

Management of brown spot of maize: Use regionally appropriate resistant and hybrids varieties. In the fall, crush the infected corn waste and chisel plow. Tillage can be done in the spring or fall based on the conservation tillage techniques endorsed for the area. Planting susceptible varieties in areas with high humidity, for example, river-bottom soils, is not recommended. Crop rotation also helps in reducing the inoculum. *Physoderma* is known to survive in the soil for almost 3 years as sporangia. According to spore-trapping statistics, sporangia can rapidly spread to nearby fields and may be dispersed or successfully carried long distances. Sporangia can also be carried into disease-free areas by infected plants that are transported for silage.

11.2.3 Downy Mildews of Maize

Downy mildews are caused by several species of the genera *Peronosclerospora, Sclerophthora*, and *Sclerospora*. Among the mildews the following are the significant diseases.

Crazy top downy mildew of maize: *Sclerophthora macrospora* (*Sclerophthora/Phytophthora macrospora*) causes crazy top downy mildew.

Symptoms: The duration of the infection and the degree of systemic colonization influence the symptoms of the crazy top. The primary signs include excessive tillering that distorts the plant and tassel proliferation, which can result in a mass of leafy structures on the tassel. Stunted plants may result from infection. There may also be excessive leaf-like structures on other plant parts, such as the ear. The diseased leaves are leathery and slender and might have extended yellow stripes. Pollens are not produced in infected plants' tassels (Premalatha *et al.*, 2012).

Disease cycle: *S. macrospora* persists in the soil as oospores. The oospores germinate in the presence of excessive moisture (mostly after flooding) and form zoospores. These zoospores are then moved with water's help toward the young seedlings and affect young leaf tissues. The infection is systemic as the fungus attacks the foliar parts, including the tassel. Since the fungus alters hormonal balance, the ear is typically absent. The crazy top disease is more likely to occur in areas with ponding, heavy rain, and protracted floods. This pathogen can also attack wild grasses, which is why diseased corn plants can be observed at the periphery of a cornfield (Trdan and Celar, 2000).

Management: Control grassy weeds to lessen the floods or ponding, improve drainage, and prevent the accumulation of inoculum.

11.2.3.1 Brown Stripe Downy Mildew

Pathogen: Its causal organism is *S. rayssiae* var. *Zeae*. It is an obligatory parasite and cannot grow in artificial culture.

Symptoms: Only the lesions on the leaves are formed by *S. rayssiae* var. *Zeae* (Payak and Renfro, 1967). At the early infection stage, yellowish or yellowish stripes are formed on leaves ranging in width from 3 to 7 mm. While in some other corn varieties, these stripes may be reddish to purple. The lesions stretch parallel to and are delimited by the veins of the leaves and have clearly defined margins. The union of nearby lesions results in more blotching and striping. First, the disease symptoms will be seen on the leaf's lower surface, which may have the most severe striping and appear burnt and pale brown. Severely damaged leaves may also shed early. Reduced size of the seed has been associated with *S. rayssiae* var. *zeae* infection (Lal and Prasad, 1989). Severe disease that develops before flowering can cause early plant death. Brown stripe downy mildew does not cause deformation of vegetative or floral tissues, unlike other downy mildews of maize. The infection does not affect the plant systemically (Singh, 1971). On both the upper and lower surfaces of lesions, sporangial growth appears as a fuzzy growth that is grayish-white when considerable moisture is present. Sporangia diminish when the lesions turn necrotic. Oospores do not occur in vascular tissue; they only grow in necrotic tissues, mesophyll, or below stomata.

Disease cycle: The emphasis is also placed on the pathogen's seed-borne form of survival, which allows it to persist through oospores and collateral hosts. From the infected crop residues, oospores are transferred to soil and can persist in soil for up to 3 years. The significance of *Digitaria sanguinalis*, a pathogen's collateral host, has been highlighted by several plant pathologists. It has also been observed that *S. rayssiae* var. *zeae* can be found in the embryo of seeds (as mycelium) from diseased plants. The pathogen's perennating structures sprout during the host's growth season, and the initial infection starts. The production of sporangia *D. sanguinalis* can lead to a primary infection. The primary infection manifests as tiny flecks on the leaves that develop into brown streaks after oospores germinate by germ tubes and generate sporangia. Sporangia formed by streaks of primary infection cause the secondary infection. Between 12 and 4 in the morning is the peak time for sporangia development, when they germinate and produce zoospores (Chaithra *et al.*, 2022). The ideal temperature for *S. rayssiae* var. *zeae* sporangia development is 22–25°C, whereas the ideal temperature for zoospore development is 20–22°C (Singh *et al.*, 1970).

Management: The disease incidence on the crop is decreased by destroying affected plants as soon as possible. Lime and zinc sulphate (0.5%) sprayed on the plants significantly reduced the disease. After 10 days of sowing, four to six sprays of dithane M-45 (0.3%) or dithane Z-70 at intervals of 7 days are highly effective. Seeds treated with Ridomil 25 WP (metalaxyl) at a rate of 4 grams per kg completely eradicate the infection. The disease is thought to be controlled by sowing resistant variety, even though no resistant strains of this disease have yet been introduced in our country.

11.2.3.2 Philippine Downy Mildew of Maize

Pathogen: Philippine downy mildew (PDM) of maize is caused by the *Peronosclerospora philippinensis* (Cueva *et al.*, 2022).

Symptoms: On the stems and leaves of corn plants, PDM symptoms can be seen. The disease affects the ability of the plant to produce healthy cobs, resulting in low production and, ultimately, the low commercial value of the crop. Disease severity depends on the stage of plant growth at the time of infection and environmental conditions. Young, infected seedlings become unable to produce fertile ears. While the older infected plants mature and show stunted growth, few grains are produced in deformed cobs. After 3–6 days of infection, pale yellow to whitish discolorations appears on the leaf blade. On the leaves, elongated chlorotic streaks are formed with downy growth of conidiophores and conidia. Then, deformation of tassels occurs, and ears also get aborted. The infected plant becomes feeble and stunted at the severity of the disease and can die in a month. Infected plants mature when the attack is moderate but form tiny, malformed ears (Weston, 1920; Dalmacio and Raymundo, 1972).

Disease cycle: Conidia, produced by *P. philippinensis*, from the infected plant material and can be grown on crops like sorghum, corn, weed species, or oats are subsequently spread between plants by an airborne method. PDM transmission is little influenced by temperature, with larger conidia forming at higher temperatures, whereas the infection rate was considerably reduced between 10°C and 16°C (Bonde *et al.*, 1992). According to experiments on field maize, the generation of spores was controlled by a small amount of moisture or dew on the surfaces of infected leaves. Sporulation always occurred at 90% relative humidity or above (Dalmacio and Raymundo, 1972). *P. philippinensis* spores germinate on the host's leaves, with the germ tube entering the leaf through the stomata and then invading the mesophyll cells. Hyphae are formed in one of two ways: thin, sparsely branched, long, unevenly sized, and with irregular branches. Oospores are produced by *P. philippinensis* in plant tissue, although it is unknown how they contribute to the disease's spread. Damp seeds transmit the disease but not through seeds with moisture content below 14% (Weston, 1920).

Management: High nitrogen applications make plants more vulnerable to attack, but this was not proved experimentally in resistant verities (Yamada and Aday, 1977). PDM resistance is polygenic and primarily controlled by additive gene effects (Leon *et al.*, 1993). Several resistant cultivars have been developed using these resistance genes (Schmitt and Freytag, 1977; Exconde, 1976; Raymundo and Exconde, 1976). Mostly resistant cultivars inhibited the infection in the area, showing slower systemic infection and thus decreasing the overall negative effect. Many fungicides are also used for PDM management, like metalaxyl, maneb, and fentin hydroxide (Exconde, 1976; Exconde and Raymundo, 1976; Exconde, 1975). Seed-borne transmission is decreased or eliminated when seeds are dried to a less than 14% moisture level. Soil-borne infection can be decreased by controlling alternate grass hosts within the paddock and rotating maize or sorghum crops for over 3 years.

11.2.4 Maize Rusts

Two types of rust are commonly reported on maize crops, that is, polysora and common rust.

11.2.4.1 Common Rust of Maize

Pathogen: Its casual organism is *Puccinia sorghi* (Delate, 2009).

Symptoms of common rust of maize: Several rust lesions are common throughout the growing season in the field. These lesions develop on any part of the plant aboveground, although they are most common on the leaves. Symptoms usually appear after tasseling. We can quickly differentiate it from another disease because of the formation of dark, reddish-brown pustules (uredinia) on the lower and upper surfaces of the infected leaves. Pustules are covered by the epidermal layer of the leaf where they have penetrated, usually measure less than 1/4 in. in length, and are elongated to oval in shape. Pustules may appear in bands on the leaf's surface if infections occur when the leaves are still in the whorl (Ramirez-Cabral *et al.*, 2017).

Disease cycle and epidemiology: Unlike most other foliar corn diseases, this rust fungus does not overwinter in crop residue. During mid-June to mid-July (the growing season), spores should be transported from tropical and subtropical regions northward, where it survives on the alternative host (wood sorrel or corn). In general, older leaves are less susceptible than younger leaves. Long durations of cool temperatures (60–74°F) and high relative humidity are conducive to rust spread and development. In these circumstances, Symptoms usually appear after 7 days of infection on susceptible varieties of sweet corn, and the severity of infection in corn hybrids usually leads to death and chlorosis of leaf sheaths and leaves. The infection spreads to new fields, plants, and leaves when the wind disperses the uredospores (produced during the season). Pustules on the maturing corn plant produce dark-pigmented telia that take the place of uredinia and produce teliospores, turning the pustules brownish black. In temperate regions, the fungus does not infect the alternate host, *Oxalis* species (wood sorrel), like in some states of the US Corn Belt and Ohio, and the teliospore does not contribute to the disease cycle (no real epidemiological significance). In contrast, in tropical regions, the teliospores infect the wood sorrel (Guerra *et al.*, 2019).

Management: Although rust is usually seen on corn in Ohio, fungicide applications have been relatively rare. An earlier fungicide spray may be required to efficiently control infection during seasons with significant rust on the lower leaves before silking (Utpal and Ritika, 2015).

11.2.4.2 Polysora Rust of Maize

Pathogen: Polysora rust is caused by *Puccinia polysora*, also known as southern rust, particularly devastating to late-planted or late-maturing hybrids. The disease is more severe in early-infected plants and when the disease spreads to leaves above the ear (which contribute most to grain filling). In contrast to common rust, the warm growing season favors the development of polysora rust. In temperate areas, polysora rot can significantly reduce yield (Sun *et al.*, 2021). For each 10% of the total infected leaf area, an 8% reduction in yield is recorded (Ramirez-Cabral *et al.*, 2017).

Symptoms of polysora rust of maize: Earlier, leaf lesions tend to be small, round to oval, and frequently with a noticeable light green to yellow color. These lesions grow into elevated, circular to oval, light orange to cinnamon-red pustules that range in size from 0.2 to 2 mm and are widely dispersed on the upper leaf surfaces. Pustules are more common on the upper than the lower leaf surface. The pustules turn dark brown to black as the plant matures and starts to produce teliospores. These pustules eventually burst through the epidermis, exposing masses of powdery spores, and then secondary disease cycles start and give the leaves a rusty color. These spores are produced in concentric rings that surround the initial pustules. At the disease severity, leaves can prematurely senesce and turn chlorotic; stalk lodging may happen. Additionally, lesions are formed on the husk, leaf sheath, and stalk (Halvorson *et al.*, 2021).

Disease cycle and epidemiology: In hot, humid climates with ideal infection-favoring temperatures of 24–28°C, polysora rust is a common problem. Polysora rust is more prevalent during high-humidity periods. Therefore, maize that was planted and matured later is more susceptible. On infested residues that have been left in the field, *Puccinia polysora* cannot live. During the growing season, infection is caused due to disseminated spores. The main source of inoculum at the beginning and during the growing season is urediospores. A large quantity of initial disease inoculum will result from the continued production of maize and the existence of diseased corn at the start of successive seasons. Urediospores are spread to newly planted corn by the rain splashes and wind. When favorable disease circumstances, rust disseminates quickly, with new infections developing in just 7 days. Spores can survive even when spread over distances greater than 100 km (Sun *et al.*, 2021).

Management: The most practical way to manage polysora rust is to use resistant maize hybrids. Early planting can help in reducing the disease severity. When the disease severity is expected to be high, fungicides can be used to manage polysora rust. To protect the early phases of reproductive growth, fungicides must be used after 7 weeks of planting. Depending on the weather and disease severity, more applications of fungicides can be required. Fungicides should be used when there is a small quantity of secondary inoculum, that is, when lesions are first seen on the leaves. In South Africa, no specific fungicide is registered against polysora rust, whereas the fungicide used against the common rust also provides adequate protection (Veerabhadraswamy, 2014).

11.2.5 Anthracnose Leaf Blight of Corn

Pathogen: The anthracnose leaf blight and stalk rot of corn are caused by the *Colletotrichum graminicola*, causing yield losses of 40% and more than 80% lodging have been recorded (Tesso *et al.*, 2012).

Symptoms: Young seedlings may exhibit anthracnose foliar symptoms early in the growing season. The fungus causes small, irregular to oval, red-brown to brown pustules on leaves, frequently with a yellow edge. Frequently, concentric fungal growth gives lesions a "target-like" appearance. A 20–30× hand lens can be used to see fungus bodies inside lesions encircled by black hairs, and a microscope can be used to see spores. Foliar symptoms are difficult to distinguish because they resemble other foliar diseases of corn in the field. Because of the buildup of protective compounds in leaves after the V6 stage, some hybrids' foliar phase of the disease can be significantly diminished. Some hybrids may exhibit a second flush of foliar symptoms on the upper leaf following tasseling. If leaf blighting develops earlier than 6 weeks after tasseling, it reduces yield. A hand lens or a microscope can be used to see the fungal structures (Jirak-Peterson and Esker, 2011).

Disease cycle: The pathogen overwinters on maize residue. The fungus releases spores on residue in the spring during warm, rainy conditions, and these spores can infect corn seedlings' foliage and roots. If the disease spreads below the soil's surface, the pith and root get infected, and symptoms might not appear until later in the growing season. Primary infections are brought on by spores that are rain-splashed onto seedlings. Spores can be continuously formed on foliar lesions if the weather is wet, cloudy, and warm, which can cause severe foliar symptoms in some hybrids (Ma *et al.*, 2022). Infections in the leaf sheath may spread to the stem if the plant has a mechanical injury. When the plant reaches maturity, *C. graminicola* in the stalk starts to break down the pith for nutrition. This greatly reduces the stalk's integrity and causes the plant to die (Venard *et al.*, 2005).

Management: Resistant hybrids, stress reduction, residue management, and crop rotation are the best ways to manage anthracnose. Crop losses are frequently caused by lodging from stalk infections likely to start from root infections. Therefore, it is not advised to use fungicides to prevent anthracnose stalk rot. To lessen the effects of lodging-related yield loss, the field should be prepared for the earliest harvest if stalk rot incidence is greater than 10% (Venard *et al.*, 2005).

11.2.6 Southern Corn Leaf Blight

Pathogen: Southern corn leaf blight (SCLB) disease is caused by *Bipolaris maydis*. In its teleomorph state, it is also called *Cochliobolus heterostrophus*. This ascomycete fungus can spread through ascospores or conidia (Agrios and George Nicholas, 2005). *B. maydis* has 3 races: Race C, Race T, and Race O; the symptoms of SCLB differ depending on the race of the infecting pathogen. Race T is contagious to Texas male sterile cytoplasm-containing corn plants (cms-T cytoplasm maize), and this sensitivity led to the SCLB pandemic in the United States from 1969 to 1970 (Ullstrup, 1972). Race T is of particular significance because of this. Although SCLB grows best in moist and warm environments, it can be found in many of the world's maize-growing regions (Bucheyeki,

2012). Breeding for host resistance, using fungicides, cultural control, and growing resistance verities are common management strategies.

Symptoms: The types of symptoms produced depend on the pathogen's specific race Singh, R., and Srivastava, R. P. (2016). The occurrence of lesions on the leaves is a characteristic sign of SCLB (Bucheyeki, 2012). In the case of Race O, lesions have tan and buff-brown borders; they start as small diamond-shaped lesions and enlarge and become rectangular as they extend within the veins. These lesions are present in the maize plant's leaves (Bucheyeki, 2012). The dimensions of a lesion are 3–22 mm in length and 2–6 mm in width (Aregbesola *et al.*, 2020). Race T lesions have green to yellow or chlorotic haloes and are tan. Later, lesions develop red-dark brown margins and can spread to other plant parts aboveground, such as the ear, stem, and sheath. Lesions can be larger than those generated by Race O, measuring 6–27 mm in length and 6–12 mm in width, and can be elliptical or spindle-shaped (Bucheyeki, 2012). Seedlings with Race T infection begin to wilt, eventually dying after 3–4 weeks. Race C causes 5 mm long necrotic lesions. They also cause wilt (Wei *et al.*, 1988). *B. maydis* can infect the husks, sheaths, stalks, leaves, ears, and cobs. The ear may die prematurely if the infection develops early, resulting in ear drops. SCLB-affected kernels will have a black mold, with a felty covering to them, which could lead to cob rot. Race T on maize of cms-T cytoplasm results in more severe ear rot (Calvert and Zuber, 1973). Infected seedlings may wilt and die within a few weeks of the planting date (Agrios, 2005).

Disease cycle: *C. heterostrophus* discharges sexual ascospores or asexual conidia to infect maize plants as part of its cyclical disease cycle. The asexual cycle, which is well known to exist in nature, is the main issue. Conidia (the major inoculum), when present, are released from lesions on an infected corn plant and spread to neighboring plants by splashing rain or wind. A healthy plant's sheath or leaf will allow the conidia of *B. maydis* to germinate on the tissue via polar germ tubes. The germ tubes enter through the leaf itself or a stomata-like natural opening. The fungus's mycelium invading the parenchymatous leaf tissue causes the cells to turn brown and collapse. These lesions result in conidiophores formation, which, under the appropriate situation, can either spread the infection to other parts of the original host plant (leaves, stalks, husks, or kernels) or release conidia to infect other plants in the area (Agrios, 2005). According to the definition of "favorable conditions," there must be water on the leaf surface, and the surrounding temperature must be 60–80°F. Spores develop and enter the plant under these circumstances within 6 hours (the University of Illinois Extension, Common Leaf Blights and Spots of Corn). In the form of spores and mycelium, the fungus overwinters in the corn residue, waiting for the right conditions to reappear in the spring (Agrios, 2005). As already stated, *B. maydis* also has an ascospore-producing sexual stage; however, this has only been seen in lab cultures. The ascocarp *Cochliobolus*, a rare species of perithecium, contains its ascospores (inside asci). Therefore, conidial infection is the primary asexual route of SCLB infection (Bucheyeki, 2012).

Management: Using resistant varieties is the most effective method of controlling SCLB. There are both single-gene and polygene origins of resistance have been discovered. Race O is more common because normal cytoplasm maize can withstand Races T and C. There may be flecking in certain resistant hybrids, but this is merely a response to resistance and will not have any negative economic effects. The spread of all races can be stopped via alternative control methods. In order to prevent *B. maydis* from overwintering in agricultural debris, it is crucial to manage crop residues during the growing period. Any leftover material can be broken down with the help of tillage. Contrary to minimum tillage, which can leave residue on the soil surface, it has been discovered that burying residues by plowing has reduced the incidence of SCLB crop rotation with non-host crops is another method of cultural control used to lessen SCLB. Foliar fungicides are another method used for the management of corn leaf blight. From 14 days before tasseling to 21 days following, foliar disease control is essential when leaf blight damage is most likely to occur. When lesions become visible on plants infected with SCLB, fungicides should be used. During the growing season, reapplications might be required depending on the environmental conditions (Sumner and Littrell, 1974; Bucheyeki, 2012).

11.2.7 *Fusarium* and *Gibberella* Stalk Rots of Maize

Pathogens: The maize stalk rot is reported to cause by *Fusarium graminearum*, (previously also named *Gibberella zeae*) (Zhang *et al.*, 2016).

Symptoms: Until spore-producing structures are visible, the symptoms caused by these diseases cannot be distinguished from those caused by *Stenocarpella* or *Cephalosporium*. Wilted plants are still standing after drying, and the lowest internodes begin to show tiny, dark-brown lesions. Dark brown phloem appears on the splitting of the diseased stalk, and tissues generally show a noticeable browning. As the disease progresses, the pith is shredded, and the surrounding tissues get discolored (Shin *et al.*, 2014). Most often, it occurs during dry and hot years. The bottom regions of the stalk are often yellow toward the dent (stage R5), and the leaves often have a dull green color. When pressed, stalk tissue may readily degenerate and collapse. In extreme circumstances, plants may separate at the lower nodes. The fungus is extremely widespread and can be found on healthy stalks as well. Under suitable conditions, the infection will only lead to stalk rot disease (Shin *et al.*, 2014).

Management: Avoid low-potassium and high-nitrogen soils, ensure proper field drainage, avoid rotating wheat and corn, and prevent plants from being stressed or injured.

11.2.8 Anthracnose Stalk Rot of Maize

Pathogen: One of the most prevalent stalk rots is anthracnose stalk rot caused by *Colletotrichum graminicola* (Cota *et al.*, 2012; Belisário *et al.*, 2022).

Symptoms: Just before natural senescence, symptoms start to appear. Primarily wounds caused due to insect feeding, mechanical trauma, hail, or any other reason. Shiny black spots on the bottom internodes of the stalk are the most noticeable sign of the stalk rot phase (Cota *et al.*, 2012). The stalk easily collapses when pinched, and discoloration can be seen within the pith. With a hand lens, tiny, black, whisker-like fungal appendages (setae) can be observed on diseased tissue. The topmost portion of the plant appears bleached during the top die-back phase, while the below plant tissue is still green. When the leaf sheaths are peeled off, black lesions could be visible on the stalks in the top internodes. In contrast to other stalk rots where the stalk breaks closer to the ground, lodging caused by anthracnose stalk rot may cause the stalk to break higher up on the stalk (Jirak-Peterson and Esker, 2011). High humidity with cloudy days favors disease spread.

Management: It can be managed by reducing plant wounding and stress by using resistant varieties, and in crop rotation, avoid rotating to sorghum (Cota *et al.*, 2012).

11.2.9 Head Smut of Maize

Pathogen: This disease occurs in most parts of the world and is caused by *Sphacelotheca reiliana* (Konlasuk *et al.*, 2015).

Symptoms: Typically, symptoms can be seen on tassels and cob. The ear and tassel are replaced with large smut sori. The tassel can rarely be transformed into smut sorus whole or partially. The smutted plants have reduced growth and yield and remain green in color than the other plants. Smut spores, which are numerous, thick-walled, spherical, finely spined, and reddish brown to black, are formed (Yanfeng *et al.*, 2014).

Disease cycle and epidemiology: *S. reiliana* (*Sorosporium reilianum*, also known as *Ustilago reiliana*) is a fungus that lives in the soil for at least 5–7 years as spores. Spores can travel great distances in the wind. Maize pickers are the most likely means of propagation in Oregon's Willamette Valley. Only corn has been affected by the fungus in the Pacific Northwest. There are two strains of *Sphacelotheca reiliana*. One attack on Sudan grass and sorghum and the other on corn only. According to reports, infection strikes at the seedling stage; nevertheless, tests conducted in greenhouses show that disease can cause many weeks after maize has been sown. After infection, the pathogen spreads systemically. Infection occurs more rapidly at soil temperatures from 68–86°F and gradually at 59°F. The fungus can be transported on the seed surface but is not seed-borne (Juroszek and Tiedemann, 2013).

Management: Crop rotation with pulses, field sanitation, and seed treatment with Thiram or Captan at a rate of 4 g per kg found effective in controlling this disease. Field sanitation and equipment sterilization is also effective for disease control (Singh and Chawla, 2012).

11.2.10 Penicillium Ear Rot

Pathogen: The *Penicillium* species, such as *Penicillium oxalicum*, a prevalent and harmful storage fungus, can proliferate in grain when the moisture content is above 18% (Logrieco *et al.*, 2003).

Symptoms: Powdery green or blue fungal growth between and on kernels, most frequently near the tip of the corn ear, is the most common sign of infection on ears damaged by insects or mechanical means. If the virus invades maize kernels stored in high moisture conditions, they may become bleached or streaked, and discoloration of the embryo or "blue eye" may occur (Jeffers, 2004).

Management: By preventing injury to the ears, we can reduce the disease occurrence.

11.2.11 Common Smut of Corn

Pathogen: Common corn smut, caused by the fungus *Ustilago maydis*, can be recognized by tumors like galls that develop on actively growing tissues and contain masses of dark, sooty teliospores (Pope and McCarter, 1992).

Symptoms: Teosinte (*Zea mexicana*) and maize (*Zea mays*) are the hosts that *U. maydis* infects. The sooty teliospore masses give rise to other smut diseases, including the common smut of maize. Galls that resemble tumors and range in size from 0.4 to 12 in. (1–30 cm) in diameter are the most noticeable symptoms. Any meristematic tissue can get infected. Galls are commonly found on stalks, leaves' midribs, nodal shoots, tassels, ears, and other plant parts (Mohan *et al.*, 2013). The infection is localized (i.e., the host is not colonized systemically); regardless, the plant's various aboveground components can develop galls. When a young plant's apical meristem is infected, galls can occasionally form under the soil's surface. Smut galls are made up of host and fungus tissues. Young galls are hard, white, and covered in semi-glossy periderm. Galls develop semi-fleshy internal tissues as they develop, and as teliospores form, black tissue streaks appear. The periderm ruptures when the galls mature, releasing a mass of powdery teliospores (Jackson Ziems, 2014). Usually, between 10 and 14 days of the disease, galls become visible. Three weeks following the infection of the ovaries, ear galls reach maturity. Three to 6 days following infection, kernels with minor discoloration or malformation are visible (Pataky and Snetselaar, 2006; Jackson-Ziems, 2014; Frommer *et al.*, 2019).

Disease cycle and epidemiology: The fungus overwinters in soil or crop residue as diploid teliospores, that can survive for many years. The wind or splashing rain can disperse teliospores directly, or they can develop and go through meiosis to form haploid sporidia, which the rain splash or wind can also disperse. As they reproduce and create dikaryotic infection hyphae, sporidia bud-like yeast. Each infection is localized. Any part of the plant that is aboveground can become infected, especially young tissues that are actively growing (meristematic tissues) (Pataky and Snetselaar, 2006; Yadav and Sharma, 2022). There is no consensus on the weather patterns, which are most conducive to common smut, but most studies demonstrate that the disease tends to be more common after humid and wet weather. Injury occurring during detasseling can also cause infection. Following violent thunderstorms with strong winds, galls on the stem and leaves of seedlings are frequently seen, particularly where plants get injured via blown soil. Naturally, rain and wind would favor the germination and spread of teliospores (Yadav and Sharma, 2022). Therefore, anything which promotes rapid elongation and localized cell division, like an injury, could make one more vulnerable to smut infection. Factors that restrict or inhibit pollination increase the prevalence of common smut ear galls as ovaries are adequately protected against infection by *U. maydis* shortly after fertilization, most likely because silks linked to fertilized ovaries die and therefore do not remain vulnerable to infection. As an illustration, hot and dry conditions frequently result in asynchronous pollen generation and silk development, which leads to ineffective pollination. Common smut ear galls could be common if *U. maydis* is easily transmitted to the stigmas of unfertilized ovaries during or after drought-like conditions (Pataky and Snetselaar, 2006). Thus, some researchers link the presence of ear galls to dry and hot conditions, but this leads to an increase in the number of unpollinated ovaries with quickly developing silks that directly alter the occurrence of ear galls.

Management: Crop rotation, seed treatments, sanitation, fertility adjustment, foliar fungicide application, and biological controls are a few approaches to managing common smut (El-Fiki *et al.*, 2003). Host resistance is the only feasible way to manage common smut where *U. maydis* is common. However, no corn line is resistant to infection by *U. maydis* (Pataky and Snetselaar, 2006).

11.2.12 Diseases Caused by Bacteria

11.2.12.1 Bacterial Stalk Rot of Maize

Pathogen: *Erwinia chrysanthemi* pv. *zeae* (Sabet) is a gram-negative, motile, rod-shaped bacterium that causes stalk rot in maize (Thind and Payak, 1985).

Symptoms: The primary symptom is the stalk and leaf sheath discoloration at a node. Lesions form on the sheaths and leaves as the disease develops. The infection then starts in the stalk and quickly moves into the leaves. An unpleasant odor can be recognized as decay occurs, and the plant's top can be separated from the plant. Complete stalk rot causes the top to collapse (Patandjengi *et al.*, 2021). Bacterial stalk rot can infect the plant at any node, from the soil's surface to the tassels and ear leaves. Severe infections may affect tasseling and, subsequently, pollination. The bacteria often do not transfer to adjacent plants until they are vectored by an insect (Patandjengi *et al.*, 2021). Stalk splitting exposes soft, slimy rot and internal discoloration, usually appearing at the nodes (Sinha and Prasad, 1977). Infected plants are frequently dispersed over the field since the bacteria typically do not transfer from one plant to another. However, several insect vectors have been reported to transmit diseases from plant to plant. High relative humidity and temperatures favor top rot and bacterial stalk growth. It can be an issue in regions with excessive rainfall or where overhead irrigation is employed, for example, water pumping via a pond, a slow-moving stream, or a lake (Kumar *et al.*, 2015).

Management: Preventing excessive irrigation and fall cultivation to incorporate crop debris is all aspects of controlling and bacterial stalk rot.

11.2.12.2 Stewart's Wilt Rot of Maize

Pathogen: Stewart's wilt is a serious bacterial disease of corn caused by the bacterium *Erwinia stewartii.*

Symptoms: Stewart's wilt on corn has two stages. When young plants are systemically infected, the wilt stage of seedlings starts. Plants infected after the seedling stage enter the leaf blight phase. The corn flea beetle (*Chaetocnema pulicaria*), an insect, injures plant tissues while feeding on them (Pataky, 2003). The bacterium that causes Stewart's wilt, *Erwinia stewartii*, overwinters as a vector on the corn flea beetle. In both stages of the infection, the initial symptoms are the same. Feeding wounds' around the leaf tissue first become wet. Parallel to the leaf veins, yellow to pale green linear stripes with wavy borders form. On vulnerable cultivars, these lesions, with time, may stretch the full length of the leaf and turn necrotic. Plants with systemic infection show symptoms on newly emerging leaves and may develop holes in the stalks near the soil surface. When plants are infected systemically, bacteria can frequently damage grains and propagate within the plant's vascular system. Various corn varieties respond to Stewart's wilt in different ways. Systemic infection occurs frequently (Pataky, 2003) or never occurs in resistant cultivars, where symptoms are typically restricted to the area around flea beetle feeding wounds that are 1–2 in. (2–3 cm) in diameter. The primary stalks of seedlings may be damaged by infection if it happens when they are sprouting, which could cause a tremendous growth of tillers. Main stalk death, although not experimentally shown, is most likely caused by *E. stewartii* infection at the main growth point carried on by flea beetles feeding on developing coleoptiles (the tissue covering the first true leaf) or premature seedlings. The foliar symptoms of the seedling wilt and leaf blight phases are identical. Depending on the cultivar's susceptibility, necrotic or chlorotic tissues may extend throughout the leaves or only a few inches of the leaf. Stewart's wilt may cause premature leaf death, making the plant more vulnerable to stem rot and lower yields (Pataky *et al.*, 2000).

Disease cycle and epidemiology: The two main hosts for *E. stewartii* are *Chaetocnema pulicaria*, a corn flea beetle, and corn. Nearly all the bacteria are spread by maize flea beetles. There are no known cases of Stewart's wilt occurring frequently or for an extended period without this insect (Ammar *et al.*, 2014). Over 28,500 insect specimens from 94 species and 76 genera were examined by F. W. Poos and Charlotte Elliot for the presence of *E. stewartii*; they concluded that *C. pulicaria* was the only species significant for retaining the bacterium and dispersing the disease (Pataky, 2003). Although the bacteria has frequently been identified from *Chaetocnema denticulata*, other insects are significant field vectors (Pataky, 2003). In the intestinal tract of corn flea beetles, *E. stewartii* overwinters. When the soil surface temperature is reached nearly 65–70°F (18–21°C), these insects come out of hibernation in the spring (Pataky, 2003). As they feed on corn, flea beetles spread *E. stewartii*. Without the vector, the bacteria cannot transmit from one plant to another. Even while *C. pulicaria*'s feces and regurgitated material include bacteria, and the most likely source of infection seems to be fecal contamination of feeding wounds. Flea beetles can produce two or more summer generations during the growing season. The peak of the first summer generation occurs around the middle of June, and succeeding generations start to appear after 4 more weeks. The dynamics of the flea beetle population correlate with two cycles of Stewart's wilt infection. The initial cycle of Stewart's wilt causes the most damage because seedlings become infected with *E. stewartii*, which is spread by the flea beetle generation that overwinters. When the bacteria are spread by the first summer generation of the insect, the second cycle of infection commences (Ammar *et al.*, 2014).

In field corn, the second cycle of infection often results in the leaf blight stage of Stewart's wilt, while it has been known to result in the seedling wilt stage in late-planted sweet corn. The bacterium is acquired by succeeding generations of flea beetles from diseased plants, and they then become the population that overwinters (Pataky, 2003). Seed transmission has little impact on the epidemiology of Stewart's wilt in North America, even though *E. stewartii* can be spread by seeds. Similarly, when the seed is produced on resistant inbred seed parents, there is almost little chance of *E. stewartii* spreading by seed to other parts of the world. The parent plant's susceptibility significantly impacts whether a seed carries *E. stewartii*. Seed may retain *E. stewartii* infection if seed parent plants are systemically affected during the disease's seedling wilt stage. Seed infection is highly uncommon if Stewart's wilt leaf blight stage occurs but the parent plant of the seed is not systemically infected. Because the effects of the seed parent plants' reactions or the presence of vectors were not considered in the early study on *E. stewartii* seed transmission in the first half of the 20th century, seed transmission rates were overstated for nearly 60 years. According to recent studies, the probability of spreading *E. stewartii* by seed is approximately 1 in 50,000 for plants that are vulnerable to the disease and have a systemic infection and 1 in 20,000,000 for plants that are resistant to the disease and have leaf blight signs.

Management: Stewart's wilt can be managed using disease control strategies, such as eradication, host resistance, and exclusion (Pataky, 2003).

11.2.12.3 Bacterial Leaf Streak of Corn

Pathogen: The bacterium *Xanthomonas vasicola*, which causes bacterial leaf streak, has been identified on field corn, sweet corn, and popcorn. There is no cure for the bacterial pathogen, which can infect a plant in the field and spread the infection to nearby plants (Korus *et al.*, 2017).

Symptoms: symptoms can appear at any growth stage, although they are often first seen on the lower leaves, spreading to the middle and higher part of the crop canopy after flowering (Korus *et al.*, 2017). The first signs of the condition start as transparent, wet streaks among veins and advance to longer yellowish necrotic stripes that can combine to produce bigger patches of symptomatic tissue. Bacterial exudates on the leaf's surface can appear as dry, tiny, and yellowish droplets. Large regions of necrosis are produced as bacterial leaf streak (BLS) lesions combine, and leaf tissue dies (Broders, 2017).

Disease Cycle and Epidemiology: The BLS pathogen is now thought to live on plant leftovers from prior infections. When bacterial exudates from diseased plants are spread to healthy plants by rain splash, wind, or overhead irrigation water, secondary infections take place. When temperatures are over 90°F, overhead irrigation is associated with a rise in disease incidence (Ortiz-Castro *et al.*, 2020). As lesions form on the leaves, the pathogen enters corn leaves through the stomata, causing a distinctive banded pattern. The infection does not enter vascular tissue and seems confined to the spaces between the major veins. As it is not systemic, it does not produce withering like Stewart's or Goss's wilt. Despite foliar lesions covering approximately 40% of the leaf area, no subsequent reduction in yield has been recorded (Ortiz-Castro *et al.*, 2020). The host plants include grass crops and weeds, such as yellow nudge sedges, oats, rice, green foxtail, shatter cane, timothy, Johnson-grass, and downy brome.

Management: Crop rotation with the non-host crop, such as alfalfa, soybeans, or wheat, eliminates volunteer corn and weeds, which can contain the bacterium over a long period. Several grass weed hosts could serve as substitute hosts, and fall tillage can hasten the breakdown of crop residue and decrease inoculum level, but this control method must be balanced against the risks of soil erosion (Ortiz-Castro *et al.*, 2020).

11.2.13 Diseases Caused by Viruses

11.2.13.1 Maize Chlorotic Dwarf Virus

Pathogen: The virus that causes chlorotic maize dwarf is called the maize chlorotic dwarf virus (MCDV). The disease has been documented in 19 states in the United States and Mexico, from Texas to Ohio and Missouri. Between the 1960s and 1970s, the disease severely damaged the Midwest and southern US economies. The virus and its vector are still prevalent in Ohio, and warmer temperatures increase the vector's ability to transmit the virus, even though significant yield reductions in the field (dent) corn due to MCDV have not yet been observed. There have been few systematic studies to evaluate yield losses in agricultural areas. Losses brought on by MCDV in vulnerable hybrids have ranged experimentally from 5 to 91%. The Johnsongrass plant, which serves as the virus's overwintering host, and the maize leafhopper *Graminella nigrifrons*, the virus's primary vector, are most prevalent in areas where MCDV is present (Wayadande and Nault, 1993).

Symptoms: Symptom severity varies depending on the maize hybrid, viral isolate, and stage of plant development at the time of infection. The viral strain and the hybrid both show different symptoms. MCDV can produce various symptoms, such as shortening of the upper internodes, yellowing or reddening of the leaves, plant stunting, chlorosis, and twisting and tearing (Jones *et al.*, 2004). Infected leaves may also become "corduroy" in texture rather than shiny and smooth. Vein banding is this virus's most readily identifiable sign (Jones *et al.*, 2004). Older plants or those that have minor strain infections may not show any signs of vein chlorosis.

Disease cycle: The black-faced leafhopper *Graminella nigrifrons* is the main carrier of MCDV. Since the leafhopper prefers to feed on grasses like barnyard grass (*Echinochloa crusgalli*), crabgrass (*Digitaria sanguinalis*), ryegrass (*Lolium perenne*), and bermudagrass, corn is not its preferred host (*Cynadon dactylon*). The virus cannot be spread until the leafhopper must feed on an infected plant for a minimum of 15 minutes and a maximum of several hours. The virus can instantly spread from the infected insect for up to 4 days. Only a few gramineous species, such as wheat and sorghum, are susceptible to MCDV infection, and these plants do not show any symptoms of infection. The virus overwinters in the rhizomes of *Sorghum halepense* (Johnsongrass), the only known perennial host. MCDV cannot be spread through rub-inoculation or seed (Morales *et al.*, 2014).

Management: The best method to prevent the disease is to control Johnsongrass. Early in the growing season, use herbicide to remove Johnsongrass in and around maize fields. Johnsongrass rhizomes, the virus's primary host, allow MCDV to survive the winter. It is not advised to control the vector with foliar pesticides. The best results are obtained when seeds are treated with systemic insecticides to prevent leafhoppers and other pests (Stewart *et al.*, 2014).

11.2.13.2 Maize Streak Virus

Pathogen: The maize leaf hopper (*Cicadulina mbila*) is the primary vector of the maize streak virus (MSV), but some other species of leafhopper, including *C. arachidis*, *C. storeyi*, and *C. dabrowski*, have also been reported in the transmission of the virus (Harkins *et al.*, 2009). The sucking mouth parts of the leafhopper allow them to enter plant cells with the help of the gut, salivary enzymes, and mechanical force. Plants can die early due to the severity of the infection. The termination of ear formation and stunted plant, grain filling, and development in diseased plants are the main cause of reduction in yield.

Symptoms: It has been noticed that severely infected maize is stunting (Shepherd *et al.*, 2010). Very few rounded spots appear in the youngest leaves as the first disease symptoms a week after infection. The number of spots expands parallel to the leaf veins as the plant grows. Fully extended leaves change from the dark green color of typical foliage to chlorosis with scattered yellow stripes along the veins.

Disease cycle and epidemiology: Existence of insect transmission vectors among host plants. The insects that may spread the maize streak virus from plant to plant belong to the *Cicadulina* spp. The virus and vectors overwinter mostly in grasses (*Brachiaria lata* and *Setaria barbata*) and irrigation regions where maize can be cultivated during the dry period (Martin and Shepherd, 2009). Alternate hosts or crops vulnerable to disease are continuously accessible during high temperatures and rainfall.

Management: Remove crop debris and avoid planting crop downwind; as leafhoppers prefer shade, maize should be planted in open spaces. Reduce the number of weeds that may contain MSV vectors, especially grasses (Martin and Shepherd, 2009). Planting should be done early to prevent the optimal temperature that allows vectors to multiply and transmit the virus. Chemical insecticides like Bullock (beta-cyfluthrin), Gaucho FS 350 (imidacloprid), and others can successfully control the vector. Utilize certified seed that has been treated to eliminate MSV vectors (Magenya *et al.*, 2008).

REFERENCES

Agrios, G. N. (2005). *Plant Diseases Caused by Fungi. Plant Pathology* (5th ed, pp. 137, 268, 467–468). Elsevier Academic, Amsterdam.

Ammar, E. D., Correa, V. R., Hogenhout, S. A., & Redinbaugh, M. G. (2014). Immunofluorescence localization and ultrastructure of Stewart's wilt bacterium *Panoea stewartii* in maize leaves and in its flea beetle vector *Chaetocnema pulicaria* (Coleoptera: Chrisomelidae). *Journal of Microscopy and Ultrastructure*, 2(1), 28–33.

Aregbesola, E., Ortega-Beltran, A., Falade, T., Jonathan, G., Hearne, S., & Bandyopadhyay, R. (2020). A detached leaf assay to rapidly screen for resistance of maize *to Bipolaris maydis*, the causal agent of southern corn leaf blight. *European Journal of Plant Pathology*, 156(1), 133–145.

Aylor, D. E., Schultes, N. P., & Shields, E. J. (2003). An aerobiological framework for assessing cross-pollination in maize. *Agricultural and Forest Meteorology*, 119(3–4), 111–129.

Belisário, R., Robertson, A. E., & Vaillancourt, L. J. (2022, September). Maize anthracnose stalk rot in the genomic era. *Plant Disease*, 106(9), 2281–2298. doi: 10.1094/PDIS-10-21-2147-FE. Epub 2022 Aug 12. PMID: 35291814.

Bonde, M. R., Peterson, G. L., Kenneth, R. G., Vermeulen, H. D., & Sumartini, B. M. (1992). Effect of temperature on conidial germination and systemic infection of maize by *Peronosclerospora* species. *Phytopathology*, 82, 104–109.

Bortiri, E., & Hake, S. (2007). Flowering and determinacy in maize. *Journal of Experimental Botany*, 58(5), 909–916.

Boutard, A. (2012). *Beautiful Corn: America's Original Grain from Seed to Plate*. New Society Publishers, Gabriola Island.

Broders, K. (2017, December). Status of bacterial leaf streak of corn in the United States. In *Integrated Crop Management Conference* (Vol. 29, pp. 111–115). Bioagricultural Sciences and Pest Management, Colorado State University.

Bucheyeki, T. L. (2012). *Characterization and Genetic Analysis of Maize Germplasm for Resistance to Northern Corn Leaf Blight Disease in Tanzania* (Doctoral dissertation).

Calvert, O. H., & Zuber, M. S. (1973). Ear-rotting potential of *Helminthosporium maydis* race T in corn. *Phytopathology*, 63(6), 769–772.

Chaithra, M., Muttappagol, M., & Patel, P. S. (2022). Brown stripe downy mildew disease of corn–*Sclerophthara rayssiae* var. *zeae* in Karnataka. *Food and Scientific Reports,* 3(2), 35–36.

Cota, L. V., da Costa, R. V., Silva, D. D., Casela, C. R., & Parreira, D. F. (2012). Quantification of yield losses due to anthracnose stalk rot on corn in Brazilian conditions. *Journal of Phytopathology*, 160(11–12), 680–684.

Cueva, F. M. dela, Castro, A. M. de, & Torres, R. L. de. (2022). Peronosclerospora philippinensis (Philippine downy mildew of maize). *CABI Compendium. CABI International*. doi: 10.1079/cabicompendium.44646.

Dalmacio, S. C., & Raymundo, A. D. (1972). Spore density of *Sclerospora philippinensis* in relation to field temperature, relative humidity and downy mildew incidence. *Philippine Phytopathology*, 8, 72–77.

Delate, K. (2009). Organic grains, oilseeds, and other specialty crops. *Organic Farming: The Ecological System*, 54, 113–136.

El-Fiki, A. I. I., Fahmy, Z. M., Mohamed, F. G., & Abdel-Wahab, A. E. (2003). Studies on the control of corn common smut disease [*Ustilago maydis*]. *Egyptian Journal of Applied Science*, 18(11B), 433–453.

Exconde, O. R. (1975). Chemical control of maize downy mildew. *Tropical Agriculture Research Series* No. 8, 157–163.

Exconde, O. R. (1976). Philippine corn downy mildew: Assessment of present knowledge and future research needs. *Kasetsart Journal*, 10, 94–100.

Exconde, O. R., & Raymundo, A. D. (1974). Yield loss caused by Philippine corn downy mildew. *Philippine Agriculturist*, 58(3/4), 115–120.

Frommer, D., Radócz, L., & Veres, S. (2019). Changes of relative chlorophyll content in sweet corn leaves of different ages infected by corn smut. *Agriculturae Conspectus Scientificus*, 84(2), 189–192.

Guerra, F. A., De Rossi, R. L., Brücher, E., Vuletic, E., Plazas, M. C., Guerra, G. D., & Ducasse, D. A. (2019). Occurrence of the complete cycle of *Puccinia sorghi* Schw. in Argentina and implications on the common corn rust epidemiology. *European Journal of Plant Pathology*, 154(2), 171–177.

Gunjan, M., Sarangdevot, Y. S., & Vyas, B. (2021). Pharmacognostical study, and pharmacological review of *Coccinia indica* fruit and *Zea mays* leaves. *Journal of Pharmaceutical Sciences and Research*, 13(6), 335–339.

Gwirtz, J. A., & Garcia-Casal, M. N. (2014). Processing maize flour and corn meal food products. *Annals of the New York Academy of Sciences*, 1312(1), 66–75.

Halvorson, J., Kim, Y., Gill, U., & Friskop, A. (2021). First report of the southern corn rust pathogen *Puccinia polysora* on *Zea mays* in North Dakota. *Canadian Journal of Plant Pathology*, 43(Sup 2), S352-S357.

Harkins, G. W., Martin, D. P., Duffy, S., Monjane, A. L., Shepherd, D. N., Windram, O. P., . . . Varsani, A. (2009). Dating the origins of the maize-adapted strain of maize streak virus, MSV-A. *Journal of general virology,* 90(Pt 12), 3066.

Jackson-Ziems, T. (2014). Smut diseases of corn. *Papers in Plant Pathology*. 439. University of Nebraska–Lincoln Extension, Institute of Agriculture and Natural Resources.

Jeffers, D. (2004). *Maize Diseases: A Guide for Field Identification*. New York, Mexico: CIMMYT.

Jirak-Peterson, J. C., & Esker, P. D. (2011). Tillage, crop rotation, and hybrid effects on residue and corn anthracnose occurrence in Wisconsin. *Plant Disease*, 95(5), 601–610.

Jones, M. W., Redinbaugh, M. G., Anderson, R. J., & Louie, R. (2004). Identification of quantitative trait loci controlling resistance to maize chlorotic dwarf virus. *Theoretical and Applied Genetics,* 110(1), 48–57.

Juroszek, P., & Tiedemann, A. V. (2013). Climatic changes and the potential future importance of maize diseases: A short review. *Journal of Plant Diseases and Protection*, 120(2), 49–56.

Khairwal, I. S., Rai, K. N., Diwakar, B., Sharma, Y. K., Rajpurohit, B. S., Nirwan, B., & Bhattacharjee, R. (2007). Pearl millet crop management and seed production manual. International Crops Research Institute for the Semi-Arid Tropics, Patancheru, Andhra Pradesh, India.

Konlasuk, S., Xing, Y., Zhang, N., Zuo, W., Zhang, B., Tan, G., & Xu, M. (2015). ZmWAK, a quantitative resistance gene to head smut in maize, improves yield performance by reducing the endophytic pathogen *Sporisorium reiliana. Molecular Breeding*, 35, 1–10.

Korus, K. A., Lang, J. M., Adesemoye, A. O., Block, C. C., Pal, N., Leach, J. E., & Jackson-Ziems, T. A. (2017). First report of *Xanthomonas vasicola* causing bacterial leaf streak on corn in the United States. *Papers in Plant Pathology*, 101, 476.

Kumar, A., Hunjan, M. S., Singh, P. P., & Kaur, H. (2015). Status of bacterial stalk rot of maize in Punjab. *Plant Disease Research*, 30(1), 97–99.

Lal, S., and Prasad, T. 1989. Detection and management of seed-borne nature of downy mildew diseases of maize. *Seeds Farms*, 15, 35–40.

Lazarovits, G., Turnbull, A., & Johnston-Monje, D. (2014). Plant health management: Biological control of plant pathogens. In Van Alfen, N. K. (ed), *Encyclopedia of Agriculture and Food Systems* (pp. 388–399). Academic Press, Oxford.

Leon, C. de, Ahuja, V. P., Capio, E. R., & Mukherjee, B. K. (1993). Genetics of resistance to Philippine downy mildew in three maize populations. *Indian Journal of Genetics & Plant Breeding*, 53, 406–410.

Logrieco, A., Bottalico, A., Mulé, G., Moretti, A., & Perrone, G. (2003). Epidemiology of toxigenic fungi and their associated mycotoxins for some Mediterranean crops. In Xiangming Xu, John A. Bailey, Michael Cooke (eds.), *Epidemiology of Mycotoxin Producing Fungi* (pp. 645–667). Springer, Dordrecht.

Ma, W., Gao, X., Han, T., Mohammed, M. T., Yang, J., Ding, J., . . . Bhadauria, V. (2022). Molecular genetics of anthracnose resistance in maize. *Journal of Fungi*, 8(5), 540.

Magenya, O. E. V., Mueke, J., & Omwega, C. (2008). Significance and transmission of maize streak virus disease in Africa and options for management: A review. *African Journal of Biotechnology*, 7(25).

Martin, D. P., & Shepherd, D. N. (2009). The epidemiology, economic impact and control of maize streak disease. *Food Security*, 1(3), 305–315.

Mohan, S. K., Hamm, P. B., du Toit, L. J., & Clough, G. H. (2013). *Corn smuts*. A Pacific Northwest Extension Publication, Washington State University, United states.

Morales, K., Zambrano, J. L., & Stewart, L. R. (2014). Co-infection and disease severity of Ohio Maize dwarf mosaic virus and Maize chlorotic dwarf virus strains. *Plant Disease*, 98(12), 1661–1665.

Naz, S., Fatima, Z., Iqbal, P., Khan, A., Zakir, I., Noreen, S., . . . Ahmad, S. (2019). Agronomic crops: Types and uses. In *Agronomic Crops* (pp. 1–18). Springer, Singapore.

Orhun, G. E. (2013). Maize for life. *International Journal of Food Science and Nutrition Engineering*, 3(2), 13–16.

Ortiz-Castro, M., Hartman, T., Coutinho, T., Lang, J. M., Korus, K., Leach, J. E., . . . Broders, K. (2020). Current understanding of the history, global spread, ecology, evolution, and management of the corn bacterial leaf streak pathogen, *Xanthomonas vasicola pv. vasculorum. Phytopathology*, 110(6), 1124–1131.

Our Badu-Apraku, B., & Fakorede, M. A. B. (2017). Advances in genetic enhancement of early and extra-early maize for sub-Saharan Africa. 1st edition, Springer Cham, 427–452.

Pataky, J. K. (2003). Stewart's wilt of corn. *APSnet Features*, 703(10.1094).

Pataky, J. K., Michener, P. M., Freeman, N. D., Weinzierl, R. A., & Teyker, R. H. (2000). Control of Stewart's wilt in sweet corn with seed treatment insecticides. *Plant Disease*, 84(10), 1104–1108.

Pataky, J. K., & Snetselaar, K. M. (2006). Common smut of corn. *The Plant Health Instructor*, 10.

Patandjengi, B., Junaid, M., & Muis, A. (2021, November). The presence of bacterial stalk rot disease on corn in Indonesia: A review. In *IOP Conference Series: Earth and Environmental Science* (Vol. 911, No. 1, p. 012058). IOP Publishing, Makassar.

Payak, M. M., & Renfro, B. L. (1967). A new downy mildew disease of maize. *Phytopathology*, 57, 394–397.

Pope, D. D., & McCarter, S. M. (1992). Evaluation of inoculation methods for inducing common smut on corn ears. *Phytopathology,* 82(9), 950–955.

Premalatha, N., Sundaram, K. M., & Arumugachamy, S. (2012). Screening and source of resistance to downy mildew (*Peronosclerospora sorghi*) in maize (*Zea mays* L.). *Electronic Journal of Plant Breeding*, 3(2), 788–793.

Ramirez-Cabral, N. Y. Z., Kumar, L., & Shabani, F. (2017). Global risk levels for corn rusts *(Puccinia sorghi* and *Puccinia polysora)* under climate change projections. *Journal of Phytopathology*, 165(9), 563–574.

Ranum, P., Peña-Rosas, J. P., & Garcia-Casal, M. N. (2014). Global maize production, utilization, and consumption. *Annals of the New York Academy of Sciences*, 1312(1), 105–112.

Raymundo, A. D., & Exconde, O. R. (1976). Economic effectiveness of resistant varieties and duter/dithane M-45 foliar spray for the control of Philippine corn downy mildew. *Philippine Agriculturist*, 60, 52–65.

Schmitt, C. G., & Freytag, R. E. (1977). Response of selected resistant maize genotypes to three species of *Sclerospora. Plant Disease Reporter*, 61, 478–481.

Shepherd, D. N., Martin, D. P., Van der Walt, E., Dent, K., Varsani, A., & Rybicki, E. P. (2010). Maize streak virus: An old and complex 'emerging'pathogen. *Molecular Plant Pathology*, 11(1), 1–12.

Shin, J. H., Han, J. H., Lee, J. K., & Kim, K. S. (2014). Characterization of the maize stalk rot pathogens *Fusarium subglutinans* and *F. temperatum* and the effect of fungicides on their mycelial growth and colony formation. *The Plant Pathology Journal*, 30(4), 397.

Singh, J. P. (1971). Infectivity and survival of oospores of *Sclerophthora rayssiae var. zeae. Indian Journal of Experimental Biology*, 9, 530–532.

Singh, J. P., Renfro, B. L., & Payak, M. M. (1970). Studies on the epidemiology and control of brown stripe downy mildew of maize *(Sclerophthora rayssiae var. zeae). Indian Phytopathology*, 23, 194–208.

Singh, V. K., & Chawla, S. (2012). *Cultural Practices: An Ecofriendly Innovative Approach in Plant Disease Management.* International Book Publishers and Distributers, Dehradun, India.

Singh, R., and Srivastava, R. P. (2016). Southern corn leaf blight-an important disease of maize: an extension fact sheet. *Indian Research Journal of Extension Education*, 12(2), 324–327.

Sinha, B., Devi, H. C., & Chanu, W. T. (2018). Disease management of nonleguminous seasonal forages Bireswar Sinha, H. Chandrajini Devi, And W. Tampakleima Chanu. In *Forage Crops of the World, Volume I: Major Forage Crops* (pp. 319–356). Apple Academic Press, Imphal, India.

Sinha, S. K., & Prasad, M. (1977). Bacterial stalk rot of maize, its symptoms and host-range. *Zentralblatt für Bakteriologie, Parasitenkunde, Infektionskrankheiten und Hygiene. Zweite Naturwissenschaftliche Abteilung: Allgemeine, Landwirtschaftliche und Technische Mikrobiologie,* 132(1), 81–88.

Stewart, L. R., Teplier, R., Todd, J. C., Jones, M. W., Cassone, B. J., Wijeratne, S., . . . Redinbaugh, M. G. (2014). Viruses in maize and Johnsongrass in southern Ohio. *Phytopathology*, 104(12), 1360–1369.

Sumner, D. R., & Littrell, R. H. (1974). Influence of tillage, planting date, inoculum survival, and mixed populations on epidemiology of southern corn leaf blight. *Phytopathology*, 64(2), 168–173.

Sun, Q., Li, L., Guo, F., Zhang, K., Dong, J., Luo, Y., & Ma, Z. (2021). Southern corn rust caused by *Puccinia polysora* Underw: A review. *Phytopathology Research*, 3(1), 1–11.

Tesso, T. T., Perumal, R., Little, C. R., Adeyanju, A., Radwan, G. L., Prom, L. K., & Magill, C. W. (2012). Sorghum pathology and biotechnology-a fungal disease perspective; Part II. Anthracnose, stalk rot, and downy mildew. *European Journal of Plant Science and Biotechnology*, 6(Special Issue 1), 31–44.

Thind, B. S., & Payak, M. M. (1985). A review of bacterial stalk rot of maize in India. *International Journal of Pest Management, 31*(4), 311–316.

Trdan, S., & Celar, F. (2000). Outbreak of maize downy mildew *(Sclerophthora macrospora)* in Slovenia in 1999. *Agriculture Scientific and Professional Review*, 6(2), 62–66.

Trowbridge, P. J., & Bassuk, N. L. (2004). *Trees in the Urban Landscape: Site Assessment, Design, and Installation*. John Wiley & Sons, Hoboken, New Jersey.

Ullstrup, A. J. (1972). The impacts of the southern corn leaf blight epidemics of 1970-1971. *Annual review of phytopathology*, 10(1), 37–50.

Utpal, D., & Ritika, B. (2015). Integrated disease management strategy of common rust of maize incited by *Puccinia sorghi Schw. African Journal of Microbiology Research*, 9(20), 1345–1351.

Veerabhadraswamy, A. L., Pandurangegowda, K. T., & Kumar, M. P. (2014). Efficacy of strobilurin group fungicides against turcicum leaf blight and polysora rust in maize hybrids. *International Journal of Agriculture and Crop Sciences (IJACS)*, 7(3), 100–106.

Veeral, D. K., & Abirami, G. (2021). Effects of liquid organic manures on growth, yield and grain quality of sweet corn (*Zea mays* convar. *sacharata* var. *rugosa*). *Crop Research,* 56(6), 295–300.

Venard, C., & Vaillancourt, L. (2005). Growth of and colonization by *Colletotrichum graminicola* inside corn stalk tissues. *Phytopathology*, 95(6), S107–S107.

Wayadande, A. C., & Nault, L. R. (1993). Leafhopper probing behavior associated with maize chlorotic dwarf virus transmission to maize. *Phytopathology*, 83(5), 522–526.

Wei, J. K., Liu, K. M., Chen, J. P., Luo, P. C., & Stadelmann, O. Y. L. (1988). Pathological and physiological identification of race C of *Bipolaris maydis* in China. *Phytopathology*, 78(5), 550–554.

Weston, W. H. (1920). Philippine downy mildew of maize. *Journal Agricultural Research*, 19, 97–122.

Yadav, K., & Sharma, S. S. (2022). Maize diseases and their effective management in India. *Chief Editor Dr. RK Naresh*, 44.

Yamada, M., & Aday, B. A. (1977). Fertilizer conditions affecting susceptibility to downy mildew disease *Sclerospora philippinensis* Weston, in resistant and susceptible materials of maize. *Annals of the Phytopathological Society of Japan (Nihon Shokubutsubyori Gakkai Ho)*, 43, 291–293.

Yanfeng, B., Long, J., Haixiang, R., Yanping, W., Xiaomei, W., Chunmei, Z., . . . Yuxin, Q. (2014). Damage characteristics and control measures of maize head smut. *Plant Diseases & Pests,* 5(5).

Zhang, Y., He, J., Jia, L-J., Yuan, T-L., Zhang, D., Guo, Y., et al. (2016). Cellular tracking and gene profiling of *Fusarium graminearum* during Maize stalk rot disease development elucidates its strategies in confronting phosphorus limitation in the host apoplast. *PLoS Pathog*, 12(3), e1005485. https://doi.org/10.1371/journal.ppat.1005485

Zibani, A., Ali, S., & Benslimane, H. (2022). Corn diseases in Algeria: First report of three Bipolaris and two Exserohilum species causing leaf spot and leaf blight diseases. *Cereal Research Communications*, 50(3), 449–461.

Index

Page numbers in *italics* indicate figures, and **bold** indicate tables.

D

E

T

U

V

W

X

Y

Z

Printed in the United States
by Baker & Taylor Publisher Services